FOUNDATIONS OF PROBABILITY THEORY,

STATISTICAL INFERENCE,

AND STATISTICAL THEORIES OF SCIENCE

VOLUME I

THE UNIVERSITY OF WESTERN ONTARIO
SERIES IN PHILOSOPHY OF SCIENCE

A SERIES OF BOOKS

ON PHILOSOPHY OF SCIENCE, METHODOLOGY,

AND EPISTEMOLOGY

PUBLISHED IN CONNECTION WITH

THE UNIVERSITY OF WESTERN ONTARIO

PHILOSOPHY OF SCIENCE PROGRAMME

VOLUME 6

FOUNDATIONS OF PROBABILITY THEORY, STATISTICAL INFERENCE, AND STATISTICAL THEORIES OF SCIENCE

*Proceedings of an International Research Colloquium held at the
University of Western Ontario, London, Canada, 10–13 May 1973*

VOLUME I

FOUNDATIONS AND PHILOSOPHY OF EPISTEMIC APPLICATIONS OF PROBABILITY THEORY

Edited by

W. L. HARPER and C. A. HOOKER

University of Western Ontario, Ontario, Canada

D. REIDEL PUBLISHING COMPANY

DORDRECHT-HOLLAND / BOSTON-U.S.A.

Library of Congress Cataloging in Publication Data
Main entry under title:

Foundations and philosophy of epistemic applications of
    probability theory.

    (Foundations of probability theory, statistical inference
and statistical theories of science ; v. 1) (The University of Western
Ontario series in philosophy of sciences ; v. 6)
    Papers and discussions from a colloquium held at the
University of Western Ontario, May 10–13, 1973.
    Includes bibliographies.
    1. Probabilities. 2. Mathematical statistics. I.
Harper, William. II. Hooker, Clifford Alan. III. Series.
IV. Series: London, Ont. University of Western Ontario. Series
in Philosophy of science ; v. 6.
  QA273.F755   519.2   75–34354
ISBN 90–277–0616–6
ISBN 90–277–0617–4 pbk.

The set of three volumes (cloth)   ISBN 90–277–0614–X
The set of three volumes (paper)   ISBN 90–277–0615–8

Published by D. Reidel Publishing Company,
P.O. Box 17, Dordrecht, Holland

Sold and distributed in the U.S.A., Canada, and Mexico
by D. Reidel Publishing Company, Inc.
Lincoln Building, 160 Old Derby Street, Hingham,
Mass. 02043, U.S.A.

Printed in The Netherlands by D. Reidel, Dordrecht

CONTENTS

CONTENTS OF VOLUMES II AND III

VOLUME II

Foundations and Philosophy of Statistical Inference

GENERAL INTRODUCTION

In May of 1973 we organized an international research colloquium on foundations of probability, statistics, and statistical theories of science at the University of Western Ontario.

During the past four decades there have been striking formal advances in our understanding of logic, semantics and algebraic structure in probabilistic and statistical theories. These advances, which include the development of the relations between semantics and metamathematics, between logics and algebras and the algebraic-geometrical foundations of statistical theories (especially in the sciences), have led to striking new insights into the formal and conceptual structure of probability and statistical theory and their scientific applications in the form of scientific theory.

The foundations of statistics are in a state of profound conflict. Fisher's objections to some aspects of Neyman-Pearson statistics have long been well known. More recently the emergence of Baysian statistics as a radical alternative to standard views has made the conflict especially acute. In recent years the response of many practising statisticians to the conflict has been an eclectic approach to statistical inference. Many good statisticians have developed a kind of wisdom which enables them to know which problems are most appropriately handled by each of the methods available. The search for principles which would explain why each of the methods works where it does and fails where it does offers a fruitful approach to the controversy over foundations. The colloquium first aimed both at a conceptually exciting clarification and enrichment of our notion of a probability theory and at removing the cloud hanging over many of the central methods of statistical testing now in constant use within the social and natural sciences.

The second aim of the colloquium was that of exploiting the same formal developments in the structure of probability and statistical theories for an understanding of what it is to have a statistical theory of nature, or of a sentient population. A previous colloquium in this series has already examined thoroughly the recent development of the analysis of quantum mechanics in terms of its logico-algebraic structure and brought out many of the sharp and powerful insights into the basic physical significance of this theory which that formal approach provides. It was our aim in this colloquium to extend the scope of that inquiry yet further afield in an effort to understand, not just one particular idiosyncratic theory, but what it is in general we are doing when we lay down a formal statistical theory of a system (be it physical or social).

Our aim was to provide a workshop context in which the papers presented could benefit from the informed criticism of conference participants. Most of the papers that appear here have been considerably rewritten since their original presentation. We have also included comments by other participants and replies wherever possible. One of the main reasons we have taken so long to get these proceedings to press has been the time required to obtain final versions of comments and replies. We feel that the result has been worth the wait.

When the revised papers came in there was too much material to include in a single volume or even in two volumes. We have, therefore, broken the proceedings down into three volumes. Three basic problem areas emerged in the course of the conference and our three volumes correspond. Volume I deals with problems in the foundations of probability theory; Volume II is devoted to foundations of statistical inference, and Volume III is devoted to statistical theories in the physical sciences. There is considerable overlap in these areas so that in some cases a paper in one volume might just as well have gone into another.

INTRODUCTION TO VOLUME I

This first volume is devoted to some problems in the mathematical and philosophical foundations of probability theory. There are several main themes in it. One theme drawn from philosophical logic that has received considerable attention in recent literature is provided by the attempt to extend probability frameworks to algebras with operators corresponding to conditionals. The papers of Adams, Giles, Harper, and van Fraassen are all devoted to this enterprise.

Another theme is the problem of rational belief change relative to personalistic interpretations of probability. The paper by May and Harper and that by Teller are explicitly directed to this problem. The papers by Adams, van Fraassen and Harper are also relevant to this theme.

The papers by Finch, Fine and Rosenkrantz are devoted to using mathematical properties of probability to illuminate various notions important to statistical inference. These papers would have been equally at home in the second volume. We put them here because the mathematical structure of probability theory plays such a salient role in them.

ERNEST W. ADAMS

PRIOR PROBABILITIES AND COUNTERFACTUAL CONDITIONALS*

I. THE PRIOR CONDITIONAL PROBABILITY REPRESENTATION OF THE PROBABILITIES OF COUNTERFACTUALS

By a *counterfactual* I shall mean any conditional expressed in the subjunctive mood. Examples of a counterfactual and what will be called its *corresponding indicative* are:

> If Alice were here, Bill would be too.

and

> If Alice is here, Bill is too.

It will not be assumed that counterfactuals entail or in any way 'implicate' the falsity of their antecedents; on the other hand, I shall not assume that counterfactuals are equivalent to their corresponding indicatives, as some have maintained.[1] An important objective of this paper is to help to clarify the logical relation between the two forms.

In previous papers I have attempted to develop a *probabilistic* 'logic' of indicative conditionals, the basic ideas of which are essential to the present work.[2] The three fundamental theses of this theory are as follows. (1) Bivalent truth values (truth and falsity) are not appropriately applicable to indicative conditionals with false antecedents, and therefore deductive connections involving them cannot be analyzed in terms of their truth conditions.[3] (2) The *probability* of an indicative conditional is appropriately measured in terms of the associated *conditional probability*.[4] (3) An appropriate *test* of the rationality of an inference pattern involving indicative conditionals is to determine whether and under what circumstances it is possible for all of the premises of the inference to be highly probable, while the conclusion of the inference is improbable.[5] Applying this probabilistic test of rationality yields an immediate explanation of the irrationality of the so-called *fallacies of material implication* (e.g., to infer 'if A then B' from 'not A'), and also brings to light surprising irrational exceptions to generally accepted rules of reasoning involving the condi-

Harper and Hooker (eds.), Foundations of Probability Theory, Statistical Inference, and Statistical Theories of Science, Vol. I, 1–21. All Rights Reserved. Copyright © 1976 by D. Reidel Publishing Company, Dordrecht-Holland.

tional such as the Hypothetical Syllogism (to infer 'if A then C' from 'if A then B' and 'if B then C').

Turning to the counterfactual, I will take it as a working hypothesis that what was true of indicatives holds for counterfactuals as well: namely, that truth values are not applicable to them under all circumstances, and so their deductive relations to other propositions cannot be analyzed in terms of truth conditions.[6] This negative hypothesis, however, leads us to ask how the *probabilities* of counterfactuals are to be represented, if the probabilistic test of rationality is to be applied to them. This is the principal problem with which this paper is concerned. No solution to it which is adequate under all circumstances will be proposed here, but in this and the following sections a partial solution will be proposed, and some of its implications discussed.

Our proposed partial solution is that in certain circumstances the probability of a counterfactual conditional is appropriately represented mathematically as a *prior conditional probability*. Illustrations occur in the context of typical problems of inverse probable inference ('reasoning from effects to causes') such as the following. Two urns, A and N, are filled with black and white balls, urn A containing 0.1% white and 99.9% black balls, and urn N containing 50% of each color. One urn is selected at random and placed before an 'observer', who draws one ball at random from it. This ball proves to be white. Following the drawing, it seems rational to maintain.

> The probability that a black ball would have been drawn if urn A had been selected is 0.999,

which has the form of a claim about the probability of the counterfactual conditional:

(C)   A black ball would have been drawn if urn A had been selected.

This probability of 0.999 is furthermore appropriately called a *prior conditional probability*, since it is equal to a ratio of prior unconditional probabilities, $p_0(A \text{ and } B)/p_0(A)$, where the subscript '0' is here used to designate prior probabilities, and 'A' and 'B' symbolize the 'events' of urn A's being selected, and a black ball's being drawn, respectively. The probability of C is furthermore equal to 1 minus the probability that a white ball would have been drawn if urn A had been selected, $p_0(W/A)$, which enters into the Baysian inference formula:

(1) $$\frac{p(A)}{p(N)} = \frac{p_0(A)}{p_0(N)} \times \frac{p_0(W/A)}{p_0(W/N)},$$

where 'W' symbolizes the newly acquired 'evidence' that a white ball was drawn.[7]

It is essential to insist on the word *prior* in describing the probabilities of counterfactuals, since these conditional probabilities must be clearly distinguished from the corresponding *posterior* conditional probabilities, which are the probabilities of the indicative counterparts to the counterfactuals. Knowing that a white ball has been drawn by the observer, the posterior conditional probability, $p(B/A) = p(B$ and $A)/p(A)$, is clearly equal to 0 (and not equal to 0.999, which is the prior conditional probability). Given that the posterior conditional probability is 0, it would be totally irrational accept or assert the corresponding indicative,

(I)   A black ball was drawn if urn A was selected.

Thus, to the extent that the prior conditional probability representation is valid, we have a theoretical explanation of the difference between the counterfactual and the indicative, as well as of the relation between them: the probability of the counterfactual is a prior conditional probability while that of the corresponding indicative is the corresponding posterior conditional probability.

Another point to which it is worth drawing attention is that, given the correctness of the proposed representation, it would follow that counterfactuals will often 'arise from' indicatives, when new information is acquired which renders a previously acceptable indicative conditional no longer acceptable. These will be cases where what are called 'prior probabilities' actually are equal to what had previously been 'posterior probabilities' before the acquisition of the new information. This is the case in our ball-drawing example. Prior to drawing the ball from the urn, it would have been rational to maintain.

> The probability that a black ball will be drawn if urn A is selected is 0.999,

which is the probability of the indicative

A black ball will be drawn if urn A is selected.

The 'mood shift' following the acquisition of 'adverse evidence' will prove important in subsequent discussions. In the next few sections we will discuss further applications of our partial representation. The final section will discuss some of its limitations.

II. NONCONDITIONAL COUNTERFACTUALS

Are there propositions of ordinary language which are appropriately describable as nonconditional counterfactuals? They should stand logically to counterfactual conditionals as ordinary 'factual' propositions stand to indicative conditionals. Furthermore, assuming the prior probability representation, their probabilities should be nonconditional prior probabilities. Also, we should expect them to arise frequently from ordinary factual propositions by a 'mood shift transformation', following the acquisition of evidence adverse to the original proposition. Certain propositions constructed using the 'non-valuational' deontic operator 'should' or 'ought to' seem to fill this bill.[8]

Consider the following example. A person expects his friend Jones to pay him a visit, arriving before 8 p.m. of a given day. Prior to this time it is rational to maintain 'Jones will be here before 8 p.m.' ('here' being at the man's house). Now 8 p.m. arrives, but Jones doesn't. It is no longer rational to maintain the factual proposition, but it would be rational to assert 'Jones ought to have been here before 8 p.m.'. The latter would be asserted on the same grounds as originally furnished the basis of the factual assertion, and so it is plausible that its probability should be equal to the probability originally attached to the factual proposition; i.e., this should be a prior nonconditional probability.

One thing significant about what I am inclined to call a nonconditional counterfactual, expressing what ought to be the case but is not, is that such claims often lead to the demand for an *explanation*. Further comments on the role of counterfactuals and prior probabilities in explanation are made in the next section.

III. PRIOR PROBABILITIES, COUNTERFACTUALS, AND EXPLANATION

It is recognized that probabilities must play an important part in a general theory of explanation, but realizing that more than one 'probability' – e.g., prior and posterior – can be attached to the same 'event' should lead us to ask which is the more immediately relevant to explanation. That propositions as to what ought to be the case but is not, whose probabilities are prior, often evoke the demand for explanation suggests that prior probabilities and their associated counterfactuals at least have an important role. This is further borne out by consideration of how such *a priori* unlikely events are explained.

A mundane example is the failure of electrical power in a house – an *a priori* unlikely event which 'ought not' to occur. A possible explanation is that one of the power lines leading to the house is down. A necessary condition for this to be a possible explanation is that it should be the case that *if the power lines were down, the power would be off*, or at least that the *a priori* probability of the power's being off if the power lines were down should be considerably higher than the likelihood of the power's being off if the lines were not down. More generally, a necessary requirement that a possible occurrence O should explain an observed phenomenon P is that it should be *a priori* probable that if O were the case then P would be, or at least more probable that if O were the case P would be than that P should be the case *simpliciter*. Furthermore, analysis of the relationships between the prior and posterior probabilities involved shows why and under what circumstances the fact that a possible occurrence O would explain phenomenon P entails that O must be more likely *a posteriori* than an alternative hypothesis which 'fails to account for P'.

A striking fact about the counterfactual conditionals entering into explanation is that they are neither equivalent to their corresponding indicatives, not do they in any sense imply the falsity of their antecedents. A person suggesting that a downed power line might explain a power failure, claiming that if the power line were down then the power would be off, would not be refuted if he learned that in fact the lines were down. On the other hand, the counterfactual conditional claim is no mere stylistic variant if the corresponding indicative, which would be 'if the power line is down then the power is off'. Given that the power is off, the *a posteriori* claim that *if* the lines are down the power is off is totally un-informative. But the *a priori*, counterfactual claim might provide important information. Of course, the prior probability representation of the probabilities of counterfactuals might have suggested directly that there is no necessary connection between the acceptability (i.e., high probability) of the counterfactual and the falsity or probable falsity of its antecedent, and the explanation examples merely corroborate this suggestion.

IV. THE INTRINSIC LOGIC OF THE COUNTERFACTUAL

Assuming the correctness of the prior probability representation it would follow that the logical laws interrelating purely counterfactual propositions should be in a sense 'isomorphic' to those interrelating the corresponding indicatives. The probabilistic 'logic' of the indicative conditional is set forth in detail in (1) and (2), and I will confine myself to a few examples by way of illustrating its transformation to the counterfactual. The inference schema of the Hypothetical syllogism is not sound in full generality, and counterexamples to it – concrete instances in which the two premises are probable while the conclusion is improbable – transform to counterexamples to the counterfactual version of the schema by a simple change of mood. Thus, a mood transformation of a counterexample discussed in (1) yields the irrational inference:

> If Jones had won the election, Smith would have retired. If Smith had died before the election, Jones would have won. Therefore, if Smith had died before the election, he would have retired.

(Smith and Jones having been candidates for a public office of which Smith was the incumbent.) An inference schema which is probabilistically sound is what I am inclined to call the *expanded hypothetical syllogism*: to infer 'if *A* then *C*' from the two premises 'if *A* then *B*' and 'if *A* and *B* then *C*'. This is valid in the indicative version, and I have been unable to find any counterexamples to it in the counterfactual. Of course, there is no reason to think that this pattern is valid where some of the propositions involved are in the indicative and others are in the counterfactual mood.

Our 'isomorphism' hypothesis can be extended to patterns of inference involving both conditional and nonconditional propositions, provided the counterfactual 'translations' of the nonconditional propositions involved are taken to be the nonconditional counterfactuals discussed in Section II. *Modus Ponens* is a schema which is sound in its indicative version, and rather unsystematic survey suggests some validity to its counterfactual transform: to infer '*B* ought to be the case' from '*A* ought to be the case' and 'if *A* were the case then *B* would be the case'. Of course, this pattern is not sound where the conditional is counterfactual but the nonconditional propositions are indicative (e.g., to infer 'Bill is here' from 'if Alice were here Bill would be here' and 'Alice is here'). *Modus Tollens* is another pattern of inference which is sound in its indicative version, and which can be analyzed in a manner similar to *Modus Ponens*. However, it turns out that the most common real life 'instances' of *Modes Tollens* involve *both* the counterfactual and the indicative, and the analysis of this raises new issues. Before turning to these, though, a comment on the logic of consistency and contradiction is in order.

A set of propositions can be described as *probabilistically consistent* if (roughly) it is possible that all propositions of the set should be simultaneously probable. The two propositions *B* and 'if *A* then not *B*' are consistent in this sense, while the propositions 'if *A* then *B*' and 'if *A* then not *B*' are inconsistent.[9] The isomorphism hypothesis would entail that the foregoing should hold, whether the propositions are in the indicative or in the counterfactual mood. Thus, in the counterfactual it would be consistent to maintain 'the power should be on' and 'if the lines were down the power wouldn't be on', but it would not be consistent to maintain 'if the lines were down the power would be on' and 'if the lines were down, the power would be off'. The inconsistency of the latter pair suggests that the form 'if *A* were the case, *B* would not be' can legitimately be called the *contradictory* of 'if *A* were the case then *B* would be'. More generally, if *A* is regarded as designating a 'possible world', then the probabilistic logic of propositions of the form 'if *A* were the case then *B* would be' is essentially that of the propositions '*B* would be true in possible world *A*'.

V. MODUS TOLLENS

The following describes a bit of real life reasoning which might be represented abstractly as a *Modus Tollens* inference for the purposes of applying logical theory. A man whom we shall call the Reasoner is about to arrive at a party, where he does not know who the other guests will be. The Reasoner does know a girl, Alice, and her boyfriend, Bill, and he thinks 'if Alice is at the party, then Bill will be too'. Now the Reasoner arrives at the party, and immediately upon arrival, before he has had a chance to see for himself who is present, he learns that Bill is not there. He then 'makes an inference' and concludes that Alice is not there. If this reasoning process were given as a textbook example, in which the problem would be to assess its soundness, it would probably be formulated as a *Modus Tollens* inference with an indicative conditional premises, thus:

> If Alice is at the party, Bill will be. Bill is not at the party. Therefore, Alice is not at the party.

The 'indicative' *Modus Tollens* formulation of the above reasoning would seriously misrepresent it. The reason is that, upon learning that Bill was not at the party, the Reasoner would *give up* his belief 'if Alice is at the party, Bill will be too', and if that proposition were no longer accepted, it could not properly be regarded as one of the 'premises' of the inference. It would seem intuitively strange if the Reasoner were to continue to insist 'if Alice is here, Bill is too' after learning that Bill was not at the party, and we shall see in a moment that it would be quite irrational to do so. In fact, rather than maintain the indicative conditional premise, the Reasoner would be most likely to replace it by the corresponding counterfactual, and reason thus:

> Bill is not at the party. If Alice were at the party, Bill would be. Therefore, Alice is not at the party.[10]

Formulated in this way, the reasoning seems to be a *Modus Tollens* inference with a *counterfactual* premise, which, assuming the non-equivalence of the counterfactual and the indicative, should be taken into account in assessing its soundness.

It would be oversimple to view the foregoing reasoning process *simply* as a *Modus Tollens* inference. Recall that the Reasoner was described as originally believing an indicative conditional and that the only *new information* he was supposed to have acquired was that the consequent of this conditional was false – i.e., that Bill was not at the party. Receipt of this information resulted in three things: (1) giving up belief in the indicative conditional, (2) 'substituting' the counterfactual for it, and (3), 'inferring' that Alice was not at the party. Steps (1) and (2) do not seem happily describable as 'inferences' at all, yet an adequate analysis of the reasoning process should be able to take all three stages into account, and not merely the final step.

I would suggest that rather than viewing the foregoing and similar reasoning processes simply as deductions or series of deductions, it is more fruitful to regard them as phenomena of *probability change*, where the receipt of new information leads to lowering the probabilities of certain propositions, and raising the probabilities of others. What we have described as 'giving up' an indicative conditional can be viewed as *lowering its probability* (possibly below some 'level of acceptability'), and what we have described as 'inferring' that Alice is not at the party can be alternatively regarded as an *increase* in the probability of that proposition, resultant on acquisition of the information that Bill is not at the party. Steps (1) and (2) are in fact very simply accounted for in terms of probability changes. Formulas of Bayesian probability change imply that the probability which should be attached to the indicative conditional 'if A is the case then B is' (where 'A' and 'B' symbolize Alice's being at the party, and Bill's being there, respectively) *after* 'not B' is learned is equal to the probability which had previously been attached to 'if A is the case and B is not, then B is the case'. The latter probability is obviously 0, and it follows that subsequent to learning that Bill is not at the party, the probability of 'if Alice is at the party then Bill is' drops to 0.[11] The substitution of the counterfactual for the indicative conditional subsequent to learning the falsity of the consequent can be accounted for on the hypothesis that the probability of the counterfactual *after* this information is learned is equal to the *prior* conditional probability of the indicative conditional. Given that the indicative conditional was sufficiently probable to be accepted *a priori* (i.e., prior to learning the falsity of its consequent), the counterfactual should be probable enough to be accepted *a posteriori*.

When the inference with which the reasoning process concludes is construed as probability change phenomenon and analyzed accordingly, a striking fact emerges: namely that inferences of this pattern are not always rational. More exactly, it is not always rational to conclude

'not A' upon learning 'not B', having previously accepted the indicative
conditional 'if A then B'. If 'not B' is treated as new information, then
Equation (2) below relates the posterior ratio of the probabilities of A and
'not A' to the prior ratio of these probabilities ('prior' being prior to
acquiring information 'not B'):

$$(2) \qquad \frac{p(\text{not } A)}{p(A)} = \frac{p_0(\text{not } A)}{p_0(A)} \times \frac{p_0(\text{not } B/\text{not } A)}{p_0(\text{not } B/A)}$$

If all we know is that prior to learning 'not B', the indicative conditional
'if A then B' was highly probable, then we know that the prior probability $p_0(\text{not } B/A) = 1 - p_0(B/A)$, which enters into the denominator
on the right in Equation (2), must be very low. But the fact that $p_0(\text{not } B/A)$
is low is obviously insufficient to guarantee that the posterior probability
ratio $p(\text{not } A)/p(A)$ is high, which is what would have to be the case if
the conclusion 'not A' were to be justified.

The theoretical possibility that it might not be rational to conclude
'not A' upon learning 'not B', given prior acceptance of 'if A then B',
is sometimes realized in practice. One situation in which the posterior
probability ratio $p(\text{not } A)/p(A)$ is not high, even though the prior conditional probability $p_0(\text{not } B/A)$ is low, is that in which the prior conditional probability $p_0(\text{not } B/\text{not } A)$ entering into the numerator in
Equation (2) is also low. This is the situation in which prior to learning
'not B' not only 'if A then B' could be affirmed, but also 'if not A then B'.
This might have been the case in our special example, if prior to arriving
at the party the Reasoner had believed not only 'if Alice is at the party,
Bill will be' but also 'if Alice is not at the party, Bill will be' – i.e., the
Reasoner might have believed 'Bill will be at the party, whether or not
Alice is there'. It is intuitively obvious that under these circumstances
the Reasoner would not have been entitled to conclude that Alice was not
at the party, upon learning that Bill was not there.

That inferences in accord with the special *Modus Tollens* pattern in
which 'not A' is concluded upon learning 'not B', given prior acceptance of
'if A then B', are not always rational does not mean that they are never
rational. All that has been shown is that if such an inference is rational,
this cannot be so merely in virtue of its being an instance of this pattern.
Certain further conditions must be satisfied, which will be discussed in
the following section as they apply to this and other inference patterns.

By way of concluding this section we shall note that it is arguable that the additional conditions for rationality *are* satisfied in our 'Alice and Bill' example, but this argument throws doubt on the correctness of the prior probability representation of the probabilities of counterfactuals.

One case in which it is irrational to conclude 'not A' upon learning 'not B', given prior acceptance of the indicative 'if A then B' is, as we have seen, that in which, 'if not A then B' would also have been accepted prior to learning 'not B'. But in circumstances in which a Reasoner initially believes both 'if A then B' and 'if not A then B', it would seem intuitively very odd if he were to affirm the counterfactual 'if A were the case then B would be' subsequent to learning 'not B'. Therefore the fact that our Reasoner did substitute the counterfactual 'if Alice were at the party, Bill would be too' upon learning that Bill was not at the party would seem to show that he (the Reasoner) could not have affirmed 'if Alice is not at the party, Bill will be' before learning that Bill was not at the party. And, assuming the Reasoner could not have made the latter affirmation prior to learning that Bill was not at the party, it should follow that he was justified in ultimately concluding that Alice was not at the party.

That the foregoing argument calls into question the prior probability representation is clear. If the probability of 'if A were the case then B would be' were in all cases equal to that of the corresponding indicative prior to learning 'not B', the acceptability of the counterfactual subsequent to learning 'not B' should not depend on whether or not 'if not A then B' could also have been affirmed prior to learning 'not B'. Our intuitive observation that persons would not affirm 'if A were the case then B would be' after learning 'not B' if they had previously accepted both indicatives 'if A then B' and 'if not A then B' appears to contradict this, however. This is our first indication of the limitations of the prior probability representation, which will be discussed at greater length in the final section. The next section digresses to discuss some of the implications of the probability change point of view as a way of construing inference.

VI. INFERENCE AS PROBABILITY CHANGE; RATIONAL INFERENCE FROM A CONTRADICTION

A significant feature of our analysis of the 'Alice and Bill' example is that it took explicit account of the fact that the second premise of the inference – that Bill was not at the party – represented information acquired by the Reasoner *after* he came to accept the first premise. Our probabilistic representation makes it possible to take theoretical account of the order of acquisition of information, since probabilities, unlike truth values, change when new information is acquired. That probabilities change, and inference can be construed as probability change, actually has nothing specifically to do with the fact that the inferences in question involve conditionals. In this section we will consider briefly how other patterns of inference besides Modus Tollens, including ones not involving conditionals at all, appear when they are treated as probability change phenomena.

Suppose we have a truth-conditionally sound inference pattern with $n+1$ premises, $\alpha_1, ..., \alpha_n$ and β, and a conclusion γ. We are interested in particular applications in which a reasoner initially accepts propositions of the forms $\alpha_1, ..., \alpha_n$, then learns a proposition of form β, and the question is whether or not he is then entitled to conclude γ. In order that probability should be applicable, we shall suppose that $\alpha_1, ..., \alpha_n$ and γ are all either non-conditional or 'simple' conditionals (not containing other conditionals as parts), and β is non-conditional. Let p_0 and p symbolize prior and posterior probabilities respectively; i.e., probabilities which it is rational for the reasoner to attach to propositions before and after receiving information β. The acceptability of the conclusion, γ, after β is learned depends on the posterior probability $p(\gamma)$, and this in turn is related to prior probabilities by the following formulas. If γ is non-conditional then

(3) $p(\gamma) = p_0(\gamma/\beta)$,

and if γ is conditional, say $\gamma = $ 'if γ_1 then γ_2', then

(4) $p(\gamma) = p_0(\gamma_2/\gamma_1 \text{ and } \beta)$.

It follows from formulas (3) and (4) that γ should be accepted after β is learned if and only if 'if β then γ' should have been accepted before β was learned, if γ is non-conditional, and if γ is the conditional 'if γ_1 then γ_2', it should be accepted after β is learned if and only if 'if γ_1 and β then γ_2' should have been accepted before β was learned. We know that prior to learning β all of $\alpha_1, ..., \alpha_n$ were acceptable (i.e., the probabilities $p_0(\alpha_1), ..., ..., p_0(\alpha_n)$ are all high), and we want to know whether this insures that the proposition 'if β then γ' or 'if γ_1 and β then γ_2' must also be acceptable before β is learned.

Consider the non-conditional case first. It might appear that if $\alpha_1, \ldots, \alpha_n$ are all acceptable before learning β, and $\alpha_1, \ldots, \alpha_n$ together with β truth-conditionally implies γ, then 'if β then γ' must also be acceptable before learning β, and therefore that γ should be accepted subsequent to learning β. One way of interpreting the Deduction Theorem (rule of Conditionalization) would seem to imply this. This appearance would be mistaken, however. In fact, if $\alpha_1, \ldots, \alpha_n$ do not include any conditionals, then a general theorem given in (1) shows that, provided β does not logically entail γ, it will always be possible for $\alpha_1, \ldots, \alpha_n$ to be 'arbitrarily highly probable' while 'if β then γ' has probability 0. Thus, in the purely non-conditional case (no conditionals in either premises or conclusion), provided γ doesn't follow from β alone, it must always be possible for $\alpha_1, \ldots, \alpha_n$ to be accepted, for β then to be learned, but for it to be irrational conclude γ.

The *Modus Tollendo Ponens* inference pattern, to infer B from 'either A or B' and 'not A', illustrates the foregoing. Suppose a reasoner initially to believe a proposition of the form 'either A or B', then to learn 'not A'. The result cited above shows that there can be situations in which it would not be rational to conclude B. A hypothetical example might be one in which a reasoner initially believes 'either it rained or it snowed in Tucson in the winter of 1960' (he believes this in 1973, on general principals that it ususally rains or snows in the winter). Now he learns that it did not rain in Tucson in the winter of 1960. Intuitively it would seem irrational for the reasoner thereupon to conclude that it did snow in Tucson in the winter in question, and probabilistic analysis supports this intuition. This example has the peculiarity that learning that it did not rain in Tucson in the winter of 1960 would probably result in the reasoner's *giving up* his original belief that it either rained or snowed there in 1960; i.e., the new information would *conflict* with the prior premise. We shall see shortly that this sort of conflict is characteristic of situations in which persons would be unjustified in inferring γ upon learning β, given prior acceptance of $\alpha_1, \ldots, \alpha_n$, in spite of the fact that $\alpha_1, \ldots, \alpha_n$ and β truth-conditionally imply γ. Before considering this sort of conflict, however, we turn briefly to patterns of inference involving conditionals.

Modus Ponens is probably the most basic truth-conditionally sound pattern of conditional inference, and, strikingly, probabilistic analysis shows that it can never be irrational to 'apply' this pattern sequentially; i.e., where a conditional 'if A then B' is initially accepted, A is then learned, and B is thereupon concluded. According to Equation (3), the probability to be attached to B subsequent to learning A is equal to that originally attached to the conditional 'if A then B', hence if the conditional was acceptable before learning A, then B alone should be acceptable after learning A. For this reason it seems appropriate to regard Equation (3) as expressing a probabilistic version of *Modus Ponens*. The other basic pattern of conditional inference is *Modus Tollens*, and we have already seen that it is not always rational to apply it in the sense of inferring 'not A' upon learning 'not B', given prior acceptance of 'if A then B'.

Among patterns of inference with conditional conclusions, two may be considered by way of illustration. One is to conclude 'if A then C' upon learning B, given prior acceptance of 'if A and B then C' – this pattern being generally regarded as sound. Probabilistic analysis supports the thesis that this sort of reasoning is generally sound, since according to Equation (4) the probability to be attached to 'if A then C' subsequent to learning B is equal to that previously attached to 'if A and B then C'. Given that the latter was high enough to be acceptable prior to learning B, 'if A then C' should be probable enough to be accepted after learning B. An inference pattern which is not always rational in sequential application is to infer 'if A then C' upon learning 'not B', given prior acceptance of 'if A then either B or C.' That this is not always rational follows from the facts that: (1) the probability to be attached to 'if A then C' after learning 'not B' is equal to the probability of 'if A and not B then C' before learning 'not B', and (2) it is possible for 'if A then B or C' to be highly probable while 'if A and not B then C' is highly improbable. Concrete examples are not difficult to construct, but will be omitted here.

Returning finally to the non-conditional case (i.e., the case in which the conclusion is non-conditional), we want to consider the nature of those situations in which it would not be rational to conclude γ upon learning β, given prior acceptance of $\alpha_1, ..., \alpha_n$, in spite of the fact that $\alpha_1, ..., \alpha_n$ together with β truth-conditionally imply γ. It follows from a theorem in (2) that these must be situations in which β conflicts with $\alpha_1, ..., \alpha_n$ in the sense that one or more of $\alpha_1, ..., \alpha_n$ would no longer be accepted after β

was learned. This would have to be the case since, if all of $\alpha_1, ..., \alpha_n$ were accepted – hence all were highly probable – after β was learned, the probability of falsehood of γ would be no more than the sum of the probabilities of falsehood of the $n+1$ propositions $\alpha_1, ..., \alpha_n$ and β, which, assuming that n is not 'too large', would have to be small.[12] This is illustrated in the irrational special case of *Modus Tollens* discussed previously, in which the Reasoner is imagined to initially believe both 'if Alice is at the party then Bill will be' and 'if Alice is not at the party then Bill will be', before learning that Bill is not at the party. Under the circumstances, learning that Bill was not at the party would certainly cause the Reasoner to retract one or other of his earlier beliefs.

The upshot of the foregoing is that where γ is non-conditional, it will be rational to conclude γ upon learning β, given prior acceptance of $\alpha_1, ..., \alpha_n$ and the fact that $\alpha_1, ..., \alpha_n$ and β truth-conditionally imply γ, except in the case in which β conflicts with $\alpha_1, ..., \alpha_n$ (and leaving aside difficulties arising when there are 'too many premises'). Thus we have a kind of converse of the traditional maxim that everything follows from a contradiction, in that certain kinds of inferences are held to be rational except when their premises conflict (where the conflict may have the form of a logical contradiction). This is not to advance the claim that inferences with contradictory premises are *never* rational, which would be as absurd an oversimplification as the traditional maxim. In fact, Equation (2) shows just how high a probability should be attached to the conclusion 'not A' upon learning 'not B', given prior high probabilities of both 'if A then B' and 'if not A then B'. What this equation makes clear, however, is that it is hopeless to look for an adequate theory of rational inference from conflicting propositions within a logical theory which takes into account only the truth-conditions of these propositions, since in the conflict case the acceptability of conclusions will be critically dependent on the degrees of probability of the premises, and not merely on whether they are 'sufficiently probable to be accepted.'

In concluding this section, I must mention two limitations on the probability change approach as a way of analyzing rational inference. One arises from the fact that there is no presently accepted way of representing the probabilities of conditionals whose antecedents are conditional,[13] and it follows from this that Equation (3) cannot be generalized to specify how probabilities should change upon receipt of new information (symbolized by β) which is itself of conditional form. The second difficulty has to do with the fact that such probability change equations as (3) and (4) tacitly assume that the newly acquired information, β, has *a posteriori* probability 1.[14] It is easy to see that if a systematic treatment of more general kinds of reasoning, involving several items of information acquired in succession, is to be possible, then this idealization cannot be maintained. In both the cases of inference from newly acquired conditional information and several items of information, therefore, we presently lack the mathematical framework within which a realistic probabilistic analysis is possible.

VII. A QUALIFICATION AND A LIMITATION; THE AMBIGUITY OF THE COUNTERFACTUAL

We have already encountered a serious difficulty with the prior probability representation of counterfactuals' probabilities which will be discussed in a moment. First, however, an important qualification must be noted. This is that the term 'prior' used to designate the sorts of probabilities which are distinguished from 'posterior' probabilities in applications of statistics to problems of inverse inference often cannot be literally identified with probabilities which at some prior *time* had been posterior probabilities, and our posited representation of counterfactuals' probabilities must be clarified to suggest that counterfactuals' probabilities may be represented by those probabilities which are *called* 'prior' in the theory of inverse inference, whether or not they are equatable with what had earlier been posterior probabilities. To illustrate, suppose that in the ball-drawing example described in Section I, the observer or reasoner (whose job it is to estimate the probabilities of the ball's having been drawn from one or other of Urns A and N) had first drawn out the white ball and *then* learned what the proportions of balls in the urns were, and that one of them was chosen at random and put before him. Under the circumstances the observer would then assign a 'prior' probability of 0.5 that the urn from which the ball was drawn was Urn A, and a prior conditional probability of 0.999 that a black ball would have been drawn if the urn selected had been Urn A (this subsequent to learning that 99.9% of the balls in Urn A are black). But neither of these 'prior' probabilities has any necessary connection with what had been posterior probabilities at some earlier time. As a matter of fact, the identification of the 'prior' conditional probability, which is a counterfactual's probability, with the proportion of black balls in Urn A (whether or not this proportion is known before a ball is actually drawn) suggests that it might be somewhat less misleading to employ the occasionally used term 'propensity probabilities' in application to them.

The foregoing qualification accounts for two things about counterfactuals. One is that in many cases the acquisition of information adverse to a previously held indicative conditional does not lead to substituting the corresponding counterfactual for it, although what had previously been the posterior probability of the conditional was high. For instance, in the Alice and Bill example discussed in Section V, the Reasoner previously thinking 'if Alice is at the party, Bill will be too' might have learned that Alice and Bill had had a falling out, and this would not have led him to substitute the counterfactual 'if Alice were at the party, Bill would be'. The acquisition of this information would instead have led the Reasoner to revise his estimates of the relevant *propensities*, in consequence of which he would not assert the counterfactual.

Identifying counterfactuals' probabilities with 'propensity' probabilities suggests how the frequently noted link between counterfactuals and natural laws and especially *dispositional concepts* might be developed in probabilistic terms. It is not inappropriate to regard the probability 0.999 of drawing a black ball from an urn containing 99.9% black balls as something like a probabilistic disposition of the urn to yield black balls. Similarly, it is plausible to identify the probabilities of certain dispositional propositions with those of counterfactuals; for instance, one might reasonably equate the probability that a certain object was brittle at some previous time with the probability that it would have broken if it had been subjected to such and such a stress at that time, which is the probability of a counterfactual conditional.[15]

The difficulty noted in Section V was that there are occasions in which it would seem intuitively odd to maintain a counterfactual, in spite of the fact that the corresponding 'prior' conditional probability is high. This remains a difficulty even where the word 'prior' is amended to read 'propensity', as such probabilities enter into the analysis of inverse inference. Our special case of the Alice and Bill reasoning illustrates this. If the Reasoner had originally believed 'if Alice is at the party Bill will be too, and if she is not there, Bill will be there anyway', he would not substitute 'if Alice were at the party, Bill would be' upon learning that Bill was not at the party even though he might still regard the probabilistic 'propensity' for Bill to attend parties at which Alice is present as high. Similarly, the intuitive feeling that 'A and not B' is inconsistent with the counterfactual 'if A were the case then B would be' shows that we do not always maintain the counterfactual where the corresponding propensity probability is high. Thus, if the Reasoner initially believing 'if Alice is at the party, Bill will be too' had then learned 'Alice is at the party and Bill is not', he would not have substituted 'if Alice were at the party, Bill would be'.[16]

The foregoing shows that the prior (i.e., propensity) representation of counterfactuals' probabilities cannot be universally valid, though the use of the subjunctive in typical descriptions of prior conditional probabilities (e.g., 'the probability that a black ball would have been drawn if Urn A had been selected') seems to me to be strong evidence for the limited validity of this representation. Rather than seeking a more general representation, I am going to argue in conclusion that in view of important ambiguities in the counterfactual, it may be more fruitful for the purpose of analyzing reasoning involving counterfactuals if the implicit probabilities are made explicit and their relations studied directly, than to attempt to construct a univocal 'logic of the counterfactual' in terms of which to assess the reasoning.

Imagine our Reasoner this time thinking about the question as to whether a friend of his, Jones, was or was not present at a particular party which the Reasoner did not himself attend. The Reasoner begins with two items of *a priori* information: (1) there is another man, Brown, who can't bear Jones at parties, and who avoids parties at which Jones is present; (2) when either Brown or Jones is not partying, he frequents a particular bar. With this background, the Reasoner now learns that Jones didn't see Brown on the evening of the party, and he takes this as a sign that Jones was at the party, reasoning as follows:

> If Jones had not been at the party there would have been a fairly good chance of his seeing Brown at the bar they both frequent, while if he had been at the party he would not have seen Brown because Brown doesn't attend parties at which

Jones is present. Therefore Jones would have been more likely to do what he actually did – not see Brown – if he had gone to the party than if he had not. Therefore Jones probably attended the party.

Analyzed in terms of the probabilities which are implicit, the foregoing reasoning is sound, but the point to be stressed here is that the Reasoner affirmed the counterfactual 'if Jones had been at the party he would not have seen Brown', because of the fact that if Jones had gone to the party Brown would not have (because he avoids parties which Jones attends). Suppose, however, that the Reasoner now learns that in fact Brown was at the party. He then concludes that Jones wasn't at the party, reasoning:

Brown was at the party. So, if Jones had been at the party he would have seen Brown. But Jones didn't see Brown on the evening of the party. Hence, Jones was not at the party.

Once again, probabilistic analysis shows the reasoning to be sound, but the point to be noted is that in this reasoning the Reasoner affirmed the contradictory of what he had previously maintained, namely 'if Jones had been at the party he would have seen Brown.' We would like to know why the Reasoner did not continue to assert that if Jones had been at the party he would not have seen Brown because Brown would not have been there.

I would suggest that it would be wrong to conclude that one of the contradictory' counterfactuals in the foregoing example, 'if Jones had been at the party he would have seen Brown' and 'if Jones had been at the party he would not have seen Brown', would have been right and the other wrong. In fact, each can be connected to an appropriate propensity probability which is plausibly implicit in the reasoning: the chances of Jones's seeing Brown at parties in general, and the chances of Jones's seeing Brown at parties at which Brown is present. The bare use of the subjunctive does not in this case determine unambiguously which propensity is meant. If this sort of ambiguity is characteristic of the subjunctive mood in general (as it is of the modal forms), this suggests that disambiguation, in which propensity probabilities may be expected to play an important role, is what is required if an adequate theory of reasoning involving the counterfactual is to be developed.

University of California

NOTES

* The present research was partially supported by National Science Foundation Grant GS-30564.

[1] For example, see Ayers [4].

[2] See my papers [1] and [2]. Brian Ellis has developed a similar approach in [6].

[3] The question as to whether truth-values apply to simple conditionals (ones not containing other conditionals) is closely bound up with that as to the applicability of probabilities to compounds containing conditionals. Van Fraassen [9] has shown how to extend 'simple' conditional probabilities to compounds containing conditionals in such a way as to satisfy the Kolmogorov axioms, but David Lewis [8] has shown that this cannot be done if the rule that probabilities change by conditionalization is also to be satisfied.

[4] The *correctness* of the conditional probability measure of conditionals' probabilities, though perhaps intuitively plausible, can itself be debated. David Lewis [8], for example, argues that conditional probabilities measure only the propriety of *asserting* a conditional, but not the rationality of *accepting* it. In my manuscript [3], I argue on pragmatic grounds that conditional probabilities provide a good measure for the *desirability* of holding and acting on certain beliefs expressible as conditionals.

[5] This vague 'probabilistic test of rationality' obviously needs precization, and a particular, if rather arbitrary, precise version is given in [1]. The details are unimportant here.

[6] This working hypothesis is perhaps more doubtful in the counterfactual case that in that of the indicative. The issue is closely bound up with that of the interdefinability of counterfactuals and dispositionals (in certain cases at least), and the appropriate applicability of truth-values to dispositional propositions. It must be conceded that we find no difficulty in applying truth-conditional logic and non-conditional probabilities to dispositional propositions, which suggests that there are many counterfactuals which can be treated 'as though' truth-values were applicable to them.

[7] The use of the expression 'prior conditional probability' to designate the probabilities $p_0(W/A)$ and $p_0(W/N)$ in Equation (1) is admittedly not standard. This is because these probabilities are usually presented as 'givens', whose own interpretation is not discussed in theoretical statistics. If probabilities are to change by Bayesian inference (conditionalization), however, these probabilities must equal ratios of prior non-conditional probabilities. The issue as to the proper interpretation of these probabilities is, however, further complicated by the possibility of regarding them as *propensities*. This will arise in Section VII.

[8] The distinction between the non-valuational or 'rational expectation' sense of 'ought' and other, 'ethical' senses, is made by Robinson [7].

[9] A doubtful special case is that in which the common antecedent of the 'contradictory' conditionals has probability 0. In this case conditional probabilities are undefined, and one does not know what to say about the logical relation between 'if A then B' and 'if A then not B'.

[10] In real life such a conclusion would be more apt to be expressed modally: 'Therefore, Alice must not be at the party'. Interconnections with such 'practical modalities' as the foregoing will not be discussed here.

[11] This assumes the standard idealization: namely that 'new information' (here that Bill is not at the party) becomes certain *a posteriori*. That this idealization is often a poor one in practice means that there will sometimes be cases in which 'not B' is learned but the indicative conditional 'if A is the case then B is' is still probable.

[12] Howard Levine and I are presently working on the problem of making clear just when

it *is* rational to infer conclusions from premises which truth-conditionally imply them, when there are very many premises, and each individual one has a small but non-zero uncertainty. Interesting mathematical problems when such inferences are looked at from the point of view of the probabilities associated with them, concerning which we hope to publish some results shortly.

[13] The result of David Lewis [8] previously cited shows that probability change resulting from receipt of information in conditional form cannot be easily described within an extension of current theories of probability change, since this would require defining the probability of a conditional with a conditional antecedent, which conformed to the law of probability change by conditionalization.

[14] Recent work of W. A. Harper aims at avoiding this idealization.

[15] That counterfactual's probabilities might be identified in certain cases with the probabilities of dispositional propositions calls into question our working hypothesis that truth values don't apply to counterfactuals. At any rate, we have no hesitation in analyzing dispositionals as though they were truth-conditional in investigating their deductive interrelations and we also attach non-conditional probabilities to them. The theoretical issues involved here have some connections with the so called problem of the *realism* of dispositional statements, as discussed by Dummett [5].

[16] He might, however, use something like the non-conditional counterfactual, and say 'If Alice is here then Bill ought to be'.

BIBLIOGRAPHY

[1] Adams, E. W., 'The Logic of Conditionals', *Inquiry* **8** (1965) 166–197.
[2] Adams, E. W., 'Probability and the Logic of Conditionals', in P. Suppes and J. Hintikka (eds.), *Aspects of Inductive Logic*, North-Holland, 1968, pp. 265–316.
[3] Adams, E. W., *The Logic of Conditionals: An Application of Probability to Deductive Logic*, D. Reidel, Dordrecht, 1975.
[4] Ayers, M. R., 'Counterfactuals and Subjunctive Conditionals', *Mind* (1965) 347–364.
[5] Dummett, M., 'Truth', reprinted in P. Strawson (ed.), *Philosophical Logic*, Oxford, 1968.
[6] Ellis, B., 'The Logic of Subjective Probability', *Brit. J. Phil. Sci.* **24** (1973) 125–152.
[7] Robinson, R., 'Ought and Ought Not', *Philosophy* **46** (1971).
[8] Lewis, D., 'Probabilities of Conditionals and Conditional Probabilities', dittoed, Princeton, 1973.
[9] Van Fraassen, B., 'Probabilities of Conditionals', this volume, p. 261.

P. D. FINCH

INCOMPLETE DESCRIPTIONS IN THE LANGUAGE
OF PROBABILITY THEORY

1. THE SIMPLE PROBABILITY MODEL

Let $\Omega = \{\omega_1, \omega_2, \ldots, \omega_k\}$ be a finite set of outcomes which occur with respective probabilities $P(\omega_j), j = 1, 2, \ldots, k$, summing to unity. For our purposes it is convenient to think of P as a function $P : \Omega \to \mathbb{R}^+$, which maps Ω into the non-negative half-line. We call the pair (Ω, P) a simple probability model with outcome set Ω and probability function P, it being understood that P does sum to unity over Ω.

For each positive integer n, we write

$$\Omega_n = \Omega \times \Omega \times \cdots \times \Omega \quad (n \text{ factors})$$

for the n-fold Cartesian product of Ω, in particular $\Omega_l = \Omega$. The set Ω_n is the set of outcomes for a situation which involves the realisation of n successive outcomes from Ω provided there are no restrictions, such as, for example, a prohibition on repetitions, which prohibit the realisation of some elements of Ω_n. If $P' : \Omega_{(n)} \to \mathbb{R}^+$ is a probability function then $(\Omega_{(n)}, P')$ is another simple probability model.

The particular probability function $P_{(n)}$ defined by

$$(1.1) \qquad P_n(\omega_1, \omega_2, \ldots, \omega_n) = P(\omega_1) \, P(\omega_2) \ldots P(\omega_n)$$

plays an important role in the study of those situations where the successive outcomes in $(\omega_1, \omega_2, \ldots, \omega_n)$ are realised independently of one another. In fact Equation (1.1) may be taken as the formal definition of independence. We refer to (Ω_n, P_n) as the n-fold product of (Ω, P).

2. MEASUREMENT AND RANDOM VARIABLES

Let (Ω, P) be a simple probability model. There are occasions where one does not observe outcomes directly but can measure only something about those outcomes. In such cases the measurement procedure in question can often be characterised as a pair $(\mathscr{X}, X)$ where $\mathscr{X}$ is a non-

Harper and Hooker (eds.), Foundations of Probability Theory, Statistical Inference, and Statistical Theories of Science, Vol. I, 23–28. All Rights Reserved. Copyright © 1976 by D. Reidel Publishing Company, Dordrecht-Holland.

empty set and $X:\Omega\to\mathscr{X}$ is a function which maps Ω into $\mathscr{X}$. The function X is often called an $\mathscr{X}$-valued random variable on Ω but we will say that it is a measurement function on Ω.

For any 'subset' S of Ω write

$$P(S)=\sum_{\omega\in S} P(\omega)$$

with the convention that when $S=\square$, the empty set, one has $P(\square)=0$. The set

$$\{(x, P(X^{-1}(x))): \quad x\in\mathscr{X}\}$$

will be called the $(\mathscr{X}, X)$-description of (Ω, P). It lists all the possible results of measurement, under the measurement procedure $(\mathscr{X}, X)$, together with the corresponding probabilities with which they occur.

Consider now the product model (Ω_n, P_n). We define the product measurement procedure $(\mathscr{X}_n, X_n)$ for Ω_n, by taking $\mathscr{X}_n$ to be the n-fold Cartesian product of $\mathscr{X}$ with itself and defining $X_n:\Omega_n\to\mathscr{X}_n$ by

$$X(\omega_1, \omega_2, ..., \omega_n)=(X(\omega_1), X(\omega_2), ..., X(\omega_n)).$$

Then the $(\mathscr{X}_n, X_n)$-description of (Ω_n, P_n) is just

$$\{(y, P_n(x_n^{-1}(y))): \quad y\in\mathscr{X}_n\}.$$

Moreover it follows from (1.1) that

$$y=(x_1, x_2, ..., x_n)\Rightarrow P_n(X_n^{-1}(y))=\prod_{j=1}^{n} P(X^{-1}(x_j))$$

so that, for each positive integer n, the $(\mathscr{X}_n, X_n)$-description of (Ω_n, P_n) can be obtained from the $(\mathscr{X}, X)$-description of (Ω, P).

Sometimes one wants to consider the product models for general n and a convenient way to do this is to consider all the n-fold products simultaneously. Thus write

$$\Omega_*= \bigcup_{n=1}^{\infty} \Omega_n, \qquad \mathscr{X}_*= \bigcup_{n=1}^{\infty} \mathscr{X}_n$$

define $P_*:\Omega_*\to\mathbb{R}^+$ and $X_*:\Omega_*\to\mathscr{X}_*$ by the requirements

$$P_* \mid \Omega_n=P_n$$

and

$$X_* \mid \Omega_n=X_n.$$

One can then consider the pair (Ω_*, P_*) but it should be noted that although $P_*: \Omega_* \to \mathbb{R}^+$ we do not have a simple probability model because P_* does *not* sum to unity over Ω_*, in fact P_* sums to unity over each of the subsets Ω_n of Ω_*. To give a name to (Ω_*, P_*) we will call it the complete sequential linkage of (Ω, P).

In an obvious sense, the pair $(\mathscr{X}_*, X_*)$ is a measurement procedure for (Ω_*, P_*) and we call the set

$$\{(x_*, P_*(X_*^{-1}(x_*))): \quad x_* \in \mathscr{X}_*\}$$

the $(\mathscr{X}_*, X_*)$-description of (Ω_*, P_*). Because of (2.1) one has

$$(2.2) \qquad P_*(X_*^{-1}(x_*)) = \prod_{j=1}^{n(x_*)} P(X^{-1}(x_j))$$

where $n(x_*)$ is the n for which $x_* = (x_1, x_2, ..., x_n)$ belongs to X_n.

In some investigations interest is centred on the $(\mathscr{X}_*, X_*)$-description because one wants to know the distribution of measurement results for an arbitrary n-fold product of a simple probability model. For instance one might be interested in the behaviour of the $(\mathscr{X}_n, X_n)$-description for large n. Equation (2.2) asserts that the description in question can be calculated from the $(\mathscr{X}, X)$-description of the underlying simple probability model and reflects the independence of the $\mathscr{X}$-values in the realisations in question.

3. SUMMARY FUNCTIONS AND SURROGATES

Suppose now that we are not in the fortunate position of being able to obtain by measurement all the individual $\mathscr{X}$-values arising from a realisation of a product of the underlying simple probability model, or that, for some other reason, we do not use all of those values but only some 'summary' of them.

In other words suppose that there is a non-empty set M and a function $\xi: \mathscr{X}_* \to M$ with domain $\mathscr{X}_*$ such that for $x_* = (x_1, x_2, ..., x_n)$ in $\mathscr{X}_*$ we have available only the corresponding point $\xi(x_*)$ in M. Thus if $(\omega_1, \omega_2, ..., \omega_n)$ is realised then the only 'datum' available is $\xi(X(\omega_1), X(\omega_2), ..., X(\omega_n))$.

In such a situation our motivating interest may still be in the $(\mathscr{X}_*, X_*)$-description of (Ω_*, P_*) and we could try to 'estimate' it in the light of the available datum by constructing a surrogate for it. We take a surrogate of the form

(3.1) $\{(x_*, Q(x_*)): \; x_* \in \mathscr{X}_*\}$

where $Q: \mathscr{X}_* \to \mathbb{R}^+$ is some function with domain $\mathscr{X}_*$ having the property that

(3.2) $\sum_{x_* \in \mathscr{X}_n} Q(x_*) = 1.$

In other words the quantity $Q(x_*)$ is a surrogate for the unknown quantity $P_*(X_*^{-1}(x_*))$ appearing in (2.2) and Q is the surrogate function for $P_* \circ X_*^{-1}$.

What sort of properties would it be useful for the surrogate function to have. Well, in the first place, it would seem plausible to retain independence. To achieve this our surrogate function Q should have the multiplicative property of (2.2), viz

(3.3) $Q(x_*) = \prod_{j=1}^{n(x_*)} Q(x_j)$

when $n(x_*)$ is the n for which $x_* = (x_1, x_2, \ldots, x_n)$ belongs to $\mathscr{X}_n$. Equations (3.3) may be looked upon as a definition of 'surrogate independence'.

On the other hand our assumption is that all we know about the observation x_* is $\xi(x_*)$ in L, thus if x_* and x'_* are two realisations for which $\xi(x_*) = \xi(x'_*)$ we have no 'datum' to distinguish between them. It seems plausible therefore to require that

(3.4) $\forall x_*, x'_* \in \mathscr{X}_*: \; \xi(x_*) = \xi(x'_*) \Rightarrow Q(x_*) = Q(x'_*).$

But if this convention is adopted we must have

(3.5) $x_* \in \operatorname{codom} X_*: \; Q(x_*) = F(\xi(x_*))$

where $F: M \to \mathbb{R}^+$ is some function which has a domain which contains the codomain of ξ.

But now substitution from (3.5) into (3.1) gives $x_* = (x_1, x_2, \ldots, x_n) \in \operatorname{codom} X_* \Rightarrow F(\xi(x_*)) = \prod_{j=1}^{n} F(\xi(x_j))$ in other words

(3.6) $F(\xi_n x_1 x_2 \ldots x_n) = F(\xi_1 x_1) F(\xi_2 x_2) \ldots F(\xi_n x_n)$

for any $(x_1, x_2, ..., x_n)$ in $\mathscr{X}_*$ and we write

$$\xi_n x_1 x_2 ... x_n \quad \text{instead of} \quad \xi(x_1, x_2, ..., x_n).$$

Let $\chi : \text{codom}\, \xi \to \mathbb{R}^+$ be any function satisfying

$$(3.7) \qquad \chi(\xi_n x_1 x_2 ... x_n) = \chi(\xi_1 x_1)\, \chi(\xi_1 x_2) ... (\xi_1 x_n)$$

the χ is said to be a non-negative character of the summary function ξ. Thus (3.6) asserts that F is a character of ξ. Thus Equation (3.5) determines Q on the codomain of X_* in terms of ξ and a character of ξ. For an x_* which does not belong to the domain of X_* it is plausible to take the corresponding value of the surrogate function to be given by

$$Q(x_*) = P_*(X_*^{-1}(x_*)) = 0.$$

Moreover if we do so we satisfy the multiplicative law (3.3) since the codomain of X_* has the property

$$(x_1, x_2, ..., x_n) \in \text{codom}\, X_* \Leftrightarrow j = 1, 2, ..., n;$$

$$x_j \in \text{codom}\, X.$$

Thus we obtain

$$(3.8) \qquad Q(x_*) = \begin{cases} F(\xi_n x_1 x_2 ... x_n), & x_* \in \text{codom}\, X_* \\ 0, & x_* \notin \text{codom}\, X_* \end{cases}$$

where $x_* = (x_1, x_2, ..., x_n)$. However to meet the requirement (3.1) we have to 'normalise' the character in question. Finally, then, we get

$$(3.9) \qquad Q(x_*) = \begin{cases} \left[\displaystyle\sum_{x \in \text{codom}\, X} \chi(\xi_1(x))\right]^{-1} \chi(\xi_n x_1 x_2 ... x_n), \\ \qquad x_* \in \text{codom}\, X_* \\ 0, \text{ otherwise} \end{cases}$$

where χ is a non-negative character of ξ.

Remark. Note that

$$\sum_{x \in \text{codom}\, X} \chi(\xi_1(x)) < \infty$$

since our assumption is that Ω is finite.

4. CONCLUDING REMARKS

At this point one can develop the ideas of Finch [1] and [2] within the framework of probability theory. Perhaps the most important point in the preceding discussion is the use of the complete linkage (Ω_*, P_*) rather than the construction of an infinite product probability space.

In my opinion Kolmogorov's famous treatment of the foundations of formal probability theory inadvertently misdirected future developments by its emphasis on probability spaces as triples consisting of a non-empty set, a σ-field of its subsets and a normed measure. In subsequent mathematical developments finite dimensional distributions were recovered within infinite dimensional spaces in terms of cylinder set and so on. But from a practical point of view one needs an algebraic structure which emphasises the role of finite segments in their own right and not just as finite parts of the infinite.

In other words one wants to deal with what we have called Ω_* rather than infinite Cartesian products such as

$$\Omega^* = \Omega \times \Omega \times \cdots \times \Omega \times \cdots.$$

Monash University, Melbourne

BIBLIOGRAPHY

[1] Finch, P. D., *The Poverty of Statisticism.*
[2] Finch, P. D., *The Macroscopic Description of Microscopic Systems.*

TERRENCE L. FINE*

A COMPUTATIONAL COMPLEXITY VIEWPOINT ON THE STABILITY OF RELATIVE FREQUENCY AND ON STOCHASTIC INDEPENDENCE**

I. INTRODUCTION

We undertake to discuss the question of the apparent stability of relative frequency (RF) and the concept of stochastic independence (SI) from the vantage point of the relatively new notion of computational complexity (CC) pioneered by Chaitin [1], Kolmogorov [2], Loveland [3], Martin-Löf [4] and Solomonoff [5] and as discussed in Fine [6]. This discussion is primarily a condensation and emendation of Fine [7, 8]. With regard to the observation, underlying the RF interpretation of probability, that many chaotic empirical phenomena yet exhibit stable RF, we will argue that this arises as a consequence of our methodological practice of first removing noticeable regularities (e.g., drifts, cycles, etc.) from the data to which we apply the RF argument. With regard to the usual definition of SI [events A, B, are SI iff $P(A \cap B) = P(A) P(B)$] we will argue that a more stringent definition is possible and that SI should no more be a probabilistic concept than is set disjointness. The vehicle for both these arguments will be a CC notion, especially as developed by Kolmogorov and Martin-Löf.

II. A NOTION OF COMPUTATIONAL COMPLEXITY

A. *Definition of Computational Complexity*

The definition of CC that we adopt is relative to an algorithm or partial recursive function ϕ and is intended to measure the difficulty of storage of a finite binary sequence $S \in \{0, 1\}^*$ or the difficulty of transmitting S and not, say, the running time or memory required to compute S. Let $|S|$ denote the length of S.

DEFINITION 1. The conditional complexity $K_\phi(S \mid I)$ of $S \in \{0, 1\}^*$ given data $I \in \{0, 1\}^*$ and a partial recursive function ϕ of two variables

Harper and Hooker (eds.), Foundations of Probability Theory, Statistical Inference, and Statistical Theories of Science, Vol. I, 29–40. All Rights Reserved. Copyright © 1976 by D. Reidel Publishing Company, Dordrecht-Holland.

is given by

$$K_\phi(S \mid I) = \begin{cases} \min\{|P| : \phi(P, I) = S\} \\ \infty \quad \text{if} \quad (\nexists P) \; \phi(P, I) = S. \end{cases}$$

We distinguish a subclass Ψ of optimal partial recursive functions corresponding to a subclass of the universal Turing machines having the additional property that:

$$\exists f : \{0, 1\}^* \times \{0, 1\}^* \rightarrow \{0, 1\}^*, \quad f \text{ recursive},$$
$$(\forall X) \, [\exists c(X)] \, (\forall Y) \, |f(X, Y)| < c(X) + |Y|,$$
$$(\forall \psi \in \Psi) \, (\forall \phi \text{ partial recursive}) \, (\exists X) \, (\forall P, I) \, \psi[f(X, P), I] =$$
$$= \phi(P, I),$$

where defined. An example of a suitable encoding function f is

$$f(X, Y) = 1_X 0 Y;$$

i.e., the concatenation of x 1's with a 0 followed by the sequence Y. The unsubscripted $K(S \mid I)$ will denote a conditional complexity measure relative to a (there are infinitely many) $\psi \in \Psi$.

B. *Elementary Properties of K*

The proof for the following theorem is easy and is available in Fine [6].

THEOREM 1.

(a) $(\forall \phi \text{ partial recursive}) \, (\exists c) \, (\forall S, I) \, K(S \mid I) < c + K_\phi(S \mid I).$

(b) $(\forall \psi, \psi' \in \Psi) \, (\exists c) \, (\forall S, I) \, |K_\psi(S \mid I) - K_{\psi'}(S \mid I)| < c.$

(c)   If I describes a subset of $\{0, 1\}^*$, then
        $(\exists c) \, (\forall S, I) \, K(S \mid S \in I) < c + \log_2 \|I\|.$

(d)   At most, the fraction 2^{-t} of all sequences in the set I satisfy
        $K(S \mid I) < \log_2 \|I\| - t.$

Property (a) partially justifies the choice of Ψ by claiming that a sequence of high conditional complexity relative to $\psi \in \Psi$ tends to be of high conditional complexity relative to any other partial recursive function or algorithm. Property (b) suggests that it is immaterial which $\psi \in \Psi$ we choose. Property (c) provides a useful upper bound to complexity. Property (d) informs us that most sequences are maximally complex.

III. THE APPARENT STABILITY OF RELATIVE FREQUENCY

A. *Introduction*

It is widely maintained (e.g., see Cramer [9], Fisz [10], Gnedenko [11], Parzen [12] that many empirical time series (a paradigm being the record of successive tosses of a coin) are surprisingly found to be unpredictable and yet to exhibit a form of regularity through the apparent stability of the sequence of RFs. We wish to suggest that the very unpredictability of the series forces the apparent stability rather than being in conflict with it. A brief examination of our data processing practices then leads us to speculate that the prevalence of such unpredictable yet stable series is an artifact of our practices and does not reflect upon 'nature'.

B. *Definition of Apparent Stability of RF*

For simplicity, we consider only 'random' experiments with $\{0, 1\}$-valued outcomes. Let $\mathscr{E}_1, \ldots, \mathscr{E}_n$ be repetitions of an experiment $\mathscr{E}$ with outcomes $x_1, \ldots, x_n$, $x_i \in \{0, 1\}$. The RF r_j of the outcome 1 on the basis of j experiments is given by

$$r_j = \frac{1}{j} \sum_{i=1}^{j} x_i.$$

A measure of the stability of $\{r_j\}$, or its apparent convergence, is given by:

DEFINITION 2. $\{r_j\}$ converges (m, ε) if

$$\max_{m \leqslant j \leqslant n} |r_j - r_n| < \varepsilon.$$

(Equivalently, we could use the Cauchy criterion

$$\max_{m \leqslant j, k \leqslant n} |r_j - r_k| < \varepsilon).$$

The smaller m and ε are, the more stringent is the test for stability. Definition 2 seems to reasonably reflect what is meant by the intuitive judgement that the sequence of RFs generated from an empirical time series $x_1, \ldots, x_n$ is stable.

C. *A Measure of Irregularity*

To measure the degree of irregularity of a series, we accept

DEFINITION 3. A sequence S is random (ψ, t) conditioned upon information that S is in a set I if

$$K_\psi(S \mid I) > \log_2 \|I\| - t.$$

From Theorem 1, we know that no sequence S has complexity greater than $c + \log_2 \|I\|$ and that 'most' sequences are random (ψ, t). The smaller t, the more stringent is the test for randomness or irregularity. A justification for the use of the term 'random' was given by Martin-Löf [4] when he showed that sequences satisfying Definition 3 also pass statistical tests for randomness. Martin-Löf's method will also be invoked in connection with our discussion of SSI in Section IV, D and E.

Another line of justification for Definition 3 would be to relate irregularity to difficulty of prediction. This question has been examined from the point of view of sequential gambling schemes by Schnorr [13] and by Cover [14]. For the purposes of a defense of Definition 3 we might note that if we have any scheme f to sequentially predict the next term in the sequence S,

$$S = x_1, \ldots, x_n, \quad f(x_1, \ldots, x_i) = \hat{x}_{i+1},$$

$$\hat{S} = \hat{x}_1, \ldots, \hat{x}_n, \quad d(S, \hat{S}) = \sum_{i=1}^{n} [1 - \delta_{x_i, \hat{x}_i}], \quad \delta_{x,y} = \begin{cases} 1 & \text{if } x = y \\ 0 & \text{if } x \neq y \end{cases},$$

then the number of errors $d(S, \hat{S})$ made by the scheme f is lower bounded in

THEOREM 2. $(\exists c) \, (\forall f, S) \, d(S, \hat{S}) \geqslant \dfrac{K(S) - K(f) - c}{1 + \log_2 |S|}$.

  Proof. One way to compute S would be to generate it sequentially by use of f and to correct $f(x_1, \ldots, x_i)$ whenever $x_{i+1} \neq f(x_1, \ldots, x_i)$. If k errors are made, then there is a program for S of length approximately

$$c + K(f) + k(\log|S| + 1),$$

where c is the length of instruction for the algorithm using f to generate S with the k corrections. However, by definition

$$K(S) \leqslant c + K(f) + k(\log|S| + 1).$$

Hence,

$$d(S, \hat{S}) = k \geqslant \frac{K(S) - K(f) - c}{\log|S| + 1}.$$

Thus, as we would expect, a simple $[K(f)$ small$]$ scheme f must make many errors in attempting to predict a highly $[K(S)$ large$]$ irregular sequence S.

D. *A Relation between Apparent Stability of RF and Irregularity*

That high irregularity (small t) insures apparent stability (small m, ε) is the content of

THEOREM 3. Let

$$I_{n,w} = \left\{ S : |S| = n, \sum_{x_i \in S} x_i = w \right\}.$$
$$(\exists c)\,(\forall n, w, m, \varepsilon)\; K(S \mid S \in I_{n,w})$$
$$> c + \log_2 \binom{n}{w} - \log(m\varepsilon^4) \Rightarrow S \text{ convergent } (m, \varepsilon).$$

Proof. See Fine [7].

Maximally chaotic sequences must have apparently convergent RFs. Conversely, an unstable RF sequence can only arise from a sequence with some degree of regularity.

E. *The Prevalence of Stable RFs and Data Processing Practices*

The argument of Section III, D is only of interest if we have reason to believe that many empirical time series are of maximal conditional complexity. We would argue that this prevalence is a consequence of our data processing practices. To oversimplify, when a scientist or engineer encounters an empirical time series, he first studies it to detect hidden regularities such as trends, cycles, or physical laws. The series is often viewed as a sequence of noisy observations on an underlying lawlike or deterministic process. It is only when the search for such laws has failed or he has the residual left after the subtraction of a regular part that he undertakes a probabilistic model of the data. Although it is not always thought of this way, an unstable sequence of RFs indicates the possibility of additional extractable regularities in the data. Thus, it should not be surprising that, insofar as the search for a lawlike part has been thorough, on those sequences to which we apply probabilistic methods, the sequence of RFs is indeed stable.

Of course, it is not usual for the search for laws to be as thorough as is required to verify that the time series is of maximal conditional complexity. As Chaitin [15] has argued, it is very difficult to verify high complexity statements, a result that is expected in view of the ineffective nature of complexity calculations (see also Fine [6], p. 129).

IV. EMPIRICAL INDEPENDENCE AND STOCHASTIC INDEPENDENCE

A. *Introduction*

We will propose a data-based concept of independence using the notion of complexity developed in Section II. This concept, called empirical independence (EI), will then be compared to the usual probabilistic definition of stochastic independence (SI) with regard to the formal properties of the binary relation of independence. The concept of SI will then be converted *via* the notion of statistical goodness-of-fit tests into a data-based concept called statistical stochastic independence (SSI). It will then be seen that EI is a more restrictive requirement than is SSI. We conclude by claiming EI to be a better formalization of our intuitive notion of independent events and one that strongly suggests that independence is not adequately definable in terms of probability.

B. *Definition of EI*

To define an empirically oriented concept of independent events to replace SI, we assume that there is a repeatable experiment $\mathscr{E}$ having among its outcomes events A, B. We recorded the occurrences of events A, B in repetitions $\mathscr{E}_1, ..., \mathscr{E}_n$ of $\mathscr{E}$ as follows:

$$O_A = x_1, ..., x_n, \qquad O_B = y_1, ..., y_n,$$

$$x_i = \begin{cases} 1 & \text{if } A \text{ occurred in } \mathscr{E}_i \\ 0 & \text{if o.w.} \end{cases}, \qquad y_i = \begin{cases} 1 & \text{if } B \text{ occurred in } \mathscr{E}_i \\ 0 & \text{if o.w.} \end{cases}$$

We interpret the empirical independence (EI) of A, B, written $A(\text{EI})B$, as meaning that knowledge of O_A is 'uninformative' about O_B and v.v. This suggests requirements of the form $A(\text{EI})B$ iff

$$\max\{|K(O_A \mid O_B, I_{n, w_A}) - K(O_A \mid I_{n, w_A})|, |K(O_B \mid O_A, I_{n, w_B}) - K(O_B \mid I_{n, w_B})|\} < t,$$

or

$$|K(O_A O_B \mid I_{n, w_A}, I_{n, w_B}) - K(O_A \mid I_{n, w_A}) - K(O_B \mid I_{n, w_B})| < t.$$

We do not adopt these criteria for EI for we are concerned that the repeated experiments $\mathscr{E}_1, \ldots, \mathscr{E}_n$ also be unlinked. Dependence between the outcomes of $\mathscr{E}_1, \ldots, \mathscr{E}_n$ could distort our view of the relation between the occurrence of A, B as outcomes of the single experiment $\mathscr{E}$. To avoid this problem, we accept

DEFINITION 4. $A(\text{EI}) B$ if for preassigned threshold t

$$K(\langle O_A, O_B \rangle \mid I_{n, w_A}, I_{n, w_B}) > \log \binom{n}{w_A} \binom{n}{w_B} - t,$$

where $\langle O_A, O_B \rangle$ denotes the unordered pair of occurrence records, i.e., $\langle O_A, O_B \rangle$ is generated by a program P if it yields either $O_A O_B$ or $O_B O_A$.

C. *Elementary Properties of EI*

The following properties of the binary relation EI were previously established in Fine [8].

THEOREM 4.
- (a) $A(\text{EI}) B$ implies $B(\text{EI}) A$.
- (b)   There is a $c > t$ such that all but a fraction $2^{-(c-t)}$ of occurrence records O_A imply $A(\text{EI}) \Omega$ (Ω the certain event).
- (c)   All but at most a fraction 2^{-t} of pairs of occurrence records (O_A, O_B) imply $A(\text{EI}) B$.
- (d)   There is a c such that all but at most a fraction 2^{-c} of those pairs (O_A, O_B) yielding $A(\text{EI}) B$ do not also yield $A(\text{EI}) \bar{B}$ ($\bar{B}$ is the complement of B).
- (e)   There is a ε such that $A \supset B$, $A(\text{EI}) B$ implies $w_A > n - \varepsilon \sqrt{n}$ or $w_B < \varepsilon \sqrt{n}$, where

$$n = |O_A| = |O_B|, \qquad w_A = \sum_{x_i \in O_A} x_i, \qquad w_B = \sum_{y_i \in O_B} y_i.$$

Property (a) is also a property of SI. For SI, we have that $(\forall A) A(\text{EI}) \Omega$. However, for EI, (b) indicates the possibility of some exceptions to this property of SI. Property (c) has no direct counterpart for SI. For SI, $A(\text{SI}) B \Rightarrow A(\text{SI}) \bar{B}$. This property could have been built into the definition of EI. However, with Definition 4 we only have the slightly weaker property (d). The counterpart of (e) for SI is that $A \supset B$, $A(\text{SI}) B \Rightarrow$

$\Rightarrow P(A)=1$ or $P(B)=0$. If one adopts a limit of RF viewpoint, then the discrepancies of $0(\sqrt{n})$ are asymptotically negligible.

A more significant discrepancy between the formal properties of SI and EI is that

$$(*)\qquad B\cap C=\emptyset,\qquad A\,(\mathrm{SI})\,B,\qquad A\,(\mathrm{SI})\,C\Rightarrow A\,(\mathrm{SI})\,B\cup C,$$

does not generally hold for EI. That this is to the credit of EI follows from an undesirable consequence of $(*)$. If $(*)$, then there exists, say, O_{A^*}, O_{B^*} with $A^*\subset B^*$, $A^*(\mathrm{EI})\,B^*$, $w_{A^*}=\frac{3}{4}n$, $w_{B^*}=\frac{1}{4}n$. To verify this, note that by Theorem 4c most pairs (O_A, O_B) with specified weight w_A, w_B are judged EI. Hence, at least $\frac{1}{2}$ of all pairs (O_A, O_B) with $w_A=1$, $w_B=\frac{3}{4}n$ must be independent. Choose B^* for O_{B^*} with $w_{B^*}=\frac{3}{4}n$ and O_{B^*} independent of at least $\frac{1}{4}$ of the sequences O_{A_i} satisfying

$$w_{A_i}=1,\qquad A_i\subset B^*,\qquad A_i(\mathrm{EI})\,B^*.$$

Let

$$A^*=\bigcup_{i=1}^{n/4} A_i.$$

By the hypothesis of $(*)$, $A^*(\mathrm{EI})\,B^*$, although

$$w_{A^*}=\frac{n}{4},\qquad A^*\subset B^*,\qquad w_{B^*}=\frac{3}{4}n,$$

contradicts Theorem 4e.

D. *Statistical Stochastic Independence*

To further explore the relation between EI and SI, we need an empirical or statistical version of SI. The concept of SI is probabilistic and can best be referred to occurrence data through a statistical goodness-of-fit test for the probabilistic hypotheis of SI.

1. *The probabilistic hypothesis of SI.* We formulate the general concept of SI more specifically in connection with a probabilistic model for the occurrence records O_A, O_B of independent events A, B. We do not assume that the repeated experiments $\mathscr{E}_1,\ldots,\mathscr{E}_n$ are necessarily independent, but do assume that they are at least exchangeable. To be precise, let μ be a probability measure on the unit square $[0,1]^2$, $\{\lambda_i\}$ a discrete probability mass function, w_i^A, w_i^B integers in $[0,n]$, $n=|O_A|=|O_B|$,

$O_A = x_1, ..., x_n, O_B = y_1, ..., y_n$. We assume that

$$(**) \quad P(O_A, O_B) = \lambda_0 \int_0^1 \int \mu(dp_A \, dp_B) \times$$

$$\times \left\{ p_A \sum^n x_{i_{(1-p_A)}} n - \sum^n x_i p_B \sum^n y_{i_{(1-p_B)}} n - \sum^n y_i \right\} +$$

$$+ \sum_i \lambda_i \left[\binom{n}{w_i^A} \binom{n}{w_i^B} \right]^{-1} \delta_{\sum\limits_{j=1}^n x_j, w_i^A} \delta_{\sum\limits_{j=1}^n y_j, w_i^B} .$$

The form assumed for P is that derived by De Finetti and Hewitt-Savage [16] for the general exchangeable or symmetric measure. The special case of independent repetitions corresponds to $\lambda_0 = 1$, μ degenerate with full support at a single point.

2. *Goodness-of-fit tests.* A goodness-of-fit test $\mathcal{T} = \{T_\delta\}$ for a probabilistic hypothesis (collection of probability measures) $\{P_\theta, \theta \in \Theta\}$ is a collection of rejection regions $T_\delta \subset \Omega$ (sample space) satisfying

DEFINITION 5. $\mathcal{T} = \{T_\delta\}$ is a test for $\{P_\theta, \theta \in \Theta\}$ iff

(a) $\qquad \delta > \delta' \Rightarrow T_\delta \subseteq T_{\delta'}$,

(b) $\qquad \max\limits_\theta P_\theta(S) = 0 \Rightarrow (\forall \delta) \, S \in T_\delta$

(c) $\qquad \sup\limits_\theta P_\theta(T_\delta) \leqslant 2^{-\delta}$

(d) $\qquad \mathcal{T}$ is recursive.

This notion was first discussed in connection with complexity by Martin-Löf [4]. T_δ is a rejection region at level $2^{-\delta}$; i.e., if $S \in T_\delta$, then we reject $\{P_\theta, \theta \in \Theta\}$ at level $2^{-\delta}$. The critical level $2^{-m_{\mathcal{T}}(S)}$ of the test $\mathcal{T}$ is given by

$$m_{\mathcal{T}}(S) = \sup\{\delta : S \in T_\delta\}.$$

The function $m_{\mathcal{T}}$ completely describes the results of applying test $\mathcal{T}$.

To relate the general notion of a test to our specific probabilistic hypothesis, we claim that a test $\mathcal{T}$ is a test for the family of probabilities described by $(**)$ iff

$$\Omega = \bigcup_{n=1}^{\infty} (\{0, 1\}^n \times \{0, 1\}^n),$$

$$(\forall \delta)\ \| T_\delta \cap \{(O_A, O_B): O_A \in I_{n, w_A}, O_B \in I_{n, w_B}\} \| \leq$$

$$\leq 2^{-\delta} \left[\binom{n}{w_A} \binom{n}{w_B} \right].$$

Verification of this claim is accomplished by verifying that Definition 5c is satisfied when the supremum is taken over the family $(**)$.

We now define the empirical version of SI denoted SSI (statistical stochastic independence) through

DEFINITION 6. If $\mathscr{T}$ is a test for independence, then $A\,(\text{SSI})\,B$ relative to $\mathscr{T}$, a threshold τ, and data (O_A, O_B) if $m_{\mathscr{T}}(O_A, O_B) < \tau$.

Note that we allow countably many tests $\mathscr{T}$ for independence or, equivalently, admit countably many versions of statistical stochastic independence (SSI). This latitude in the formalization of the frequentist interpretation of the hypothesis $A\,(\text{SI})\,B$ is desirable for our purposes. $A\,(\text{EI})\,B$ will be seen to imply satisfaction of all the versions of $A\,(\text{SSI})\,B$. Our argument does not hinge upon more specific features of a statistical test for SI.

E. *Relations Between EI and SSI*

The fundamental relation between EI represented through $K(\langle O_A, O_B \rangle \mid I_{n, w_A}, I_{n, w_B})$ and SSI represented through $m_{\mathscr{T}}(O_A, O_B)$ is given by

THEOREM 5.
    (a) $(\forall \mathscr{T}\ \text{test for SI})\ (\exists c)\ (\forall O_A, O_B)$

$$K[\langle O_A, O_B \rangle \mid I_{n, w_A}, I_{n, w_B}] + m_{\mathscr{T}}(O_A, O_B) < c +$$

$$+ \log \left[\binom{n}{w_A} \binom{n}{w_B} \right].$$

    (b) $(\forall \mathscr{T})\ (\exists c)\ (\forall n, w_A, w_B)\ (\exists O_A \in I_{n, w_A}, O_B \in I_{n, w_B})$

$$K[\langle O_A, O_B \rangle \mid I_{n, w_A}, I_{n, w_B}] < c$$

and

$$m_{\mathscr{T}}(O_A, O_B) = \min \{m: O'_A \in I_{n, w_A}, O'_B \in I_{n, w_B}, m = m_{\mathscr{T}}(O'_A, O'_B)\}.$$

Proof. See Fine [6], p. 162.

The implication of Theorem 5a is that

$$A(\text{EI})\ B \Rightarrow m_{\mathscr{T}}(O_A, O_B) < t + c.$$

Hence, if $t + c < \tau$, then also $A(\text{SSI})\ B$. The implication of Theorem 5b is that there exist data sequences yielding minimal values of $m_{\mathscr{T}}$ and hence suggesting $A(\text{SSI})\ B$ for which the conditional complexity is so low that false $A(\text{EI})\ B$. While EI implies SSI, the converse is false.

F. *Conclusions*

EI is a stronger concept of independence than is any of the versions (different tests $\mathscr{T}$) of SSI. Furthermore, EI while hewing closely to our intuitive requirements for events to be independent, leads to different formal properties of the binary relation of independence than possessed by SI. The fact that EI implies SSI and SSI is a test for the usual probabilistic definition of SI $[P(A \cap B) = P(A)\ P(B)]$, suggests the necessity of the usual probabilistic definition of SI. The only possible modification of the probabilistic definition would be to *a priori* restrict the possible measure to which we apply it. For some measure P' and events A, B for which $P'(A \cap B) = P'(A)\ P'(B)$, we would decide that false $A(\text{SI})\ B$. However, this would conflict with EI through the possibility of observing occurrence records O_A, O_B with $w_A/n = P'(A)$, $w_B/n = P'(B)$, and high conditional complexity.

These observations lead us to conclude that at root independence can not be defined within probability theory. There is a suggestive analogy with disjointness wherein

$$A \cap B = \emptyset \Rightarrow P(A \cup B) = P(A) + P(B),$$

but disjointness cannot be defined in terms of probability.

In defense of the usual definition of SI, we might note its apparent simplicity (insofar as any statement involving probabilities can be deemed simple) and our expectation that it is only a small fraction of pairs (O_A, O_B) for which EI and SI or SSI conflict.

Cornell University, Ithaca

NOTES

* On leave, 1972–1973, Department of Electrical Engineering, Stanford University, Stanford, California 94305.
** Prepared with partial support from NSF Grant GK 26054 X1.

BIBLIOGRAPHY

[1] G. J. Chaitin, 'On the Length of Programs for Computing Finite Binary Sequences',
    J. Assoc. Comput. Mach. **16**, 145–149, 1969.
[2] A. N. Kolmogorov, 'Three Approaches to the Quantitative Definition of Informa-
    tion', *Problemy Peredacii Informatsii* **1**, 4–7, 1965.
[3] D. Loveland, 'A Variant of the Kolmogorov Concept of Complexity', *Information
    and Control* **15**, 510–526, 1969.
[4] P. Martin-Löf, 'The Definition of Random Sequences', *Information and Control* **9**,
    602–619, 1966.
[5] R. J. Solomonoff, 'A Formal Theory of Inductive Inference', *Information and Control*
    7, 1–22, 224–254, 1964.
[6] T. L. Fine, *Theories of Probability: An Examination of Foundations*, Academic Press,
    New York, pp. 118–164, 1973.
[7] T. L. Fine, 'On the Apparent Convergence of Relative Frequency and Its Implica-
    tions', *IEEE Trans. on Information Theory* **IT-16**, 251–257, 1970.
[8] T. L. Fine, 'Stochastic Independence and Computational Complexity', *Information
    Processing* **71**, North-Holland, Amsterdam, 1972.
[9] H. Cramer, *Mathematical Methods of Statistics*, Princeton University Press, Princeton,
    pp. 141, 1957.
[10] M. Fisz, *Probability Theory and Mathematical Statistics* (3rd ed.), Wiley, New York,
    p. 5, 1963.
[11] B. Gnedenko, *Theory of Probability* (4th ed.), Chelsea, Bronx, pp. 54–55, 1968.
[12] E. Parzen, *Modern Probability Theory and Its Applications*, Wiley, New York, p. 2,
    1960.
[13] C. P. Schnorr, *Zufalligkeit und Wahrscheinlichkeit*, Springer-Verlag, Berlin, New
    York, 1971.
[14] T. M. Cover, private communication.
[15] G. J. Chaitin, 'Information-Theoretic Computational Complexity', *IEEE Trans. on
    Information Theory*, **IT-20**, 10–15, 1974.
[16] E. Hewitt and L. J. Savage, 'Symmetric Measures on Cartesian Products', *Trans.
    Amer. Math. Soc.* **80**, 484, 489, 1955.

ROBIN GILES

A LOGIC FOR SUBJECTIVE BELIEF*

ABSTRACT. We regard logic as a language for communicating beliefs, and adopt the principle that a proposition is defined not by assigning objective conditions for its truth but *by associating with its assertion a definite commitment*. For a prime proposition, the commitment is to pay $1 should a 'trial' yield a negative outcome. (The possibility of trial is assumed, but no truth value is postulated and we do not assume that repetitions of a trial always yield the same outcome.) With prime propositions alone it is not possible to express *partial* beliefs. This possibility enters with the definition of the logical connective → (implies): following Lorenzen, we say the assertion of $P \rightarrow Q$ entails a commitment to assert Q if the 'opponent' asserts P. If P and Q are prime this expresses the belief that Q is 'at least as true as' P; more complicated propositions represent more elaborate beliefs. The other connectives are defined in a similar way. The assertion of a proposition by the 'proponent' $\mathbb{P}$ now leads, via a dialogue, to a *position* in which each speaker is committed to certain prime propositions. A certain set of these positions is regarded as acceptable by $\mathbb{P}$; making weak assumptions of rationality of $\mathbb{P}$ we arrive at an axiom concerning this set and thence construct a *risk function* which uniquely characterizes it. In general, the risk function is subadditive (on a certain group) but, in the case of a "confident" speaker (when the set of acceptable positions is maximal and the risk function is minimal) it is additive and directly determines a probability assignment on the prime propositions. Using these constructions, we show finally that *every* rational speaker behaves (in a certain exact sense) *as if* he believes that every prime proposition has a definite objective probability of which he has only partial information. Thus we obtain a justification, in subjective terms, for a belief in objective probability.

1. INTRODUCTION

In certain cases there is no difficulty in using formal logic for the expression of beliefs. For instance, if A denotes the proposition 'it is raining' and B denotes 'the sun is shining' I express a definite belief by asserting one of the following: A, B, $\neg A$ (not A), $A \vee B$ (A or B), or, less plausibly, $A \wedge B$ (A and B), or even $A \rightarrow B$ (A implies B) which means (somewhat surprisingly) $B \vee \neg A$. In fact, by using combinations of the connectives $\neg$, $\rightarrow$, $\vee$, $\wedge$, I can express, in a single formula, *any* definite belief concerning the truth or falsehood of any finite number of prime propositions (propositions without logical symbols) A, B, ...

However, in practice we are usually concerned with *degrees* of belief. I may wish to say 'it is probable that A' or 'A is more likely than B' or 'I am

Harper and Hooker (eds.), Foundations of Probability Theory, Statistical Inference, and Statistical Theories of Science,
Vol. I, 41–72. All Rights Reserved. Copyright © 1976 by D. Reidel Publishing Company, Dordrecht-Holland.

almost sure that $\neg (A \wedge B)$'. The classical propositional calculus does not provide facilities for such assertions. Of course, one can supplement it by introducing, *ad hoc*, new primitive terms – 'probably', 'possibly', etc. – or even by introducing quantitative probability as a primitive term. However, if this is done it is necessary not only to give formal rules for the manipulation of the new terms but, if the theory is to be of any *use*, also to explain precisely the *interpretation* of these terms.

Now, it is an obvious general principle that for the utmost precision of a theory the number of primitive terms should be kept to a minimum and the terms themselves should be chosen so that their interpretation can be given in as simple and clear (and therefore, hopefully, unambiguous) language as possible. To the extent that this aim is attained, it may be possible to show that the formal rules or *axioms* of the theory are necessary consequences of the meanings of the terms involved, in which case the theory is clearly placed on a much more secure foundation than if the axioms were simply postulated.

In particular, it follows that the introduction of new primitive terms in order to deal with degrees of belief should not be undertaken if it can be avoided. I shall show here how it is possible to achieve the desired aim instead by a change in the interpretation of the logical connectives and of the notion of a prime proposition itself.

In the next section we shall discuss much more deeply what is meant by a prime proposition and, in particular, what is involved in the *assertion* of such a proposition. In Section 3 the new meanings of the logical connectives are explained. Due to limitations in space the resulting logic is not developed here. (It is outlined in Giles (1972) and developed more systematically in Giles (1973).) Instead we examine in Section 4 the kind of partial beliefs that can be expressed in the new language and show in Section 5 how the concept of probability arises.

2. PRIME PROPOSITIONS

We shall be interested in the beliefs of various agents called *speakers*, but only in so far as they are *expressed* in communications between the speakers. We assume the speakers make *assertions* employing a language which contains, among other things, logical terms. Fundamental to the whole development is the notion of a *tangible meaning* for an assertion:

DEFINITION 2.1. A *tangible meaning* for an assertion consists of an exact description of some obligation undertaken by the speaker.

This is to be contrasted with a common or *avowed* meaning, by which we denote any meaning that involves no definite commitment on the part of the speaker, but is rather of the nature of an explanation which the speaker might offer to an enquirer. The avowed meaning corresponds to the sales-talk of a salesman, while the tangible meaning is the analogue of the contract or guarantee. As the analogy suggests, for the formal development we shall be concerned only with the tangible meaning of an assertion. This will be taken to define it completely: two assertions may be identified if they incur the same commitment.

There are two advantages in this approach. First, it is eminently practical: one is going to treat more seriously the assertions of a speaker who has something to offer and something at stake. Secondly, if an assertion incurs no obligation on the speaker we can give no *logical* reason why it should not be made indiscriminately. If there is a commitment, however, we may be able to deduce, from the particular nature of the obligation, conditions under which (a) there can be no harm in making the assertion, or (b) it would be clearly unwise to make it. In this way we have the possibility of deducing, *from the meanings of the assertions*, the laws which must govern their use by any rational speaker. In the absence of tangible meanings such laws can only be either deduced in a relatively obscure way from avowed meanings or, more commonly, simply postulated ad hoc as axioms; in either case the foundation is relatively insecure.

This approach can be extended. Any term used in assertions by the speakers can (and should) be defined by explaining its role in determining the tangible meaning of the assertions in which it appears; we call this a *tangible definition* for the term and say that the term has been given a *tangible meaning*. In particular, this applies to the logical symbols. For instance, $\rightarrow$ is to be defined by assigning a tangible meaning to $P \rightarrow Q$ in terms of those of P and Q. In this way our development of logic becomes independent of the notion of truth.

In particular, we can accept as prime propositions (i.e. propositions with no logical symbols) statements that have no definite truth values in the conventional sense. Instead of viewing a prime proposition as 'a statement that is either true or false' we – bearing in mind the need to impose an obligation on him who asserts it – regard it as a statement that can be *tested*: i.e., in principle, we associate with it a procedure that results in a definite *outcome*, 'yes' or 'no'. However, we must be realistic; this procedure has to be specified, but no procedure can be described in every detail. So, to be more precise, we say rather that, associated with any prime proposition A, there is a *document* giving instructions for carrying out an experimental procedure that results in an outcome, 'yes' or 'no'. We say that the document specifies an *elementary experiment* [1] and refer to an experimental execution of these instructions as an (admissible) *trial* of A. The proposition A is well defined only in so far as these instructions are clear in the following sense: they need not by any means specify one particular experimental procedure but they must be such that whenever a purported trial has been carried out the speakers involved are always in agreement as to (a) whether it really was an admissible trial, and (b) if so, what the outcome was. In order to ensure this, it is necessary that the documents should employ only concepts that are very directly related to experience, preferably so directly that they require no explanation, but can simply be demonstrated. For a further discussion of how this can be done and of the internal structure of elementary experiments in general see Giles (1972).

With this understanding of the significance of a prime proposition we can now assign a tangible meaning to every prime proposition as follows:

DEFINITION 2.2. He who asserts a prime proposition A promises to pay $\$1$ should a trial of A yield the outcome 'no'.

The choice of '$\$1$' here is simply for typographical convenience; any other unit of 'goods' could be substituted. However, it is important that the same 'stake' be associated with each prime proposition.

It is to be understood that the obligation incurred in the assertion of A is discharged when the indicated trial (and payment if required) has been made. Thus the assertion incurs at most a loss of $\$1$. (This 'principle of limited liability' persists throughout the whole logic.)

Two differences from classical logic are already apparent. First, whereas in classical logic it makes no difference whether a proposition is asserted once or twice, here this is not so: the second assertion incurs a new obligation, effectively doubling the stake. Secondly, whereas in classical logic it is required (or assumed) that every proposition is either true or false and the existence of a testing procedure, while welcome, is not insisted on, at least in orthodox expositions,[2] in the new logic the existence of a testing procedure (trial) is fundamental, and we do not need to assume that the proposition is either true or false: more precisely, we need not assume – what is tacitly supposed in classical logic – that the proposition is *dispersion-free* in that repeated trials will always give the same outcome. This is a virtue of the new approach for two reasons: (a) Except in the case of propositions of a mathematical nature (e.g. $27 + 18 = = 45$), where it can be seen to be a consequence of the rules for carrying out a trial that the result will always be the same, it is not possible to be *sure* that any proposition is dispersion-free. (b) There are many well-defined elementary experiments in which, owing to quantum and/or thermal phenomena, repeated trials do not usually give the same result.[3] Each such elementary experiment defines a prime proposition and these *dispersive* prime propositions can be treated as easily in the new logic as the dispersion-free propositions of classical logic.

3. INTERPRETATION OF THE LOGICAL SYMBOLS

Propositions are built out of prime propositions by means of the logical connectives → (implies), ∨ (or), ∧ (and), and ¬ (not) in the usual way:[4]

DEFINITION 3.1. An expression is a *proposition* if it is either a prime proposition or of the form $(P \to Q)$, $(P \vee Q)$, $(P \wedge Q)$, or $\neg P$, where P and Q are propositions.

This establishes the syntax, which is the same as in the classical propositional calculus. As was explained in Section 2, the *meaning* of each logical symbol will now be given by a definition which assigns a tangible meaning to any assertion in which it is used.

The term 'opponent' used in these definitions is to be understood in terms of the following 'rules of discourse', which are proposed primarily in the interests of simplicity. When a speaker, who may in this connection be called the *proponent*, makes an assertion it is to be regarded in the first instance as an *offer* to assume the indicated commitment. Any other speaker may respond to this offer; he says, perhaps, 'I accept'. Thereupon the commitment becomes effective and a 'debate' ensues between this *opponent* and the proponent, in which no other speaker may intervene.

In the following, P and Q denote arbitrary propositions. For $\neg$ we adopt:

DEFINITION 3.2. He who asserts $\neg P$ offers to pay \$1 to his opponent if he will assert P.

That this definition is intuitively reasonable is clear, at least in the case when P is a prime proposition. For the proponent will then recover this money from the opponent iff (= if and only if) the trial of P yields 'no'. The justification of the definition of $\vee$ (or) is even simpler:

DEFINITION 3.3. He who asserts $P \vee Q$ undertakes to assert either P or Q at his own choice.

We now come to the connective $\rightarrow$ (implies). We could adopt the classical definition, introducing $P \rightarrow Q$ as an abbreviation for $Q \vee \neg P$; this leads to a structure equivalent to classical logic. However, we obtain a much richer language (and, in particular, one in which it is possible to express *degrees* of belief) and at the same time we accord at least as closely with the intuitive meaning of 'implies' with the following definition:

DEFINITION 3.4. He who asserts $P \rightarrow Q$ offers to assert Q if his opponent will assert P.

In clarification of this definition I add the following remark: If, during any debate, either speaker asserts a proposition of the form $P \rightarrow Q$ the other speaker may choose to *admit* this assertion, in which case it is simply annulled. Alternatively, he may *challenge* it by asserting P; the original speaker must then fulfill his obligation by asserting Q, whereupon the original assertion of $P \rightarrow Q$ is annulled – an assertion may be challenged only once.

The definition of $\rightarrow$ is a crucial feature of the new logic. A similar definition was introduced by Lorenzen (1955) in giving a dialogue interpretation of intuitionistic logic.[5] The resulting logics are quite different, however, mainly because Lorenzen did not employ Definition 2.2 (there are other differences as well). In the present context, Definition 3.4 leads instead to the many-valued logics $Ł_n$ ($n = 2, 3, \ldots$ and ∞) introduced by Łukasiewicz (1930) around 1920. (See Giles 1972.)

If we add to the formal language the symbol F to stand for the particular prime proposition whose tangible meaning is given by:

DEFINITION 3.5. He who asserts F agrees to pay his opponent $\$1$. then $\neg P$ may be identified with $P \to F$: these assertions incur the same commitment. The device of replacing $\neg$ by F is technically convenient. We therefore ignore Definition 3.2 and introduce instead:

DEFINITION 3.6. $\neg P$ is an abbreviation for $P \to F$.

In selecting a definition for $\wedge$ (and) the obvious choice would be to let the assertion of $P \wedge Q$ be equivalent to the assertion of both P and Q. However, there are several reasons why this is not suitable here. Firstly, $P \wedge P$ would not be equivalent to P. Secondly, the analogous definition of $\forall x P(x)$ (see 3.8 below) is clearly not viable. Finally, such a choice would mean abandoning the principle of limited liability. So, instead, we adopt the following obvious analogue of Definition 3.3:

DEFINITION 3.7. He who asserts $P \wedge Q$ undertakes to assert either P or Q at his opponent's choice.

Observe that the speaker is not obliged to assert *both* P and Q. He must, however, be prepared to assert either, since he cannot tell which his opponent may choose.

For completeness, I include here definitions of the quantifiers $\forall$ and $\exists$, although we shall not use them. Suppose that, for every n in some set (for instance the natural numbers $\mathbb{N}$), $P(n)$ is a proposition. Then:

DEFINITION 3.8. (a) He who asserts $\exists x P(x)$ undertakes to assert $P(n)$ for some n of his own choice.

(b) He who asserts $\forall x P(x)$ undertakes to assert $P(n)$ for some n of his opponent's choice.

We have now given interpretations to all the logical symbols in such a way that, by recursion, a tangible meaning is assigned to every proposition. He who asserts any proposition incurs, in so doing, a definite obligation. The general nature of this obligation is clear. Suppose that one speaker asserts a compound proposition, say $P \to Q$. His opponent may, if he wishes, challenge it by asserting P whereupon the first speaker must assert Q, his original assertion then being annulled. If P and Q are compound propositions the debate will continue, in accordance with the rules, with new propositions being asserted and previous assertions being annulled. At any time during the debate a *position* will obtain in which each speaker is committed to a finite collection of assertions which we shall call his *tenet* (repetitions of assertions may occur in a tenet). As the debate continues the position changes, the propositions asserted becoming simpler, until eventually a *final* position is reached in which all propositions asserted are prime. The appropriate trials are then carried out and each speaker pays the other in accordance with 2.2 and 3.5.

We assume that every assertion must in due course be 'dealt with' in accordance with the definitions. We make no ruling here as to the *order* in which the assertions are dealt with, since it turns out, for our present purposes, to make no difference. However, one point needs clarification. We allow the *opponent* complete freedom in arranging the trials of the prime propositions occurring in the final position. He may use what procedures he likes; all that is required is that these procedures are in fact admissible trials, as specified in the corresponding documents. The reason for this apparent favoritism is easy to see: it was the proponent who, in his initial assertion, first mentioned these propositions. He had at that time, in the process of defining them, the chance of laying down any conditions he wished. The opponent's strategy during the debate has been determined by the form of these conditions, and it would not be fair to impose further conditions *a postiori*.

To illustrate the above definitions let us consider a simple example. For convenience in the discussion the proponent and opponent will be referred to as 'I' and 'you' respectively, and any position will be denoted by an expression consisting of my tenet on the right and yours on the left of a vertical stroke.

Consider the classically equivalent propositions $A \rightarrow B$ and $B \vee \neg A$, where for simplicity we shall assume that the propositions A and B are prime. In classical logic these propositions 'have the same meaning': each is true iff B is true or A is false. However, in the new logic they cannot be identified since their assertions represent different commitments. Indeed, if I assert $A \rightarrow B$ we shall (according to whether you challenge or not) reach one of the final positions $\emptyset \mid \emptyset$ or $A \mid B$.[6] Thus in asserting $A \rightarrow B$ I commit myself to be placed in one (at your choice) of these final positions. On the other hand, if I assert $B \vee \neg A$ I am obliged either to assert B, thus reaching the final position $\emptyset \mid B$, or $\neg A$ which, if you challenge, yields the final position $A \mid F$: i.e. I commit myself to be placed in one of the final positions $\emptyset \mid B$ or $A \mid F$, at my choice.

It is clear that, formally, these commitments are quite distinct. To see that they are actually so, suppose that I believe more strongly that an outcome 'yes' will result from a trial of B than from a trial of A.[7] Then I may reason that, in the position $A \mid B$, I am more likely to gain than to lose and so be willing to assert $A \to B$ whereas, unless I am *sure* of the outcome of at least one of A and B, I would expect to lose on average from the assertion of $B \vee \neg A$. Indeed, the argument shows that the commitment involved in asserting $A \to B$ is strictly weaker than that for $B \vee \neg A$.

This example brings out two points: firstly, the new definitions of the connectives yield a syntactically identical but semantically richer language than that of classical logic; secondly, there is in this language a possibility of expressing relative degrees of belief. In fact, it can be shown (though we shall not get that far in this paper) that an arbitrary quantitative declaration of (relative or absolute) degree of belief can be expressed with arbitrary accuracy by a suitable proposition in the language.

Further illustrative examples will be found in Giles (1972).

Let us now consider the proposition $A \to A$, where A is a prime proposition. Clearly, I (the proponent) will be prepared to assert $A \to A$ iff I consider the position $A \mid A$ acceptable. What does this entail? You (the opponent), who arrange all trials, may if you wish use different procedures (so long as both accord with the instructions on the relevant document) for the trials of the two instances of A. If you are able to use for *your* assertion a procedure which gives a lower probability of outcome 'no' than that given by the procedure you use for *my* assertion then you can cause me to lose (on average). More significantly, iff I *believe* that no such possibility is open to you then I will consider the position $A \mid A$ acceptable (and should be willing to assert $A \to A$). We then say that *in my opinion* or *for me* the proposition A is *perfect*:

DEFINITION 3.9. A prime proposition A (and the corresponding elementary experiment) is *perfect for a speaker* S iff the position $A \mid A$ is acceptable to him.[8] If a prime proposition A is perfect for every speaker we propose to consider then we say A is *perfect*.

Roughly speaking, he who asserts that a prime proposition is perfect expresses his belief that its probability of yielding outcome 'yes' will not be affected by any additional conditions imposed on the procedure for trial (i.e. additional to those given on the document defining the proposition). For example, the drawing of a ball from an urn containing black and white balls, with outcome 'yes' for a white ball drawn, is (with the usual understandings – no looking, etc.) a perfect elementary experiment. The elementary experiments dealt with in physical theories – in contrast to those which arise in the laboratory – are perfect (see Giles, 1972).[9]

Of course, if a speaker believes a proposition A to be dispersion-free, in that every trial will yield the same result, he will consider $A \mid A$ acceptable, and so deem A perfect. This applies, in particular, to the 'truth-definite' propositions treated in classical logic. Since the propositions arising both in the physical sciences and in classical logic are perfect, and since the theory of imperfect propositions is not a straightforward generalization we shall make the:

STANDING ASSUMPTION 3.10. Henceforth, all prime propositions considered will be assumed to be perfect.

4. THE RISK FUNCTION

In this section we shall investigate the beliefs of a rational observer and expose a structure analogous to that referred to as 'coherence' in discussions of subjective probability.

Since the systematic development is non-trivial I offer first a very simple example which provides a review of some of the concepts in the last section and a preview of others to come.

EXAMPLE 4.1. An observer $\mathbb{S}$ is given the following information by someone he trusts. U and V are urns, each containing ten balls, some black and the rest white. There are six white balls in U, while V contains at least one and at most eight white balls. Prime propositions A, B, C are defined as follows. A trial of A consists in drawing a ball from U, the outcome being 'yes' if it is white. A trial of B is similar, but using the urn V. A trial of C consists in drawing a ball from U and a ball from V, the outcome being 'yes' if both are white. (Each ball must be replaced afterwards.)

We notice first that each of these propositions is perfect for $\mathbb{S}$; indeed, surely the positions $A \mid A$, $B \mid B$, and $C \mid C$ are all acceptable for $\mathbb{S}$ since his expected loss in each case is zero. However, none of the propositions is dispersion-free.

Next note that S might be expected to assign a subjective probability of 0.6 to (an outcome 'yes' for) *A*. Such an assignment is represented tangibly in the following familiar way: S is willing to pay up to $0.6 for the right to get $1 should a trial of *A* yield 'yes' and *also* to pay up to $0.4 for the right to get $1 should a trial of *A* yield 'no'. *A* is an example of a prime proposition which is *probability-definite* for S. In the case of *B*, S might well refuse to assign *any* subjective probability. Indeed, he may agree to pay up to $0.1 (or $0.2) for the right to get $1 should a trial yield 'yes' (or 'no') but surely he is not being unreasonable if this is as far as he is willing to go. We might say that he assigns *lower* and *upper* probabilities 0.1 and 0.8, respectively, to *B*. *B* is an example of a prime proposition which is not probability-definite. The same applies to *C*. Notice, however, that in spite of this 'indefiniteness' in *B* and *C* the position $C \mid B$ is acceptable to S: he can be sure that he won't lose (on average) in the resulting trials. On the other hand, *neither* of the positions $A \mid B$ and its 'reverse', $B \mid A$, is acceptable for S: in each he might possibly gain, but his information is consistent with an average loss of $0.5 and $0.2 in $A \mid B$ and $B \mid A$, respectively.

Let us now forget these introductory remarks and begin a systematic treatment. Consider a speaker S. We may expect that there will be certain propositions that S is willing to assert. Since, as we saw in Section 3, the assertion of a proposition results eventually in one's finding oneself in some final position, we are led to consider the final positions that are 'acceptable' to S. If (as we shall always assume) S is rational, it is clear that this set of positions is by no means arbitrary – for instance, surely $\emptyset \mid F$ is not acceptable. We shall now determine its properties.

To facilitate a systematic treatment we introduce some definitions and notation. Our first definition concerns the notion of a *tenet*, introduced earlier (after 3.8) as a term for a collection of propositions. For mathematical purposes it is more convenient to regard it as the formal *sum* of the propositions concerned. It will be convenient also to generalize the notion of a tenet, so that those spoken of in Section 3 will now be known as *positive* tenets:

DEFINITION 4.2. Let $\mathscr{P}$ denote the set of all propositions and $\mathscr{P}_p$ the subset consisting of prime propositions. Let $\mathcal{T}$ be the free abelian group spanned by $\mathcal{P}$. An element of $\mathcal{T}$ will be called a *tenet*. Thus a tenet is a formal linear combination $\alpha = n_1 P_1 + \cdots + n_r P_r$, where $P_1, \ldots, P_r$ are distinct propositions and $n_1, \ldots, n_r$ are integers, and addition of tenets is defined in the obvious way. If $n_i \geqslant 0$ for $1 \leqslant i \leqslant r$ we call α a *positive* tenet and write $\alpha \geqslant 0$. The *length* of the tenet α is $|\alpha| = \Sigma |n_i|$. If $|\alpha| = 0$, α is called the *empty* tenet and denoted by $\emptyset$. Any tenet α can be expressed uniquely as the difference $\alpha = \alpha^+ - \alpha^-$ of two positive tenets with $|\alpha| = |\alpha^+| + |\alpha^-|$; α^+ and α^- are called the *positive* and *negative parts* of α. If P_i is prime for $1 \leqslant i \leqslant r$ then α is called a *prime* tenet. The set of all prime tenets is denoted $\mathcal{T}_p$. We regard $\mathcal{P}$ as a subset of $\mathcal{T}$ in the obvious way: by regarding a proposition as a positive tenet of length 1. Suppose α is a positive tenet. Then to *assert* α means to make n_1 assertions of P_1, n_2 assertions of P_2, and so on ($|\alpha|$ assertion in all).

The 'positions' introduced in Section 3 can best be defined formally as follows. For future use we introduce also a natural rule for *addition* of positions:

DEFINITION 4.3. An ordered pair (α, β) of positive tenets determines a *position* denoted $\alpha \mid \beta$. If α and β are prime then $\alpha \mid \beta$ is a *prime* position (or *final* position). The *reverse* of $\alpha \mid \beta$ is the position $\beta \mid \alpha$. The *sum* of two positions $\alpha \mid \beta$ and $\gamma \mid \delta$ is the position $\alpha + \gamma \mid \beta + \delta$. The set Π of all positions and the set Π_p of all prime positions are abelian semigroups with zero $\emptyset \mid \emptyset$.

We next give a more precise definition of the notion of an *acceptable* position:

DEFINITION 4.4. Let $\mathbb{S}$ be a speaker and α, β two positive tenets. We write $\mathbb{S} \models \alpha \mid \beta$ and say the position $\alpha \mid \beta$ is *acceptable* for $\mathbb{S}$ if $\mathbb{S}$ is willing, in return for an arbitrarily small fee, to assert β provided his opponent at the same time asserts α. $\mathbb{S} \models \emptyset \mid \beta$ is written in the shorter form $\mathbb{S} \models \beta$ (in particular, if β is a single proposition P, in the form $\mathbb{S} \models P$) and we then say that β (or P) is *acceptable* or *true* for $\mathbb{S}$. We say P is *false* for $\mathbb{S}$ if $\mathbb{S} \models \neg P$.

In particular, a prime proposition is true (false) for $\mathbb{S}$ iff $\mathbb{S}$ is *sure* that the outcome of a trial will be 'yes' ('no'). In general, of course, a proposition will be neither true nor false for $\mathbb{S}$.

The 'arbitrarily small fee' is introduced for technical reasons. For example, it allows us to regard the position $2F \mid 5A$, where A is the proposition defined in Example 4.1, as acceptable. In fact, S's attitude to $2F \mid 5A$ may be expected to be neutral – on average he should neither gain nor lose – but for a small fee he should be willing to be placed in this position.

We now introduce and justify an axiom concerning the set of positions acceptable to a (rational) speaker. Since we shall for the present be concerned only with one speaker S we shall write '$\models$' as an abbreviation for '$S\models$'. It is not the case that '$\models$' will be denoted by '$\not\models$'.

AXIOM 4.5. Let $\alpha, \beta, \gamma, \delta$ be prime positive tenets and A be any prime proposition. Then, for any speaker S,

   (a)   If $\models \alpha \mid \beta$ and $\models \gamma \mid \delta$ then $\models \alpha + \gamma \mid \beta + \delta$.
   (b) $\models F \mid A$.
   (c) $\not\models F$ (i.e. $\not\models \emptyset \mid F$).
   (d) $\models A \mid \emptyset$.
   (e)   (The *archimedian property*) If the relation $\models n\alpha + F \mid n\beta$ holds
            for arbitrarily large [10] positive integers n then $\models \alpha \mid \beta$.
   (f) $\models \alpha + A \mid \beta + A$ if and only if $\models \alpha \mid \beta$.

The justification of Axiom 4.5 is as follows:

(a) By hypothesis, S is willing (for a small fee) to assert β if his opponent asserts α and to assert δ if his opponent asserts γ. But then his opponent may take advantage of both these offers: if he asserts both α and γ (and pays an arbitrarily small fee) then S will assert both β and δ. Now, since $\alpha, \beta, \gamma, \delta$ are prime,[11] the situation is exactly the same as if S had been placed in the position $\alpha + \gamma \mid \beta + \delta$. Consequently, S can have no rational grounds for considering $\alpha + \gamma \mid \beta + \delta$ unacceptable.

(b), (c), and (d) are obvious: S cannot possibly lose through being placed in either of the positions $F \mid A$ or $A \mid \emptyset$ and he is bound to lose \$1 in the position $\emptyset \mid F$.

(e) Suppose that $\models n\alpha + F \mid n\beta$ holds for arbitrarily large integers n. This means that, for any $\varepsilon > 0$, S is willing to be placed in the position $n\alpha + F \mid n\beta$ for a fee of \$$\varepsilon$, which is equivalent to saying that he will accept the position $n\alpha \mid n\beta$ for a fee of \$$(1 + \varepsilon)$. It follows that for a fee of \$$(1 + \varepsilon)/n$ he should be willing to be placed in the position $\alpha \mid \beta$. But $(1 + \varepsilon)/n$ is arbitrarily small. Hence $\models \alpha \mid \beta$.

(f) Consider the position $A \mid A$. It is clear that S cannot rationally *rely* on any financial gain through being placed in this position. Indeed, the document for the elementary experiment A specifies certain conditions of admissibility for a trial. Since the same conditions apply to both occurrences of A in the position $A \mid A$, $\mathbb{S}$ has no grounds for complaint should his opponent decide to use the *same* trial for both assertions; and in this case $\mathbb{S}$'s net gain is of course zero. Consequently, if he does not dare to assert $\alpha \mid \beta$ he will not (if he is rational) dare to assert $\alpha + A \mid \beta + A$ either. The converse follows from (a) since, in view of the standing assumption (3.10) that every prime proposition is perfect, $\models A \mid A$.

As was indicated at the beginning of this section, the beliefs of any rational speaker are determined by the set of positions which he considers acceptable. We have given in Axiom 4.5 plausible assumptions concerning the nature of this set. We shall say that any set for which these assumptions are satisfied determines a *structure* (or 'structure of rational belief', if you like):

DEFINITION 4.6. Let $\mathscr{P}_p$ be any set of prime propositions and Π_p the corresponding set (see 4.3) of prime positions. Let $\mathbb{S}$ be a subset of Π_p and, for any prime position $\alpha \mid \beta$, let us write $\mathbb{S}\models\alpha \mid \beta$, and say $\alpha \mid \beta$ is *acceptable* for $\mathbb{S}$, iff $\alpha \mid \beta$ belongs to $\mathbb{S}$. Then $\mathbb{S}$ is called a *structure* iff the conditions (a)–(f) of Axiom 4.5 hold. We write $\mathbb{S}\not\models\alpha \mid \beta$ iff it is not the case that $\mathbb{S}\models\alpha \mid \beta$. When no confusion results we abbreviate '$\mathbb{S}\models$' to '$\models$'.

According to Axiom 4.5 there is, associated with each (rational) speaker $\mathbb{S}$, a structure which we may also denote by $\mathbb{S}$. Indeed, to decide whether $\mathbb{S}\models\alpha \mid \beta$ holds we merely ask the speaker $\mathbb{S}$ whether he considers the prime position $\alpha \mid \beta$ acceptable. His answers, if rational, will conform to Axiom 4.5. (We assume that he is always prepared to answer such a question. This is no great imposition: if in doubt he can safely say 'no'.)

We can thus determine the belief structure of a speaker by asking him a sequence of questions, each answered by 'yes' or 'no'; he is never required to make any quantitative assessment.

Nevertheless, we shall now show that his beliefs are characterized by a uniquely determined *risk function* $\langle \ \rangle$ which assigns to each prime position $\alpha \mid \beta$ a numerical 'risk value' $\langle \alpha \mid \beta \rangle$ which may be positive or negative and which, roughly speaking, represents the least sum (in dollars) that he would consider as sufficient compensation for placing himself in that position. The rigorous proof of this result, together with its exact statement, will be given later. For the present I offer some intuitive arguments which suggest the properties that such a risk function might be expected to possess, and thus motivate the definition and make the theorem plausible.

First, we have the basic property that links the risk function to the notion of acceptability. Iff the risk value is non-positive then surely the position is acceptable: i.e. $\langle \alpha \mid \beta \rangle \leqslant 0$ iff $\models \alpha \mid \beta$.

Next, according to 4.5 (f), two positions of the form $\alpha + A \mid \beta + A$ and $\alpha \mid \beta$ are either both acceptable or both unacceptable. It is easy to see from this that, quite generally, the acceptability of a position $\alpha \mid \beta$ is determined by the tenet (in the general sense of Definition 4.2) $\beta - \alpha$. This suggests that the same might be true of the risk value, in which case we could extend the domain of definition of the risk function $\langle \ \rangle$ to include also the group $\mathcal{T}_p$ of all prime tenets, with the original risk value being given by $\langle \alpha \mid \beta \rangle = \langle \beta - \alpha \rangle$ (α and β are here any positive tenets).

Now consider the function $\langle \ \rangle$, restricted to the group $\mathcal{T}_p$. For any tenet α, $\langle \alpha \rangle$ denotes the risk value $\langle \alpha^- \mid \alpha^+ \rangle$ of the position $\alpha^- \mid \alpha^+$. This interpretation makes it plausible that $\langle \ \rangle$ is, on $\mathcal{T}_p$, 'sublinear' in the following sense: [12]

DEFINITION 4.7. A real-valued function $\langle \ \rangle$ on the group $\mathcal{T}_p$ of all prime tenets is *sublinear* iff, for any prime tenets α, β:

(a) $\langle \alpha \rangle < \infty$.
(b) $\langle \alpha + \beta \rangle \leqslant \langle \alpha \rangle + \langle \beta \rangle$.
(c) $\langle n\alpha \rangle = n \langle \alpha \rangle$,   for $n = 1, 2, \ldots$.
(d) $\langle \emptyset \rangle = 0$.

Indeed, the only property which is perhaps unexpected is (b) (subadditivity), where one might have anticipated additivity: $\langle \alpha + \beta \rangle = = \langle \alpha \rangle + \langle \beta \rangle$. In fact, if the speaker mentally assigns a subjective probability to each prime proposition and uses these assignments to estimate the risk values, then additivity obtains. However, we have already seen (4.1) that there are cases where a rational observer will not wish to assign any subjective probability to a prime proposition. For the general case, one may argue roughly as follows: The speaker assigns a risk value sufficient to cover the worst that can happen. Thus, since the trials for $\alpha + \beta$ consist in those for α and β taken together, $\langle \alpha + \beta \rangle$ cannot exceed $\langle \alpha \rangle + \langle \beta \rangle$. But it is possible that α and β are so related that the worst situation for α never occurs along with (i.e. in the same 'possible world' as) the worst situation for β. In this way it can happen that $\langle \alpha + \beta \rangle < \langle \alpha \rangle + + \langle \beta \rangle$.

There are certain other properties that a risk function would be expected to have. Clearly, for any prime proposition A, the risk value of the position $\emptyset \mid A$ is $\langle \emptyset \mid A \rangle \leqslant 1$, and in the case $A = F$ it is 1. Similarly, the position $A \mid \emptyset$ is acceptable, so that $\langle A \mid \emptyset \rangle \leqslant 0$. Also, the speaker would clearly be willing to *pay* up to \$1 to be placed in the position $F \mid \emptyset$: i.e. $\langle F \mid \emptyset \rangle = -1$. In view of the relation $\langle \alpha \mid \beta \rangle = \langle \beta - \alpha \rangle$, these considerations suggest the following definition:

DEFINITION 4.8. A *risk function* is a real-valued sublinear function $\langle \; \rangle$ on the group $\mathcal{T}_p$ of all prime tenets, such that $\langle F \rangle = -\langle -F \rangle = 1$ and, for every prime proposition A, $\langle -A \rangle \leqslant 0$ and $\langle A \rangle \leqslant 1$. If $\langle \; \rangle$ is any risk function, we extend its domain to include the semigroup Π_p of all prime positions by writing $\langle \alpha \mid \beta \rangle = \langle \beta - \alpha \rangle$, for any prime positive tenets α, β.

We can now state the main theorem of this section:

THEOREM 4.9. Given any structure $\mathbb{S}$ there is one and only one risk function $\langle \; \rangle_{\mathbb{S}}$ such that, for any position $\alpha \mid \beta$,

(*) $\mathbb{S} \models \alpha \mid \beta$ iff $\langle \alpha \mid \beta \rangle \leqslant 0$.

Conversely, given any risk function $\langle \; \rangle$, there is exactly one structure $\mathbb{S}$ such that $\langle \; \rangle = \langle \; \rangle_{\mathbb{S}}$.

There is thus a one to one correspondence between structures (of rational belief) in the sense of Definition 4.6 and risk functions. In the next section we shall show how in turn an arbitrary risk function can be represented in terms of probability assignments.

I conclude this section with the proof of Theorem 4.9. (If desired this proof may be omitted with little loss of continuity.)

We first extend the notion of acceptability to the group $\mathcal{T}_p$:

LEMMA 4.10. For any prime tenet α write $\mathbb{S} \models \alpha$ (or briefly $\models \alpha$) iff $\mathbb{S} \models \alpha^- \mid \alpha^+$. Then, for any prime tenets α and β and any prime proposition A,

(i)   If $\alpha \geqslant 0$ and $\beta \geqslant 0$ then $\models \alpha \mid \beta$ iff $\models \beta - \alpha$.
(ii) $\models \alpha$ and $\models \beta$ implies $\models \alpha + \beta$.

(iii) $\models A - F$.

(iv) $\not\models F$.

(v) $\models -A$.

(vi)   If $\models n\alpha - F$ holds for arbitrarily large positive integers n then $\models \alpha$.

Conversely, if any relation $\models$ is defined in $\mathcal{T}_p$ with the properties (ii)–(vi) and if $\models$ is then defined in the semigroup Π_p of prime positions by means of (i) then conditions (a)–(f) of Axiom 4.5 hold in Π_p so that $\models$ determines a structure.

Proof. (i) Let $\gamma = \beta - \alpha$. Let δ be the greatest tenet for which $\alpha - \delta \geqslant 0$ and $\beta - \delta \geqslant 0$; of course, $\delta \geqslant 0$. It is easy to show that $\beta = \delta + \gamma^+$, $\alpha = \delta + \gamma^-$. Then, by $|\delta|$ applications of Axiom 4.5 (f), we conclude $\models \alpha \mid \beta$ iff $\models \gamma^- \mid \gamma^+$: i.e. iff $\models \gamma$.

(ii) If $\models \alpha$ and $\models \beta$ then $\models \alpha^- \mid \alpha^+$ and $\models \beta^- \mid \beta^+$, so that $\models \alpha^- + \beta^- \mid \alpha^+ + \beta^+$ whence, by (i), $\models \alpha^+ + \beta^+ - \alpha^- - \beta^-$: i.e. $\models \alpha + \beta$.

(iii), (iv), (v) follow immediately from (b), (c), (d) using (i); and (vi) follows by applying (e) to the position $\alpha^- \mid \alpha^+$.

The converse is equally straightforward.

Lemma 4.10 shows that the structure set up by Axiom 4.5 has been captured in the properties (i)–(vi) of $\models$. We now define what will turn out to be a suitable risk function:

DEFINITION 4.11. For each prime tenet α let [13]

$$\langle \alpha \rangle_S = \inf \{ r/n : \mathbb{S} \models n\alpha - rF, \quad n \in \mathbb{N}, \quad r \in \mathbb{Z} \}.$$

LEMMA 4.12. Let $m \in \mathbb{N}$ and let α be a tenet. Then $\models m\alpha$ iff $\models \alpha$.

Proof. That $\models \alpha$ implies $\models m\alpha$ is immediate by (ii). Conversely, if $\models m\alpha$ then, by (ii) and (v), $\models mn\alpha - F$ for $n = 1, 2, \dots$. But then, by (vi), $\models \alpha$.

LEMMA 4.13. $\langle \ \rangle$ is a risk function: i.e., if α, α_1, α_2 are prime tenets and A is a proposition, then

(a) $\langle \alpha \rangle < \infty$.

(b) $\langle \alpha_1 + \alpha_2 \rangle \leqslant \langle \alpha_1 \rangle + \langle \alpha_2 \rangle$.

(c) $\langle m\alpha \rangle = m \langle \alpha \rangle$ for $m \in \mathbb{N}$.

(d) $\langle \emptyset \rangle = 0$.

(e) $\langle F \rangle = 1$, $\langle -F \rangle = -1$.

(f) $\langle A \rangle \leqslant 1$, $\langle -A \rangle \leqslant 0$.

Moreover, we have the following generalization of Definition 4.11:

(g) $\langle\alpha\rangle\leqslant r/n$ iff $\models n\alpha-rF$ $(n\in\mathbb{N},\ r\in\mathbb{Z})$.

Also

(h) $\langle\alpha\rangle\leqslant 0$ iff $\models\alpha$.

Conversely, if $\langle\ \rangle$ is any risk function and if the relation $\models$ is defined in $\mathcal{T}_p$ by (h) then $\models$ has the properties (ii)–(vi) of Lemma 4.10 (and consequently (g) holds too).

Proof. (In the following n, n_1, n_2, m denote positive integers and r, r_1, r_2 denote integers.)

(a) Let $|\alpha^+|=r$. Then $\models\alpha^+-rF$ by (ii) and (iii), and $\models\alpha^-$ by (ii) and (v). Thus, by (iii), $\models\alpha-rF$, whence $\langle\alpha\rangle\leqslant r$.

(b) Assume $\lambda_i>\langle\alpha_i\rangle$ $(i=1,2)$. Choose r_1,r_2,n_1,n_2 so that $\lambda_i>r_i/n_i$ and $\models n_i\alpha_i-r_iF(i=1,2)$. Then, by (ii),

$$\models n_1n_2(\alpha_1+\alpha_2)-(n_2r_1+n_1r_2)\,F$$

so that $\langle\alpha_1+\alpha_2\rangle\leqslant(n_1r_2+n_2r_1)/n_1n_2<\lambda_1+\lambda_2$.

(c) Assume $\lambda>\langle m\alpha\rangle$. Choose r,n so $\lambda>r/n$ and $\models nm\alpha-rF$. Then $\alpha\leqslant r/nm=\lambda/m$. Since this holds for every such λ, $\langle\alpha\rangle\leqslant\langle m\alpha\rangle/m$. Similarly one shows that $\langle m\alpha\rangle\leqslant m\langle\alpha\rangle$.

(e) That $\langle F\rangle\leqslant 1$ and $\langle-F\rangle\leqslant-1$ follows from 4.11 and 4.10 (iii). But $0=\langle F-F\rangle\leqslant\langle F\rangle+\langle-F\rangle$ by (ii).

(f) follows from (iii) and (v) by 4.11.

(g) Assume $\langle\alpha\rangle\leqslant r/n$. For any $m\in\mathbb{N}$ choose r_1,n_1 so that $r_1/n_1<r/n++1/mn$ and $\models n_1\alpha-r_1F$. Then, by (ii) and (iv), $\models nm(n_1\alpha-r_1F)+(n_1(mr++1)-mnr_1)(-F)$: i.e. $\models n_1(nm\alpha-(mr+1)\,F)$. Hence, by 4.12, $\models m(n\alpha--rF)-F$. But m is arbitrarily large, so by (vi) $\models n\alpha-rF$. The converse is immediate by 4.11.

(h) follows from (g) with $n=1$, $r=0$.

(d) $\langle\emptyset\rangle=\langle F-F\rangle\leqslant 0$ by (h), since $\models F-F$ by (iii). On the other hand, if $\langle\emptyset\rangle<0$ we can choose m so that $\langle\emptyset\rangle<-1/m$. Then $\models m\emptyset+F$ by (g), which contradicts (iv).

Conversely, if $\langle\ \rangle$ is any risk function and $\models$ is defined on $\mathcal{T}_p$ by (h) then (ii)–(vi) of Lemma 4.10 are deduced as follows:

(ii) If $\models\alpha$ and $\models\beta$ then $\langle\alpha\rangle\leqslant 0$ and $\langle\beta\rangle\leqslant 0$ so that $\langle\alpha+\beta\rangle\leqslant 0$, whence $\models\alpha+\beta$.

(iii) $\langle A-F\rangle\leqslant\langle A\rangle+\langle-F\rangle\leqslant 0$, by (e) and (f), so $\models A-F$ by (h).

(iv) follows at once from (h) and (e), and (v) follows from (h) and (f).

(vi) Let $n \in \mathbb{N}$. If $\models n\alpha - F$ then $1 \geqslant \langle n\alpha - F \rangle + 1 = \langle n\alpha - F \rangle + \langle F \rangle \geqslant \geqslant \langle n\alpha \rangle = n\langle \alpha \rangle$. So if $\models n\alpha - F$ holds for arbitrarily large n then $0 \geqslant \langle \alpha \rangle$ so that $\models \alpha$ by (h).

Proof of Theorem 4.9. By Lemma 4.13, $\langle \ \rangle_\mathbb{S}$, defined by 4.11, is a risk function with the property 4.9 (∗). Assume $\langle \ \rangle'$ is another such risk function and α is any prime tenet. If $\langle \alpha \rangle' < \langle \alpha \rangle_\mathbb{S}$ we can choose integers r, n with $n > 0$ such that $\langle \alpha \rangle' < r/n < \langle \alpha \rangle_\mathbb{S}$. Then, using 4.9 (∗) and the last remark in Lemma 4.12, $\langle rF \mid n\alpha \rangle_\mathbb{S} > 0 > \langle rF \mid n\alpha \rangle'$, whence $\models rF \mid n\alpha$ and also $\not\models rF \mid n\alpha$, a contradiction. Thus $\langle \alpha \rangle' \not< \langle \alpha \rangle_\mathbb{S}$; similarly $\langle \alpha \rangle' \not> \not> \langle \alpha \rangle_\mathbb{S}$. Since α is arbitrary, this means that $\langle \ \rangle' = \langle \ \rangle_\mathbb{S}$. Thus the property (∗) in 4.9 determines a *unique* risk function for any given structure. It is clear, too, that distinct structures give rise to distinct risk functions.

It remains to show that every risk function arises from some structure. Let $\langle \ \rangle$ be any risk function. Then, by the last parts of Lemma 4.13 and Lemma 4.10, the relation $\models$ defined by 4.13 (h) and 4.10 (i) determines a structure which is related to $\langle \ \rangle$ by 4.9 (∗).

5. THE ORIGIN OF PROBABILITY

We saw in Section 4 that any speaker (or structure) $\mathbb{S}$ is characterized by a risk function $\langle \ \rangle_\mathbb{S}$. We shall now discuss the interpretation of the risk function, showing how it can be described in terms of probability. We shall show that $\mathbb{S}$'s behaviour, as manifested in the set of positions that he finds acceptable, is just *as if* he believes that each elementary experiment has a definite objective probability of yielding 'yes' although he, himself, has only limited information regarding the values of these probabilities. We do not, of course, *prove* that he has such a belief; neither do we assume it. We merely show that his behaviour is consistent with the hypothesis that he has this belief. Current work, not reported here, suggests that this belief is also a *practical convenience* not only for the description but for the development of a 'structure of rational belief' (4.6). Assuming this, we seem to obtain very strong justification, on purely rational grounds, for a belief in objective probability.

Naturally, we have nothing to say on the question of how these partially known objective probabilities should be represented in $\mathbb{S}$'s picture of the

world; we are concerned only with the role they play in respect of his
behaviour as manifested in the assertions he is willing to make.

We first consider the significance of $\langle A \rangle_S$ for a prime proposition A.
For simplicity we drop the subscript S.

LEMMA 5.1. For any prime proposition A,

(a) $\quad 0 \leqslant \langle A \rangle + \langle -A \rangle \leqslant 1,$

(b) $\quad -1 \leqslant \langle -A \rangle \leqslant 0 \leqslant \langle A \rangle \leqslant 1.$

Proof. Much of this is immediate from the definition of a risk function
(4.8). Nothing in addition that $\langle A \rangle + \langle -A \rangle \geqslant \langle A - A \rangle = \langle \emptyset \rangle = 0$, we
obtain (a). By (a), $\langle A \rangle \geqslant 0 - \langle -A \rangle \geqslant 0$ and $\langle -A \rangle \geqslant \langle A \rangle \geqslant -1$, which
proves (b).

For any prime proposition A let us define the *lower* and *upper proba-
bilities* of A (for S) to be $p_l = 1 - \langle A \rangle$ and $p_u = 1 + \langle -A \rangle$, respectively. By
Lemma 5.1 we have $0 \leqslant p_l \leqslant p_u \leqslant 1$. The terminology for p_l and p_u is justified
by the following observations. By 4.13 (g), for any integers r, n with $n > 0$,

(i) $\quad \models nA - rF \quad$ iff $\quad r/n \geqslant 1 - p_l,$ and

(ii) $\quad \models rF - nA \quad$ iff $\quad r/n \leqslant 1 - p_u.$

Now, suppose that S believed that A had (objectively) some unknown
probability p (of yielding the outcome 'yes'). Then he would expect to
gain, on average, $\$(r - n(1 - p))$ in the position $rF \mid nA$, and this gain is
positive iff $r/n > 1 - p$. Thus if S had formed the opinion that $p \geqslant p_l$ then he
would certainly expect to gain (on average) if $r/n > 1 - p_l$ but might fear
loss if $r/n < 1 - p_l$. This accords exactly with (i). Similarly, (ii) corresponds
to a belief that $p \leqslant p_u$. Thus we can 'explain' (i) and (ii) by supposing that S
believes that A has some unknown objective probability p in the range
$p_l \leqslant p \leqslant p_u$. It may be that, like many physicists, S has in fact such a belief;
what we have shown, however, is simply that any rational speaker (in
the precise sense given by Axiom 4.5) will behave as if he had.[14]

The case in which $p_l = p_u = p(A)$, say, is clearly of special interest; it
may be regarded as representing the situation when S (believes that he)
knows the probability of A. We shall then say that A is *probability-definite*
for S. From the definition of p_l and p_u the condition for this is that
$\langle A \rangle + \langle -A \rangle = 0$, and $p(A)$ is then given by $p(A) = 1 - \langle A \rangle$. This condition can easily be expressed directly in terms of the concept of acceptabil-
ity. Indeed, from the remarks following 5.1, it is clear that A is proba-

bility-definite iff, for every pair of integers r, n, either $rF \mid nA$ or its reverse, $nA \mid rF$ is acceptable.

The opposite extreme to probability-definiteness arises when $p_l = 0$ and $p_u = 1$, i.e. when $\langle A \rangle = 1$ and $\langle -A \rangle = 0$. This can be regarded as representing the situation when $\mathbb{S}$ feels he can say nothing whatever about the probability of A. We shall then say that A is *completely indefinite* for $\mathbb{S}$.

It is convenient to have a quantitative measure of the degree to which ($\mathbb{S}$'s opinion of) a prime proposition A deviates from being probability-definite. We shall define the *indefiniteness* $I(A)$ of A for $\mathbb{S}$ to be $I(A) = = \langle A \rangle + \langle -A \rangle$. Thus $0 \leqslant I(A) \leqslant 1$, and in fact $I(A) = p_u - p_l$.

For an actual observer – for instance in a laboratory – we may expect that $I(A)$ is a decreasing function of time. When he first considers the elementary experiment corresponding to the proposition A he may have very little information about it (other than the procedure for carrying out the experiment) so that $I(A)$ may be large, but through the subsequent carrying out of trials (or in other ways) this indefiniteness may be expected to decrease and perhaps eventually to approach zero. (Of course, every time such an observer changes his mind [15] – generally by accepting some previously unacceptable tenet, but possible in the reverse direction – he is to be regarded formally as a different 'speaker'.)

Roughly speaking, the indefiniteness of any elementary experiment A (for an observer $\mathbb{S}$) is a measure of the extent to which $\mathbb{S}$, himself, believes that the information he has about A is incomplete. For example, in Example 4.1 the prime propositions A, B, and C have indefiniteness 0, 0.7, and 0.42, respectively. For a more physical example let A be the prime proposition corresponding to the following elementary experiment. A photon is emitted from a source S so as to reach a part-silvered glass plate G. It is then either reflected or passes through, reaching a detector D. In the latter case the outcome is 'yes'. The practical physicist may feel, after a superficial inspection of the apparatus, that the probability of outcome 'yes' lies between $p_l = 0.6$ and $p_u = 0.8$, say. After a closer examination he might set $p_l = 0.72$ and $p_u = 0.73$, say. We should not expect him ever to be prepared to give an exact probability in the tangible sense that he is willing to bet either way on this basis. Thus the probability-definite elementary experiment is a limiting case – approached, but not strictly attained, in practice.[16]

These concepts can be generalized:

DEFINITION 5.2. Let α be any prime tenet. The *indefiniteness* $I(\alpha)$ of α is given by $I(\alpha) = \langle\alpha\rangle + \langle-\alpha\rangle$. If $I(\alpha) = 0$, α is *probability-definite*.

LEMMA 5.3. If α is a probability-definite tenet then, for *any* tenet β, $\langle\alpha+\beta\rangle = \langle\alpha\rangle + \langle\beta\rangle$.

Proof. By subadditivity of the risk function, $\langle\alpha+\beta\rangle \leqslant \langle\alpha\rangle + \langle\beta\rangle \leqslant \langle\alpha\rangle + \langle-\alpha\rangle + \langle\alpha+\beta\rangle = \langle\alpha+\beta\rangle$, so that equality holds throughout.

This lemma may be interpreted as showing that the speaker may always attach to a probability-definite proposition A a risk value $\langle A\rangle = 1 - p(A)$, *even when A occurs as part of a larger tenet*. (Of course, this conforms to our picture of him as believing that A has a definite probability $p(A)$.)

The function I behaves in the group $\mathcal{T}_p$ of prime tenets rather like a seminorm in a vector space: it is non-negative and sublinear. Indeed, by 4.13, we have at once:

LEMMA 5.4. For any tenets α and β,

    (a) $0 \leqslant I(\alpha) < \infty$,
    (b) $I(\alpha+\beta) \leqslant I(\alpha) + I(\beta)$,
    (c) $I(\alpha) = I(-\alpha)$,
    (d) $I(m\alpha) = mI(\alpha)$ for $m = 1, 2, \ldots$,
    (e) $I(\emptyset) = 0$,
    (f) $I(F) = 0$.

COROLLARY 5.5. The set of all probability-definite tenets forms a group.

COROLLARY 5.6. (The *triangle inequalities*). If α and β are any prime tenets then

$$|I(\alpha) - I(\beta)| \leqslant I(\alpha\pm\beta) < I(\alpha) + I(\beta).$$

In particular, if β is probability-definite, then $I(\alpha\pm\beta) = I(\alpha)$.

Corollary 5.6 applies in particular if α and β are prime propositions A and B, say. In this case one might be inclined to suppose that something stronger would be true: if $I(A)$ and $I(B)$ are large then should not $I(A+B)$ and $I(A-B)$ be large too? An example shows that this is not necessarily so.

EXAMPLE 5.7. Let S be a weak radioactive source, say of β-particles, and let D be a detector (e.g. a geiger counter) which gives a response iff a β-particle reaches it. Let us suppose that the detector is normally inoperative, but is actuated for a period of one second when a button is pressed. Let A denote (the prime proposition associated with) an elementary experiment in which the detector D is placed to the right of S at a certain distance and the button is pressed, the outcome being 'yes' if the detector responds. To be precise, we should suppose that the document labelled by the prime proposition A gives instructions for the preparation of the source S as well as for the construction of the detector D, and moreover (owing to our standing assumption (3.10) that each elementary experiment is perfect) that these instructions are so precisely stated that no significant refinement of the elementary experiment A can be given. The simplest way to picture this situation is to imagine that the same source and detector are used for each trial of A.

Now, it is quite possible for an experimentalist to be precisely informed of the (chemical?) procedures to be used in the construction of S without having any significant information about the radioactive strength of the source so produced. In this case (as we saw in the example preceding 5.2) the prime proposition A will have for him a large indefiniteness.

We now take for B another experiment *whose instructions have something in common with those for A*: as an extreme case, let us consider an elementary experiment B' for which the instructions are exactly the same as for A except that D is this time placed on the left of S. (In spite of this relationship between the texts on the associated documents, the prime propositions A and B' are of course *formally* quite distinct.) Then it is likely that our experimentalist, in estimating the risk $\langle B' - A \rangle$ associated with the position $A \mid B'$, will consider it to be very small, since he will reckon that, whatever the probability may be that B' will yield outcome 'no' (resulting in a loss to him of $1), there will be an almost equal probability that A will result in outcome 'no', so that on the average he should gain as much as he loses.[17] He is able to base this reasoning on the symmetry of the experimental arrangements and his experience of the nature of radioactivity and detectors in general, in spite of his ignorance of the strength of the particular source S. Of course, the same applies to $\langle A - B' \rangle$. Thus $I(A - B') = \langle A - B' \rangle + \langle B' - A \rangle$ is small.

Now let B'' be the same elementary experiment as B' except that the outcome is 'yes' if the detector *does not* respond. Then a similar discussion shows that, although the indefiniteness of B'' is large, the risk attached by our experimentalist to each of the positions $F \mid A + B''$ and $A + B'' \mid F$ will be numerically small, so that $I(A + B'' - F)$ is small. But, by 5.6, $I(A + B'') = = I(A + B'' - F)$.

Taking $B = B'$ and $B = B''$ in turn, we see from this example that even when $I(A)$ and $I(B)$ are large it is possible for either $I(A + B)$ or $I(A - B)$ to be small. (Of course, it is not possible for *both* $I(A + B)$ and $I(A - B)$ to be small since $2I(A) = I(2A) \leqslant I(A + B) + I(A - B)$.)

The indefiniteness of a prime proposition A is a measure of the reluctance of the speaker to lay bets involving the outcome of the corresponding elementary experiment. As the above examples suggest, this reluctance is normally due to ignorance, resulting in a failure to associate any definite probability with the elementary experiment. As experience of trials of the experiment are accumulated so we can expect this ignorance to decrease, and as a limiting case we might contemplate the situation when it is zero for every prime proposition. We are thus led to consider, as an important special case, the following type of structure:

DEFINITION 5.8. A structure $\mathbb{S}$ is *probability-definite* if every prime proposition is probability-definite for $\mathbb{S}$.

Thus a probability-definite structure $\mathbb{S}$ determines a *probability assignment*; namely, the map which assigns to each prime proposition A the probability $p(A) = 1 - \langle A \rangle$ (see the remarks following 5.1).

By Lemma 5.3, we have:

THEOREM 5.9. In a probability-definite structure:

(a)   Every tenet is probability-definite.
(b)   The risk function $\langle \ \rangle$ is *additive*: $\langle \alpha + \beta \rangle = \langle \alpha \rangle + \langle \beta \rangle$, for any tenets α, β.

Thus $\langle \ \rangle$ is then a *prime valuation* in the following sense:

DEFINITION 5.10. A real-valued function v on the group $\mathcal{T}_p$ of all prime tenets is a *prime valuation* if

(a) $v(\alpha + \beta) = v(\alpha) + v(\beta)$ for any tenets α, β,
(b) $0 \leqslant v(A) \leqslant 1$ for every prime proposition A, and
(c) $v(F) = 1$.

By a *prime propositional valuation* we mean any function v on the set $\mathscr{P}_p$ of prime propositions to the closed interval $[0, 1]$ such that $v(F) = 1$.

The term 'prime valuation' is used because the domain consists of all *prime* tenets and because, in the further development (see Giles 1972), an extension to a 'valuation' defined for all tenets is constructed.[18] However, since we shall not here be concerned with such extensions, we shall for simplicity often refer to a prime valuation simply as a *valuation*.

From 5.10 we have at once:

THEOREM 5.11.

(a) The restriction of a prime valuation to $\mathscr{P}_p$ is a prime propositional valuation. Conversely, any prime propositional valuation has a unique extension to a prime valuation.

(b) A prime valuation is a risk function.

To appreciate the significance of valuations we need to discuss the partial ordering of structures. According to Definition 4.6 the structure corresponding to a rational speaker is the set of all prime positions which are acceptable to him. The class $\mathscr{S}$ of all structures has thus a natural partial ordering by inclusion. The set $\mathscr{R}$ of all risk functions can also be partially ordered in a natural way: given two risk functions $\langle \ \rangle$ and $\langle \ \rangle'$ we write $\langle \ \rangle \geqslant \langle \ \rangle'$ iff $\langle \alpha \rangle \geqslant \langle \alpha \rangle'$ for every prime tenet α. We now show that the natural one to one correspondence between structures and risk functions (4.9) reverses the order:

THEOREM 5.12. Let $\mathbb{S}$ and $\mathbb{T}$ be structures. Then the following conditions are equivalent:

(a) $\mathbb{S} \supseteq \mathbb{T}$,

(b) $\langle \ \rangle_\mathbb{T} \geqslant \langle \ \rangle_\mathbb{S}$.

Proof. That (b) implies (a) is immediate from 4.13 (h). On the other hand, if $\langle \ \rangle_\mathbb{T} \not\geqslant \langle \ \rangle_\mathbb{S}$ then, for some α and some integers r, n with $n > 0$, $\langle \alpha \rangle_\mathbb{T} < r/n < \langle \alpha \rangle_\mathbb{S}$. But then $\mathbb{T} \models n\alpha - rF$ and $\mathbb{S} \not\models n\alpha - rF$ so that $\mathbb{S} \not\supseteq \mathbb{T}$.

According to condition (a), $\mathbb{S}$ finds acceptable every position that is acceptable to $\mathbb{T}$;[19] one might describe this situation by saying that $\mathbb{S}$ is *at least as confident as* $\mathbb{T}$. With this terminology there is a speaker $\mathbb{C}$ (for 'cautious') who is less confident than any other: indeed, $\mathbb{C}$ considers a

position acceptable only if he cannot do otherwise without being irrational. By 5.12, to establish the existence of $\mathbb{C}$ it suffices to prove:

THEOREM 5.13. In $\mathscr{R}$ there is a greatest element $\langle\ \rangle_{\mathbb{C}}$ given by
$\langle\alpha\rangle_{\mathbb{C}} = |\alpha^+|$, for every tenet α.

Proof. In the proof of 4.13 we saw that, for any risk function $\langle\ \rangle$ and
tenet α, $\langle\alpha\rangle \leqslant |\alpha^+|$. Thus it is only necessary to verify that $\langle\ \rangle_{\mathbb{C}}$ is a risk
function. But, for any tenets α and β, $\langle\alpha+\beta\rangle_{\mathbb{C}} = |(\alpha+\beta)^+| \leqslant |\alpha^+| + |\beta^+| =$
$= \langle\alpha\rangle_{\mathbb{C}} + \langle\beta\rangle_{\mathbb{C}}$, and the other properties of a risk function are even
more immediate.

$\mathbb{C}$ is *the smallest* structure. There is no greatest structure but there exist
maximal structures, corresponding to 'confident speakers' and minimal
risk functions. Indeed:

THEOREM 5.14. A prime valuation is a minimal risk function. Equivalently, a probability-definite structure is a maximal structure.

Proof. Let v be a prime valuation and suppose that $\langle\ \rangle$ is a risk func-
tion with $\langle\ \rangle \leqslant v$. Then $\langle\ \rangle = v$, for if not then, for some tenet α, $\langle\alpha\rangle +$
$+ \langle-\alpha\rangle < v(\alpha) + v(-\alpha) = 0$, a contradiction.

The converse of Theorem 5.14 depends on a variation of the Hahn-
Banach theorem:

THEOREM 5.15. Given any risk function $\langle\ \rangle$ and any tenet α there is a
valuation v with $v \leqslant \langle\ \rangle$ and $v(\alpha) = \langle\alpha\rangle$.

For a proof, see Giles (1964) page 219.

COROLLARY 5.16. A minimal risk function is a valuation. Equivalently, a maximal structure is probability-definite.

Thus the minimal risk functions and the prime valuations coincide;
equivalently, the maximal structures and the probability-definite structures coincide. Since valuations are certainly simpler than risk functions,
it is reasonable to try to describe a given risk function in terms of valuations; this corresponds to describing a belief structure in terms of probability assignments. This is done in the following theorem:

THEOREM 5.17.
(i) Let $\mathscr{V}$ be any (non-empty) set of valuations. For each tenet α let

$\langle\alpha\rangle = \sup\{v(\alpha): v\in\mathscr{V}\}$. Then $\langle\;\rangle$ is a risk function. (In fact, this is true even when $\mathscr{V}$ is a set of risk functions.)

(ii) Conversely, every risk function $\langle\;\rangle_S$ can be written in this form by setting $\mathscr{V} = \mathscr{V}_S$, where $\mathscr{V}_S$ is the set of all valuations underlying $\langle\;\rangle_S$ (i.e. $\mathscr{V}_S = \{v: v\leqslant\langle\;\rangle_S\}$.)

Proof. (i) We have to establish conditions (a)–(f) of 4.13, Of these, all but (b) are immediate. Thus we have only to show that if α, β are any tenets then $\langle\alpha+\beta\rangle \leqslant \langle\alpha\rangle + \langle\beta\rangle$. But if $\langle\alpha+\beta\rangle > \langle\alpha\rangle + \langle\beta\rangle$, then, for some v in $\mathscr{V}$, $v(\alpha+\beta) > \langle\alpha\rangle + \langle\beta\rangle$, and this is a contradiction, since $\langle\alpha\rangle + \langle\beta\rangle \geqslant v(\alpha) + v(\beta) \geqslant v(\alpha+\beta)$.

(ii) We must show that, for every tenet α,

$$\langle\alpha\rangle_S = \sup\{v(\alpha): v\in\mathscr{V}_S\} = \langle\alpha\rangle, \text{ say.}$$

It is clear that $\langle\alpha\rangle_S \geqslant \langle\alpha\rangle$. On the other hand, for any tenet α there is, by 5.15, a valuation v in $\mathscr{V}_S$ with $v(\alpha) = \langle\alpha\rangle_S$, which means $\langle\alpha\rangle \geqslant \langle\alpha\rangle_S$.

COROLLARY 5.18. Any risk function is the sup of the valuations it exceeds. Any structure is the intersection of the probability-definite structures that contain it.

Now for the interpretation of this mathematics. Imagine an observer $\mathbb{O}$ (not necessarily a rational speaker, in the sense of 4.5) who believes that, associated with each elementary experiment (or prime proposition) A is an objective probability (of outcome 'yes') $p(A)$, these probabilities being, however, largely unknown to him. We need not enquire what his interpretation of $p(A)$ is; it may, for instance, be the frequency interpretation or some sort of propensity interpretation. We are concerned only with the role which his belief plays in governing his behaviour: i.e. in determining which positions he considers acceptable. For every prime proposition A we have, of course, $0 \leqslant p(A) \leqslant 1$. Moreover, since the elementary experiment F is sure (by definition) to produce the outcome 'no', $p(F) = 0$. For each A let $v(A) = 1 - p(A)$. Then v is a prime propositional valuation and so (5.11) has a unique extension to a prime valuation, which we shall also denote by v.

Let $\alpha = \sum m_i A_i$ and $\beta = \sum n_j B_j$ be two prime positive tenets. Then $v(\beta-\alpha) = \sum n_j v(B_j) - \sum m_i v(A_i)$. Now $\mathbb{O}$, as a believer in objective probability, will argue as follows. For each prime proposition A, $v(A)$ is, through the meaning of probability, the average loss, in dollars per trial, that he would experience in an infinite series of trials of A. Consequently, $v(\beta-\alpha)$ represents the average total loss in dollars that he would experience in the trials which result when he is placed in the position $\alpha \mid \beta$.

Now, the objective probabilities $p(A)$ are not fully known to him. Let us suppose, however, that he is able to express his state of belief regarding the values of these probabilities by laying down a set $\mathscr{V}$ of valuations to which he believes v belongs. Then his mean loss per occasion, when placed in the position $\alpha \mid \beta$, will lie between $\inf\{v(\beta-\alpha): v\in\mathscr{V}\}$ and $\sup\{v(\beta-\alpha): v\in\mathscr{V}\}$. Suppose $\mathbb{O}$ considers a final position acceptable iff he is sure that, should he be placed in that position a large number of times, he would not lose, on the average. Then he will find $\alpha \mid \beta$ acceptable exactly iff $\sup\{v(\beta-\alpha): v\in\mathscr{V}\}\leqslant 0$.

For each tenet γ, let us set $\langle\gamma\rangle_{\mathbb{O}}=\sup\{v(\gamma): v\in\mathscr{V}\}$. Then, by 5.17, $\langle\ \rangle_{\mathbb{O}}$ is a risk function, and we have just seen that $\mathbb{O}$ considers $\alpha \mid \beta$ acceptable iff $\langle\beta-\alpha\rangle_{\mathbb{O}}\leqslant 0$. This shows that he is a 'rational speaker' in the sense of Axiom 4.5, and (by 4.9) that $\langle\ \rangle_{\mathbb{O}}$ is his risk function. Thus an observer who "believes in" objective probability and behaves as though he does not know the actual objective probabilities belonging to the elementary experiments, but knows only that this assignment of probabilities is some member of a known set of such assignments, is a rational speaker in the sense of the present theory.

Conversely, let $\mathbb{S}$ be any rational speaker. In the above description of $\mathbb{O}$ set $\mathscr{V}=\mathscr{V}_{\mathbb{S}}$. Then, by 5.17, $\langle\ \rangle_{\mathbb{O}}=\langle\ \rangle_{\mathbb{S}}$ so that, by 4.9, $\mathbb{O}$ and $\mathbb{S}$ have the same set of acceptable prime positions and so are, as speakers, indistinguishable. Thus *every* rational speaker $\mathbb{S}$ behaves as if he believed in the existence of an objective probability assignment whose value is known to him only to lie in a certain set, in the manner described for $\mathbb{O}$ above.

Queen's University, Kingston, Canada

NOTES

* This work was supported by a grant from the National Research Council of Canada.
[1] Elementary experiments are of fundamental importance in connection with the foundations of physics and quantum mechanics in particular. See Giles (1970), Jauch (1968), Mackey (1963).
[2] For instance, Fermat's Last Theorem is admitted as a proposition.

[3] For examples see Giles (1972). Strictly speaking, most (if not all) elementary experiments in practice show some 'dispersion'.

[4] Quantifiers will be introduced briefly later, but will not be used in this paper.

[5] In fact the definitions of all the logical symbols are similar to those introduced by Lorenzen.

[6] $\emptyset$ denotes the *empty tenet*, in which nothing is asserted.

[7] Example 4.1 below shows that I may well have this belief even if I am not prepared to assign any definite subjective probabilities to these events.

[8] The term 'acceptable' will be explained more precisely in Section 4.4 below.

[9] To avoid confusion it should perhaps be remarked here that the procedure in quantum mechanics referred to in Giles (1970) as a *test*, in Jauch (1968) as a *yes-no experiment*, in Mackey (1963) as a *question*, and represented in the usual theory by a projection operator, should not be regarded as a proposition. At any rate, it is not perfect, since the probability of outcome 'yes' depends on things (namely, the *state* of the system under study) not specified in the definition of the procedure. A test is *part* of a (perfect) proposition, the other part being a state. (See Giles, 1972.)

[10] I.e. given any integer m there is a greater integer n for which it holds.

[11] Were they not prime, the conclusion would require a detailed examination of the rules of debate: it is not otherwise clear that to have to conduct a debate with initial position $\alpha + \gamma \mid \beta + \delta$ is equivalent to having to conduct both the debates with initial positions $\alpha \mid \beta$ and $\gamma \mid \delta$.

[12] The term, 'sublinear', is borrowed from the theory of functions on a vector space. Although $\mathcal{T}_p$ is only an abelian group it is very like a vector space and could indeed be embedded in one. (For several reasons I do not consider this embedding appropriate here.) It is surprising that the same situation, and to a large extent the same mathematics, arose also in my work on the foundations of thermodynamics (Giles, 1964).

[13] $\mathbb{N}$ denotes the natural numbers and $\mathbb{Z}$ the integers.

[14] This can be regarded as justifying the claim at the beginning of this section, in the very special case in which there is only one prime proposition. It takes a lot more work to justify it in the general case, where there may be infinitely many prime propositions and $\mathbb{S}$'s beliefs may involve correlations of arbitrary complexity.

[15] We do not attempt in any way to discuss the question of when such a change in mind is justified.

[16] One might suppose that the sort of partial ignorance represented by distinct upper and lower probabilities might be described in more detail by introducing a subjective probability distribution for the value of an assumed objective (frequency or propensity interpretation) probability p. In fact this is not so. If a speaker $\mathbb{S}$ is prepared to commit himself to *any* definite subjective probability distribution for p then the elementary experiment A involved is already probability-definite: in fact, $p_l = p_u =$ expectation value of p. For example, if $\mathbb{S}$ believes that A has some unknown objective probability p which, in his opinion, is equally likely to lie anywhere in the interval $[0, 1]$ then it is a simple exercise to show that his expected loss for each of the positions $\langle \emptyset \mid A \rangle$ and $\langle A \mid F \rangle$ is \$0.5; it follows that $p_l = p_u = 0.5$. It seems that there is no conventional way to describe the sort of partial information corresponding to the case $p_l < p_u$.

[17] Of course, in the mental process we have just described, the experimentalist argues in terms of an unknown objective probability. I have chosen to present the argument in this way since this is in fact the natural way in which a physicist would reason. However, it is not necessary to use the concept of objective probability to reach the desired conclusion. A more naive observer might just remark that whenever, in the past, he had experience of

trials of a similar pair of experiments to A and B' (using a source other than S) he could without loss have adopted from the beginning (i.e. even before anything was known about the strength of the source) a *small* positive value for $\langle B' - A \rangle$. So, reasoning by induction; he expects the same to apply again.

[18] There is a slight difference in terminology between Definition 5.10 and the 1972 preprint: if v is a valuation in the present sense then $-v$ is a valuation in the earlier sense. I use Definition 5.10 here since it has the simpler property (b) and fits in well with the intuitively appealing concept of a risk value.

[19] We assume that corresponding to any structure there is a (possible) speaker and use the same symbol for each.

BIBLIOGRAPHY

Giles, R.: 1964, *Mathematical Foundations of Thermodynamics*, Pergamon, Oxford, England.

Giles, R.: 1970, 'Foundations for Quantum Mechanics', *J. Math. Phys.* **11**, 2139–2160.

Giles, R.: 1972, 'A Non-Classical Logic for Physics', Preprint. (A shortened version of this paper has since appeared in *Studia Logica* **33**, part 4, (1974). See also 'A Pragmatic Approach to the Formalization of Empirical Theories', to appear in *Proceedings of the Conference for Formal Methods in the Methodology of Empirical Sciences*, Warsaw, June 1974, and 'Łukasiewicz Logic and Fuzzy Set Theory', to appear in *Proceedings of the 1975 International Symposium on Multiple-Valued Logic*, Bloomington, Indiana, May 1975.)

Giles, R.: 1973, *Physics and Logic*, duplicated lecture notes.

Jauch, J. M.: 1968, *Foundations of Quantum Mechanics*, Addison-Wesley, Reading, Massachusetts.

Lorenzen, P.: 1962, *Metamathematik*, Bibliographisches Institut, Mannheim.

Łukasiewicz, J. and Tarski, A.: 1930, 'Investigations in the Sentential Calculus', translation in J. Łukasiewicz, *Selected Works*, Reidel, Dordrecht (1970), and in A. Tarski, *Logic, Semantics, Metamathematics*, Clarendon Press, Oxford, England (1956).

Mackey, G. W.: 1963, *The Mathematical Foundations of Quantum Mechanics*, Benjamin, New York.

DISCUSSION

Adams: What is the connection between belief and assertion in this approach? Is there, for example, a rule which says that a man must assert whatever he believes?

Giles: I would take *assertion* as a technical term of the theory and *belief* as merely being used in an informal way; there is no attempt here to attach any precise meaning to the notion of belief.

Adams: Then why make any assertions at all, isn't the best thing to do simply to say nothing?

Giles: Certainly you will not lose by saying nothing but you can gain by speaking. If you assert a particular proposition then it may be that your opponent will challenge this and then you can get something from him.

Adams: There seems to be no way within your theory to criticize and change the documents – they seem to be given prior to any of the subsequent debate.

Giles: It is true that I did not treat of documents in detail here, but in another paper of mine, 'A Non-Classical Logic for Physics' (see Bibliography to paper), I have gone into the structure of documents in some detail and it is there that this issue should be joined.

Brockhouse: Isn't this a peculiar game? It concerns 'putting your money where your mouth is' – the condition that others may respect your expressed opinion – and yet it concerns your own interior belief states.

Giles: Yes, it is exactly that. The meaning of any assertion is given in terms of a commitment incurred by him who asserts it. This interpretation then determines which sets of beliefs one can hold and still be rational; each rational speaker is characterized by some risk function.

Brockhouse: If we are to deal with subjective aspects of beliefs, must we not admit that some assertions are more important than others?

Giles: Yes, but you can take this into account if you wish – you simply attach differing dollar values to differing sentences and, for example, if you attach a value of $3.00 to an assertion A then you assert A not once only but three times.

Marin: How do you treat cases of partial information – for these are, after all, by far the commonest cases. Normally, we only have some evidence that some assertion or other was made and this evidence is usually incomplete. Again we often only wish to assert sentences with certain probabilities given the information at hand, rather than being certain that we wish to assert them.

Giles: It is certainly assumed here that each speaker always knows what the other has asserted. The notion of incomplete information is embodied in the selection of a risk function, rather than appearing in the form of a probability assignment. For example, the weakest risk function – the risk function corresponding to what I call the cautious observer, who does not assert anything unless he is absolutely sure of it – represents a very high degree of uncertainty, for one hardly asserts anything.

[Here a remark by Marin has been omitted.]

Giles: Yes, I think so. But reflect on this: In the case of a definite subjective probability $p(A)$ (of getting 'yes' for A) the risk in asserting A is $\langle \emptyset \mid A \rangle = 1 - p(A)$, while the expected gain if the opponent asserts A is the same: $-\langle A \mid \emptyset \rangle = 1 - p(A)$. Thus here $\langle \emptyset \mid A \rangle + \langle A \mid \emptyset \rangle = 0$. But this is only the case in my system when A is 'probability-definite'; in general $\langle \emptyset \mid A \rangle + \langle A \mid \emptyset \rangle \geqslant 0$, and the person who is not at all sure about A will have this sum being greater than zero. In the probability-definite case the risk function corresponds to a unique assignment of subjective probabilities to the prime propositions in the Bayesian manner; but in the general case, as I have shown, the risk function is represented instead by a *set* of allowed probability assignments.

WILLIAM L. HARPER

RATIONAL BELIEF CHANGE, POPPER FUNCTIONS AND COUNTERFACTUALS*

ABSTRACT. This paper uses Popper's treatment of probability and an epistemic constraint on probability assignments to conditionals to extend the Bayesian representation of rational belief so that revision of previously accepted evidence is allowed for. Results of this extension include an epistemic semantics for Lewis' theory of counterfactual conditionals and a representation for one kind of conceptual change.

I. ORTHODOX BAYESIAN PROBABILITY

1. *Preliminaries*

In the orthodox Bayesian tradition of Ramsey, De Finetti and Savage rational belief functions are represented by sharp probabilities.[1] This representation has been defended in a number of ways, but the arguments most characteristic of the tradition turn on the role of belief in guiding decisions.[2]

Given some fundamental assumptions about preference and some idealizations and conventions about belief functions, the representation falls out of the Bayesian analysis of rational decision making.

One of the idealizations is that the objects for which the belief function is defined are closed under boolean operations. If $P(A)$ and $P(B)$ exist then so do $P(\bar{A})$, $P(AB)$, etc. The domain of a belief function P will be a boolean algebra $\mathscr{E}$ of propositions. It will be convenient to consider propositions as sets of possible worlds so that $\mathscr{E}$ is a field of subsets of the necessary proposition T. Thus, AB and $\bar{A}$ are the ordinary set operations $A \cap B$ and $T - A$. The possible worlds in T correspond to Savage's (Savage [49] pp. 8–12) possible states of the world and the propositions correspond to his events.[3]

The values of the belief function are real numbers in the interval $[0, 1]$ with the convention that full belief in A is represented by $P(A) = 1$.

Ramsey and Savage provide axiomatic characterizations of rational preference according to which a rational agent acts as though his decisions were guided by maximizing expected utility relative to a utility

* This research was partially supported by Canada Council Grant 1302-C70-300.

Harper and Hooker (eds.), Foundations of Probability Theory, Statistical Inference, and Statistical Theories of Science, Vol. I, 73–115. All Rights Reserved. Copyright © 1976 by D. Reidel Publishing Company, Dordrecht-Holland.

function U and a probability function P. The utility function is determined up to positive linear transformations. The probability function is determined uniquely and represents the agents belief function.

Suppose that $A_0 \ldots A_{n-1}$ is a partition of the propositions relevant for a decision among acts $a_0 \ldots a_{k-1}$. The acts do not influence the subjective probability of the propositions.[4] Let $U(a_j A_i)$ represent the utility to the agent of doing act a_j when proposition A_i is the case. The expected utility $\underline{E}(a_j)$ of each act a_j is the sum over the A_i's of $U(a_j A_i) \cdot P(A_i)$.

$$(1) \qquad E(a_j) = \sum_{i < n} U(a_j A_i) \cdot P(A_i).$$

The preference ordering among the acts for a utility maximizing agent conforms to their expected utilities.

There are various ways of construing the objects of the utility function U. Richard Jeffrey would have $U(a_j A_i)$ attach directly to the proposition that the agent performs act a_j when proposition A_i is the case (Jeffrey [1] pp. 63–81). Savage has utility defined primarily for acts construed as functions mapping possible worlds into consequences, and derivatively for consequences (Savage [49] pp. 17–26).

There are other variations. The differences are interesting and important for the problem of axiomatizing preference. (A very nice summary of the field together with the most up to date treatment is to be found in Krantz *et al.* [28] pp. 369–422). For our purposes, however, all that needs attention is a kind of invariance with respect to finer partitions that most treatments share.

Suppose, as before, we have partition $A_0 \ldots A_{n-1}$ of T and acts $a_0 \ldots a_{m-1}$. For each A_i let $B_0^i \ldots B_{k_i-1}^i$ be a partition of A_i such that the acts do not influence the subjective probability of the B_j^i's.

$$(2) \qquad \sum_{i < n} U(a A_i) \cdot P(A_i) = \sum_{i < n} \sum_{j < k_i} U(a B_j^i) \cdot P(B_j^i).$$

What I shall call the finer partitions principle is that the agents acts do not conflict with (2).

2. Bets and Coherence

If an agent satisfied Savage's or Ramsey's axioms and his utilities were linear with the stakes in decisions between buying and selling bets then his belief function would determine the prices at which he would buy and sell. He would buy a bet for A at positive stake s when the price is less than $P(A) \cdot s$, sell when the price is more and be indifferent when the price equals $P(A) \cdot s$. De Finetti (De Finetti [12] pp. 103–104) showed that the direct assumption of this connection between bets and beliefs requires that P be a probability function if the agent is to avoid having book made against him. Kemeny (Kemeny [26] pp. 268–269) showed that under this bets-belief assumption having P a probability is sufficient to prevent having book made against the agent. These and later variations by many writers have come to be called dutch book arguments for representing rational belief as a probability function.

A belief function is understood to be coherent as a guide to rational decision making just in case making the bets-beliefs assumption for it does not lead to any system of bets on which the agent faces a total net loss on every outcome. Let $0 < x < s$. Any act with utility $s - x$ in outcomes where A holds and $-x$ for outcomes in $\bar{A}$ constitutes buying a bet for A at stake s and price x. Similarly any act with utilities $x - s$ in A and x for outcomes in $\bar{A}$ can be regarded as selling the bet at the same price and stake. The bets-beliefs assumption characterizes the obvious way belief ought to guide choices between acts with such utilities. The dutch book arguments show that this assumption is consistent with expected utility maximizing just in case P is a probability function.

In order to award the utilities properly, actual bets require some procedure for deciding if the proposition is true or false. Brian Ellis has investigated betting systems relative to decision procedures where some propositions may remain undecided (Ellis [11] pp. 131–136). A proposition that may remain undecided is called *semidecidable*. When A is not decided bets for A are called off. They are won or lost as A is decided to be true or false. Ellis shows that in such a framework if arbitrary semidecidable propositions are allowed then strictly coherent betting ratios cannot be probabilities.[5]

As Ellis sees it the classical dutch book arguments only apply to systems of propositions all of which are decidable, and his result shows that they cannot be generalized to cover semi-decidable propositions. As there are many interesting propositions for which no decision procedures exist Ellis considers his result to be a serious objection to the dutch book justification of the probability representation of rational belief.

In the dutch book arguments, we have been presenting, the acts of buying and selling bets have been defined simply in terms of having appropriate utilities for the outcomes. That actual bets require some method of determining what outcome obtains in order to award utilities properly is beside the point. The idea of coherence is that the bets-beliefs assumption can hold safely under all possible combinations of utilities and outcomes. Ellis' objection just doesn't apply. The classical dutch book arguments had nothing to do with decidability and never were restricted to decidable propositions.

The difficulties with actual bets do not undercut the force of the dutch book argument and Bayesians need not assume that the degree of belief $P(A)$ is behavioristically defined as the critical rate at which he would buy and sell actual bets on A. These difficulties do indicate, however, that measuring degrees of belief is not as straight-forward as one might hope.

Ramsey's treatment suggests a way to use gambling devices and prizes to measure degrees of belief (Ramsey [46] pp. 77–79). Suppose that the outcomes of some gambling devices are themselves value neutral to the agent and that he assigns equal subjective probability to the designed equi-probable outcomes. Let c be a prize that the agent desires and consider the choice between (a) and (b).

(a)   receive c, if A and nothing if $\bar{A}$
(b)   receive c, if gambling device comes up with any one of m out of the n outcomes, and nothing if it comes up with one of the rest.

The agents degree of belief is measured as close as one wants by considering his preferences between (a) and (b) in various choices of this sort. Since Ramsey's time many such procedures have been proposed, some of which are more sophisticated than this (cf. Krantz, et al. [28] pp. 900–901 for discussion and further references). All of the available procedures, however, share with the one we sketched that sometimes the choices will involve gambles on propositions for which no convenient method of verification exists.

Many interesting propositions have no convenient procedures for verification. Popper points out that most general scientific hypotheses have special problems. According to him, there is no method for finding out that they are true, but there may be a method for finding out that they

are false (Popper [43] pp. 27–48, [45] pp. 1–30 and many further publica-
tions referenced in [45]). If A cannot be verified then any attempt to build
explicit tests for deciding A into the choice between (a) and (b) is doomed.
Considerations of this sort have led Abner Shimony to give up the expected utility justification for inductive probability (Shimony [51] pp.
103–104).

I think it is important to see that when A is not conveniently verified
the option (a) is considerably different than a bet on a horse race where
the outcome will be disclosed without fail. The decision between (a) and
(b) is a kind of thought experiment of what one would do under the hypothetical assumption that choosing (a) would without fail result in pay
off (c) if A and nothing if $\bar{A}$. I do not, however, see why such hypothetical
choices cannot be made in all seriousness. Moreover, the degrees of belief
that they reveal are exactly those that would operate to guide decisions
where the agent thinks that the truth of A matters.

3. Acceptance and Strict Coherence

If P is a classical probability on $\mathscr{E}$ then $\Delta p = \{A : P(A) = 1\}$ has some important characteristics. For all $A, B \in \mathscr{E}$

(1)   If $A, B \in \Delta p$ then $A \cap B \in \Delta p$
(2)   If $A \in \Delta p$ and $A \subseteq B$ then $B \in \Delta p$
(3) $A \cap \bar{A} = \emptyset \notin \Delta p$.

A subset Δ of $\mathscr{E}$ satisfying these requirements is a proper filter of $\mathscr{E}$ and
corresponds to a consistent set of propositions closed under semantical
consequence. The semantical consequence relation is that of truth preservation. A proposition A is true at world W just in case $W \in A$. Clearly 1
preserves truth in that whenever both A and B are true in W so is $A \cap B$.
Similarly for 2. Constraint 3 is consistency.

We have introduced a convention that $P(A) = 1$ represents full belief
in A. Robert Stalnaker considers a belief function P to represent an
idealized possible state of knowledge where every A such that $P(A) = 1$
is known by the agent (Stalnaker [52] p. 66). The concept of knowledge
is idealized in that what the agent knows is closed under semantical consequence just as in Hintikka's analysis (Hintikka [21] pp. 8–39). When P
is so considered the set $K(P) = \cap \Delta p$ consists in exactly those worlds that
are epistemically possible relative to P in Hintikka's sense of epistemic

possibility (*Ibid.*, and Hintikka [22]). An agent who knows every A in Δp in effect knows that the actual world is in $K(P)$. I interpret $P(A)=1$ as the agent accepts A, and interpret $K(P)$ as a proposition that expresses the total content of what he accepts. Stalnaker's heuristic of a possible state of knowledge is a good one, however, because the agent will usually regard himself as knowing just those propositions he accepts.

Let h be a function mapping $\mathscr{E}$ into $K(P)$ so that for each A in $\mathscr{E}$

$$h(A)=A\cap K(P)$$

and let $\mathscr{E}_p$ be the range of h. $\mathscr{E}_p$ is a field of subsets of $K(P)$ isomorphic to the quotient algebra of $\mathscr{E}$ modulo the filter Δp, and h is a homomorphism of $\mathscr{E}$ onto $\mathscr{E}_p$. Where we have a decision D relative to $\mathscr{E}$ with partition $A_0 \ldots A_{n-1}$ of T and acts $a_0 \ldots a_{k-1}$ let $h(D)$ be the corresponding decision in $\mathscr{E}_p$ with partition $h(A_0) \ldots h(A_{n-1})$ of $K(P)$ and expected utility $E_h(a_j)=\sum_{i<n} u(a_j h(A_i)) \cdot P(h(A_i))$ for each act a_j.

Remark 2.1. D and $h(D)$ are equivalent in that for every a_j

$$\underline{E}(a_j)=\underline{E}_h(a_j).$$

Proof. Break $\underline{E}(a_j)$ down into a sum over $h(A_i)$'s and a sum over $(A_i-h(A_i))$'s. The later sum is zero. ■ Every decision relative to partitions of T in field $\mathscr{E}$ is equivalent to its corresponding decision relative to partitions of $K(P)$ in field $\mathscr{E}_p$. The expected utility framework does not distinguish between $U(a_j A_i)$ and $U(a_j A_i \cap K(P))$.

There are two opposed ways of dealing with this point and they represent two quite different approaches to the Bayesian analysis of rational belief. On the one hand, we may say that no rational agent would accept contingent propositions because this would be tantamount to ignoring what could happen in some of the possible outcomes. On the other hand, we may allow that rational agents do accept contingent propositions and hold that for a rational agent only those outcomes consistent with what he knows are relevant for making his decisions. The second position sees the Bayesian belief function as an extension of our ordinary ideas of belief and knowledge. In addition to representing what the agent regards himself as knowing it also assigns degrees of belief to those propositions his knowledge does not decide. On this view the agent may be quite rational to accept any of the propositions we would usually regard him as knowing. The first view is the strict-coherence position. On it the

ordinary notions of belief and knowledge are inoperative for decision
making. They are replaced by the more adequate notion of rational partial
belief (see Richard Jeffrey [24] for an excellent defense of this view).
Those cases where an agent would normally be regarded as accepting
some contingent propositions are really just cases where his degree of
belief is close to but not equal to 1 (see Teller [55] p. 240).

An agent who thinks he rationally accepts some contingent truths
might argue:

> "I am sure that my hand is before me on the page, and that
> the population of the United States is greater than that of
> Canada."

The strict coherence advocate replies:

> "Then you should be indifferent between paying $1000 for
> a chance to lose it if you are wrong and get it back if you are
> right, and paying nothing for a chance to get $1000 if you
> are wrong and nothing if you are right."

If the agent refuses to see the light, the strict coherence advocate simply
increases the amount. Sooner or later the agent breaks down and admits
that he is just a little bit unsure.

At first glance the case for strict coherence seems strong. But, the situation is less simple than it may seem. Try the Ramsey degree of belief
measurement. Consider a choice between

(a)   Receive $1000 if my hand is really there on the page and
        nothing if not.
(b)   Receive $1000 if the random device comes up on any but m
        of the n possible outcomes and nothing if not.

No matter how small m/n is made, so long as there is at least one un-
favorable outcome, I will prefer (a) to (b). This also holds for the proposition that the population of Canada is less than that of the United States.

If the agents degree of belief is 1 why isn't he indifferent between the
choices offered by the strict coherence advocate? One answer might be
that the utilities of receiving a net of zero dollars can differ with the choice
context. If the agent breaks down at one stake, but not at another then
one has prima facie evidence that the stake can change the relative de-

sirabilities of zero dollar net gain in different contexts. Even if the choice offered by the strict coherence advocate could avoid all difficulties of this sort a result in his favor would be a puzzle for decision theory rather than an unambiguous defense of his position.

Strict coherence is usually defined and defended relative to the bets-belief assumption. A belief function is strictly coherent just in case the bets-belief assumption does not lead to any system of bets where the agent suffers a net gain on no outcome and a net loss on some outcome. From Kemeny's result we have that a belief function is strictly coherent just in case it is coherent and $P(A) = 1$ just for those propositions that are true in every *relevant possible outcome*. A bet having utilities matching the money offered in the strict coherence advocate's choice would lead to a violation of strict coherence if the agent accepts a contingent proposition and the relevant outcomes are all the worlds in T. Stalnaker suggests the obvious way to have acceptance of contingent propositions not violate strict coherence (Stalnaker [54] p. 68). The relevant possible outcomes are just those in $K(P)$.

Strict coherence was originally introduced by Shimony as a constraint on confirmation functions (Shimony [50] pp. 9–12). A confirmation function $\mathscr{C}$ on $\mathscr{E}$ maps $\mathscr{E} \times \mathscr{E} - \{\emptyset\}$ into the reals and satisfies the following basic axioms:

(1) $0 \leqslant \mathscr{C}(H/E) \leqslant 1$
(2) $\mathscr{C}(E/E) = 1$
(3) $\mathscr{C}(H/E) + \mathscr{C}(\bar{H}/E) = 1$
(4)   If $E \cap H \neq \emptyset$ then $\mathscr{C}(H \cap J/E) = \mathscr{C}(H/E) \cdot \mathscr{C}(J/E \cap H)$.

I have used Carnap's axiomatization (Carnap [7] p. 38), because confirmation functions have been so closely identified with Carnap's program of inductive logic. The motivation for the constraints on confirmation is that $\mathscr{C}$ is to be a conditional probability adequate to play the following role:

> If an ideally rational agent were to have exactly E as his total evidence then his degree of belief in proposition H ought to be $\mathscr{C}(H/E)$.

Carnap's program may be regarded as the attempt to put as many constraints on $\mathscr{C}$ as can be justified by this role.[6]

When Shimony applied the strict coherence argument to confirmation functions he made $\mathscr{C}(H/E)$ determine betting rates for bets on H when the relevant possible cases are restricted to just those in E. On this procedure a confirmation function is strictly coherent just in case it is regular. Regularity is the constraint that

$$\mathscr{C}(H/E)=1 \quad \text{only if} \quad E \subseteq H.^7$$

The heuristics of confirmation indicate that regularity is a natural constraint on confirmation functions. Suppose $\mathscr{C}(B/A)=1$ but $A \not\subseteq B$. Thus, proposition $\bar{A} \cup B$ is contingent (i.e. $\bar{A} \cup B \neq T$). For any E and H $\mathscr{C}(H/E)=\mathscr{C}(H/E \cap (\bar{A} \cup B))$. If $E \neq E \cap (\bar{A} \cup B)$ then the heuristic of confirmation is violated since $\mathscr{C}(H/E)$ corresponds to having $E \cap (\bar{A} \cup B)$ as evidence rather than just E.

When $\mathscr{C}$ is regular the corresponding absolute probability function,

$$\mathscr{C}(H)=\mathscr{C}(H/T),$$

is strictly coherent in the strict sense. This does not, however, imply that the belief function of a rational agent who guides his beliefs by $\mathscr{C}$ would also be strictly coherent in the strict sense. Carnap represents rational belief of an agent with background evidence as a credence function. Where K is the total content of the agent's background evidence his rational credence function $\mathscr{C}_{(K)}$ conforms to $\mathscr{C}$ so that

$$\mathscr{C}_{(K)}(H/E)=\mathscr{C}(H/E \cap K)$$

and

$$\mathscr{C}_{(K)}(H)=\mathscr{C}(H/K).$$

Clearly $\mathscr{C}_{(K)}$ need not be strictly coherent in the strict sense.[8] Moreover, $\mathscr{C}_{(K)}$ need not even be regular. If $K \subseteq H$ then $\mathscr{C}_{(K)}(H/E)=1$ even if $E \not\subseteq H$.

This difference between $\mathscr{C}_{(K)}(H/E)$ and $\mathscr{C}(H/E)$ corresponds to the following different heuristics for conditional probability. The confirmation conditional probability $\mathscr{C}(H/E)$ is what the rational agent would assign to H if his total evidence were reduced to nothing but the proposition E. The credence conditional probability $\mathscr{C}_{(K)}(H/E)$ is what the rational agent would assign to H if E and nothing further were added to what he now accepts.

4. *Conditionalization and Learning from Experience*

One of the most striking features of the orthodox Bayesian tradition is the representation of learning from experience by conditionalization. A change from rational belief function P_0 to P_1 is by conditionalization on A just in case $P_1(B)$ is the conditional probability $P_0(B/A)$ for every B. Let us suppose that $P_0(A) > 0$ and examine the claim that $P_0(AB)/P_0(A)$ is the appropriate new degree of belief in B for a rational agent who has just altered his belief function P_0 by learning A and nothing more.

The most extensive discussion of such claims is by Paul Teller (Teller [55, 56]). Under assumptions that come to the specification that $P_0(A) > 0$ and that accepting A is the total direct epistemic input from the learning experience Teller suggests that if $P_0(B) = P_0(C)$, $B \subseteq A$ and $C \subseteq A$, then it ought to be that $P_1(B) = P_1(C)$ (Teller [55] pp. 233–238). He shows that this qualitative assumption about rational belief change is equivalent to conditionalization under fairly normal structural assumptions about belief functions (Teller [55] pp. 223–230). Teller, also, reports a rather ingenious dutch book argument by David Lewis (Teller [55] pp. 222–225).

I shall give a dutch book argument based on an idea that can be used to help extend the representation of rational belief so that conditionalization on propositions of zero probability is allowed. The idea can also be used to defend Savage's conditional expected utility argument for conditionalization from an objection Teller makes against it (Teller [56] p. 18).

Suppose that $P_0(A) > 0$ and P_1 is the appropriate new belief function when the change from P_0 is to learn A and nothing more. The new set of propositions accepted $\Delta(P_1)$ ought to be generated by $\Delta(P) \cup \{A\}$ and the content of what is accepted $K(P_1)$ ought, therefore, to be $K(P_0) \cap A$. The basic idea is that the shift from P_0 to P_1 ought to be as minimal as is required for accepting A. One obvious principle governing minimality here is that one not give up any proposition he already accepts unless he needs to. Since $P_0(A) > 0$; $K(P_0) \cap A \neq \emptyset$ and A is compatible with everything the agent accepts. Thus the agent need give up none of the propositions in $\Delta(P)$ when he accepts A.

When the bets-beliefs assumption is applied to P_1 the relevant outcomes are just those in $K(P_1)$. A conditional bet for B on A is one that is called off in relevant outcomes not in A and that is won or lost as usual for relevant outcomes in A. Since $K(P_1) = K(P_0) \cap A$ any bet for B at stake s and price $\gamma \cdot s$ relative to $K(P_1)$ outcomes has exactly the same net return on every outcome as a conditional bet for B on A at the same price and stake with $K(P_0)$ relevant outcomes. If the bets-beliefs assumption holds for both P_0 and P_1 then $P_1(B)$ must be the same as the critical price ratio for conditional bets for B on A relative to $K(P_0)$ outcomes. De Finetti (De Finetti [12] pp. 108–109) has shown that the P_0 critical price ratio for conditional bets for B on A must be $P_0(AB)/P_0(A)$ if the agent is to avoid dutch books.[10]

I think that implicit assumptions like those I make explicitly here account for the fact that some Bayesian writers (e.g. De Finetti) were content to limit their argument for conditionalization to showing that the ratio is the appropriate betting rate for conditional bets.

If all rational learning from experience is by conditionalization on new evidence and belief functions are only classical probabilities, then no way is provided for revising previously accepted evidence on the basis of new inputs. Suppose P_1 arises from P_0 by conditionalization on A. Then, $P_1(A) = 1$, $P_1(A/C) = 1$ for every C such that $P_1(C) > 0$, and no conditional probability $P_1(A/C)$ exists for any C such that $P_1(C) = 0$. Clearly all revision of the assignment $P_1(A) = 1$ blocked. Any hypothetical new evidence C that is not already rejected will continue to support A, and any hypothetical C that is rejected cannot play a new evidence role because the relevant conditional probabilities do not exist.

Richard Jeffrey (Jeffrey [23] pp. 153–164) has proposed a generalization of conditionalization according to which $P_1(A)$ need not be 1. A change from P_0 to P_1 originates in the partition $A_0 \dots A_{n-1}$ just in case for $i < n$ $P_1(B/A_i) = P_0(B/A_i)$ for every B. When this happens the coherence constraints on P_1 generate the rule:

$$P_1(B) = \sum_{i < n} P_0(B/A_i) \cdot P_1(A_i).$$

Jeffrey argues that there can be cases where one rationally responds directly to experience by shifting $P_0(A)$ to some new value $P_1(A)$ without accepting A or anything else as new evidence. He claims that in such cases the partition $\{A, \bar{A}\}$ should be an origin for the shift from P_0 to P_1.

Isaac Levi, a defender of rational acceptance, rightly saw that Jeffrey's

rule would provide for a representation of learning from experience
where no contingent evidence would ever need to be accepted (Levi [34]).
His attack on Jeffrey's rule, however, was defective (Harper and Kyburg
[20], Levi [35]). Teller has now provided a quite adequate defense of
Jeffrey's rule (Teller [55] pp. 243–257). Levi's fears have been realized.
The strict-strict coherence approach can handle learning from experience.
In fact it handles it better than the ordinary framework of conditionalization and acceptance because any change is open to correction on the
basis of future observations.

I think that the standard notions of acceptance and bodies of evidence
are too useful to give up. But, if one is to accept contingent evidence of
the usual sort some provision must be made for revision of previously
accepted evidence. A first step toward this is to allow $P(B/A)$ to be defined
even when $P(A)=0$.

II. EXTENDING THE REPRESENTATION OF RATIONAL BELIEF TO POPPER FUNCTIONS

1. *Popper's Probability Functions*

Popper provides an axiomatic treatment of probability in which conditional probability is primitive and exists everywhere.[11] Suppose **F** is a
minimal algebra with a binary operation AB and unary operation $\bar{A}$.[12]
Nothing specific about the algebraic properties of these operations is assumed. The following axioms characterize one version of a Popper
probability function P mapping $\mathbf{F} \times \mathbf{F}$ into the reals.[13] For all A, B and
C in **F**,

    a1. $0 \leqslant P(B/A) \leqslant P(A/A) = 1$
    a2.   If $P(A/B) = 1 = P(B/A)$ then $P(C/A) = P(C/B)$
    a3.   If $P(C/A) \neq 1$ then $P(\bar{B}/A) = 1 - P(B/A)$
    a4. $P(AB/C) = P(A/C) \cdot P(B/AC)$
    a5. $P(AB/C) \leqslant P(B/C)$.

Popper adds the additional constraint that there be some C and D in **F**
such that $P(A/B) \neq P(C/D)$. I shall call functions satisfying these requirements Popper functions.

In the classical mathematical treatment probability is defined as a non-
negative additive set function normalized to 1. Suppose that $\mathscr{E}$ is a field

of subsets of some non-empty set T. A function M mapping $\mathscr{E}$ into the reals is a classical probability just in case [14]

(i) $M(T)=1$, and

(ii)   If $A\cap B=\emptyset$ then $M(A\cup B)=M(A)+M(B)$.

Sometimes the treatment is generalized to have $\mathscr{E}$ an arbitrary boolean algebra. In this case T will be the maximum of $\mathscr{E}$ and in place of set intersection and union we will have boolean meet and join.

Classical conditional probabilities are introduced by definition as ratios of absolute probabilities.

(iii) $M(B/A)=M(AB)/M(A)$,

provided $M(A)>0$. If $M(A)=0$ then no classical probability $M(B/A)$ exists. In a Popper function conditional probability is primitive, but absolute probability is easily represented. For Popper function P the absolute probability $P(A)$ is conditional probability relative to $T=A\bar{A}$, so that

$$P(A)=P(A/T)$$

for A in $\mathbf{F}$. Where $P(A)>0$ the Popper conditional probability $P(B/A)$ is a ratio of absolute probabilities.

$$P(B/A)=P(AB)/P(A)$$

just as classical conditional probability. The most salient difference between a Popper function and classical probability is that $P(B/A)$ exists even when $P(A)=0$.

Popper's extension of conditional probability to all pairs of elements has some mathematical advantages. Chief among these is that P induces an interesting boolean algebra of equivalence classes on the minimal algebra $\mathbf{F}$. For elements A, B of $\mathbf{F}$ define,

d1.3. (i) $A \underset{\tilde{p}}{\sim} B$ iff $P(A/C)=P(B/C)$ for all C in $\mathbf{F}$

    (ii) $[A]p=\{C\in\mathbf{F}:A \underset{\tilde{p}}{\sim} C\}$

    (iii) $F/P=\{[C]p:C\in F\}$.

When $A \underset{\tilde{p}}{\sim} B$ we say that A is P-equivalent to B. The subset $[A]p$ of $\mathbf{F}$ is the equivalence class of A under P, and $\mathbf{F}/P$ is the set of equivalence classes induced by P on $\mathbf{F}$. The following operations are defined on $\mathbf{F}/p$

(iv) $[A]p \wedge [B]p = [AB]p$
(v) $\overline{[A]}p = [\bar{A}]p$

to form the algebraic structure $\mathbf{F}//p$. Popper proves the following theorem.

t1.1. $\mathbf{F}//p$ is a boolean algebra with
          $[A]p \wedge [B]p$ as meet and $\overline{[A]}p$ as
          complement for A, B in $\mathbf{F}$.

This theorem shows that the constraints on P are sufficient to impose boolean behaviour on the unstructured operations of $\mathbf{F}$.

In the classical treatment $\mathscr{E}$ must already be a boolean algebra, before M can be defined on it. As Popper points out, the algebraic structure of $\mathscr{E}$ is an additional assumption buried in the classical characterization of probability. The $\mathbf{F}$ of a Popper function need only be a minimal algebra, and the algebraic properties used in probability reasoning are generated by the explicit constraints on P.

The introduction of $\mathbf{F}//p$ allows us to formulate some further connections between Popper functions and classical probabilities. Let P_A be defined on $\mathbf{F}$ so that $P_A(B) = P(B/A)$ and $P_{[A]}$ be the corresponding function on $\mathbf{F}//p$ so that $P_{[A]}([B]) = P_A(B)$. We have the following remarks

r1.2. (i) If $P(\bar{A}/A) \neq 1$,   then $P_{[A]}$ is a classical probability on $\mathbf{F}//p$
      (ii) If $P(\bar{A}/A) = 1$,   then P_A (hence $P_{[A]}$) is the incoherent constant function assigning 1 to every element.

The absolute probability $P(A)$ is simply $P_T(A)$ and corresponds to the classical probability $P_{[T]}$ on $\mathbf{F}//T$.

Two elementary properties of the other extreme value for Popper functions are also of interest.

r1.3. (i) $P(A/\bar{A}) = 1$ iff $P(A/C) = 1$ for all C
      (ii) $P(A/\bar{A}) \neq 1$ iff $P(A/C) = 0$ for some C.

The first of these has the effect that the maximum of $\mathbf{F}//p$ is the set of all A such that $P(A/\bar{A}) = 1$. We shall say that A is P-valid just in case $P(A/\bar{A}) = 1$. The second remark is that whenever A is not P-valid there is some C such that $P(A/C) = 0$. This will be useful in showing P-validity by indirect proof.

2. *Stalnaker's Representation Theorem*

Robert Stalnaker has constructed a dutch book representation theorem for Popper functions as coherent extended conditional belief functions (Stalnaker [52] pp. 70–74). The representation is based on his idea that the relevant outcomes for strict coherence depend on what the agent accepts. The present treatment differs from Stalnaker's in formulation and in the details of the proof. The most important difference is the explicit emphasis on the way the constraints on $K(P_A)$ justify axioms 2 and 4 for Popper functions.

Let us extend the representation of rational belief so that $P(B/A)$ is defined for every pair A, B in $\mathscr{E}$. We want $P(B/A)$ to represent the degree of belief that would be rational for the agent to assign to B were he to accept A as his total new input from experience. We shall think of P_A (the function defined on $\mathscr{E}$ so that $P_A(B) = P(B/A)$) as the absolute belief function the rational agent would have were he to minimally revise his beliefs to accept A. In light of this motivation certain general constraints on rational extended conditional belief functions seem warranted.

Let
$$\Delta(P_A) = \{B : P(B/A) = 1\}$$
and
$$K(P_A) = \cap \, \Delta(P_A).$$

Just as with classical conditionalization $\Delta(P_A)$ is to be the set of propositions the agent would accept if his absolute belief function were P_A and $K(P_A)$ is the set of worlds where all these propositions hold. Since P_A is to be a belief function where A is accepted it is required that $P(A/A) = 1$. This gives the constraint

(1) $K(P_A) \subseteq A.$

We shall express the constraints in terms of $K(P_A)$ where possible.

Another obvious constraint is that $K(P_A)$ be nonempty for at least some A.

(2) (i) $K(P_A) \neq \emptyset,$ for some $A.$

If $K(P_A)$ is empty then A is regarded as absurd in that there is no world consistent with all the propositions the agent would be committed to were he to minimally revise his beliefs to accept A. It is convenient to have a convention for cases where $K(P_A)$ is empty.

(2) (ii)   If $K(P_A)=\emptyset$,   then $P(B/A)=1$ for all B.

This convention corresponds to the idea that anything follows from something that commits you to a contradiction.

The main constraint is that P_A be coherent when the possible cases are restricted to $K(P_A)$.

(3) P_A is coherent relative to $K(P_A)$ if $K(P_A)\neq0$.

The justification for this is the obvious one that P_A is to be a belief function appropriate to guide the agent's decisions relative to partitions of $K(P_A)$.

Our motivation that P_A be a minimal revision to accept A and the classical conditionalization property that $K(P_A)=K(P)\cap A$ when $P(A)>0$ suggest a further constraint.

(4) $K(P_{AB})=K(P_B)\cap A$,   provided $K(P_B)\cap A\neq\emptyset$.

If $K(P_B)\cap A$ is non-empty then a minimal revision of $K(P_B)$ to accept A is simply to add A to what one already accepts.

The justification of this last constraint corresponds to the principle applying to minimal revisions of belief functions: When revising your beliefs in order to accept A do not give up anything you already accept unless you need to.

We shall take any function from $\mathscr{E}\times\mathscr{E}$ into the interval $[0,1]$ which satisfies 1–4 to be a suitable representation for an extended conditional belief function (ebf).

THEOREM 2.1: If P is an extended conditional belief function then P is a Popper function.

Proof. The plan of the proof is to show that a violation of any of the axioms for Popper functions will also violate one of the constraints on extended conditional belief functions. Usually this will consist in showing a violation of coherence by means of one of the betting systems used in John Kemeny's version of the dutch book argument (Kemeny [26] pp. 263–266).

Axiom 1 for Popper functions,

a1. $0 \leqslant P(B/A) \leqslant P(A/A) = 1$,

is trivially required by constraint 1 on extended conditional belief functions.

Consider axiom 3.

a3.   If $P(C/A) \neq 1$ then $P(\bar{B}/A) = 1 - P(B/A)$.

When the hypothesis is satisfied for some C, constraint (2) (ii) on conditional belief functions requires that $K(P_A)$ be non-empty. Therefore, constraint 4 non-trivially requires coherence relative to $K(P_A)$. Unless $P(\bar{B}/A) = 1 - P(B/A)$ coherence relative to $K(P_A)$ will be violated.

The situation with axiom 5

a5. $P(AB/C) \leqslant P(B/C)$

is similar. If $K(P_C) = \emptyset$ then both sides of the inequality are trivially 1. If $K(P_C) \neq \emptyset$ then coherence with respect to $K(P_C)$ is sufficient to guarantee that the inequality holds. The axioms so far considered all fall out directly from the coherence requirement that P_A be a classical probability function on $\mathscr{E}/P_A$.

Axioms 2 and 4 involve relations between different $K(P_A)$'s. Let us deal with a4 first.

a4. $P(AB/C) = P(A/C) \cdot P(B/AC)$.

There are two cases to consider.

Case I. $K(P_C) \cap A = \emptyset$. When $K(P_C) \cap A$ is empty then both sides of a4 must be zero. Since A is false in every world in $K(P_C)$ both $P(A/C)$ and $P(AB/C)$ equal zero.

Case II. $K(P_C) \cap A \neq \emptyset$. When $K(P_C) \cap A$ is non-empty then condition 4 on conditional belief functions requires that $K(P_{AC}) = K(P_C) \cap A$. This has the effect that a conditional bet on B relative to A in $K(P_C)$ at odds $r : 1 - r$ and stake S has exactly the same outcomes as a straight bet on B at the same odds and stake in $K(P_{AC})$. Given this, bets on B in $K(P_{AC})$ can be represented as conditional bets in $K(P_C)$ so that all the degree of belief values can be represented in a system of bets all relative to the same partition of possible outcomes. If a4 is not satisfied the incoherence of P will show up in the bets used by Kemeny to show the corresponding law for confirmation functions (Kemeny [1] pp. 265–266). Though axiom 2 is not as familiar to students of probability as the other axioms it does follow from the constraints on extended conditional belief functions.

a2.   If $P(B/A)=1=P(A/B)$ then $P(C/A)=P(C/B)$.

Assume $P(B/A)=1=P(A/B)$. By condition 4 we have $K(P_{AB})=K(P_A)\cap B$ if $K(P_A)\cap B\neq\emptyset$, and $K(P_{AB})=K(P_B)\cap A$ if $K(P_B)\cap A\neq\emptyset$. Case I: $K(P_A)\cap B\neq\emptyset$ and $K(P_B)\cap A\neq\emptyset$. Here both $K(P_A)$ and $K(P_B)$ equal $K(P_{AB})$ so that a2 follows easily by coherence constraints on $K(P_{AB})$. Case II: $K(P_A)\cap B=\emptyset$. Here $K(P_A)$ is empty, since by the hypotheses of the theorem $P(B/A)=1$. Therefore, by constraint (2) (iii) $P(\bar{A}/A)=1==P(\bar{A}/\bar{A})$.

The general constraint on Popper functions

r1.3. (i) If $P(D/\bar{D})=1$ then $P(D/E)=1$ for all E

follows from axioms a1 and a3-a5 and constraint 1-4 on ebf's.[15] Since these already established axioms are sufficient for it, we may use r1.3 (i) to justify a2. Using this remark we have $P(\bar{A}/B)=1$, since $P(\bar{A}/\bar{A})=1$.

We now have $K(P_B)$ is empty, since both $P(A/B)=1$ and $P(\bar{A}/B)=1$. Thus, $K(P_B)=K(P_A)$ and both are empty. Case III $K(P_B)A=\emptyset$ is symmetrical with what we showed for case II. This completes showing the validity of a2 and completes the proof of the theorem. ∎

We say that Popper function P on field $\mathscr{E}$ is *compact* just in case for all A

$$K(P_A)=\emptyset \quad \text{only if} \quad P(\bar{A}/A)=1.$$

If $\mathscr{E}$ allows filters Δ such that $\cap\Delta=\emptyset$, but $\emptyset\notin\Delta$ then there can be Popper functions on $\mathscr{E}$ that fail to be compact. Such Popper functions will violate condition 2ii on extended conditional belief functions.

THEOREM 2.2: If P is a compact Popper function on $\mathscr{E}$ then P, is a suitable representation for an extended belief function on $\mathscr{E}$.

Proof. Constraints 1–2ii are trivially met. Constraint 3 follows from the fact that P_A is a probability function whenever $K(P_A)\neq\emptyset$ together with Kemeny's result (p. 223 above). Constraint 4 follows by manipulation from the Popper function axioms.[16] ∎

Stalnaker constructed extended belief functions and Popper functions on sentences. Suppose **S** is a set of sentences closed under syntactical operations **ab** and **ā**. Let T be the set of maps v from **S** into $\{0, 1\}$ such that $v(\mathbf{ab}) = v(\mathbf{a}) \cdot v(\mathbf{b})$ and $v(\mathbf{\bar{a}}) = 1 - v(\mathbf{a})$ for **a**, **b** in **S**. This is the set of truth valuations on **S** with **ab** as conjunction and **ā** as negation. For each **a** let A be the set of v in T such that $v(\mathbf{a}) = 1$. The set $\mathscr{E}$ of A such that **a** is in **S** is a field of subsets of T.

A function P taking $\mathbf{S} \times \mathbf{S}$ into the reals is suitable as an extended conditional belief function on **S** just in case the corresponding function P' on $\mathscr{E}$ such that

$$P'(B/A) = P(\mathbf{b/a})$$

satisfies 1–4 with respect to $\mathscr{E}$. On this formulation there is a representation theorem.

THEOREM 2.3: P is suitable as an extended conditional belief function on **S** iff P is a Popper function on S.

Proof. Just as in theorems 1 and 2 except that compactness now follows from compactness of truth functional logic. ■

Stalnaker's result is a general representation theorem for Popper functions. Any minimal algebra **F** can have a truth valuation put on it. The truth valuations provide a compact field within which to construct $K(P'_A)$. From Stalnaker's theorem we see that putting 1–4 on these $K(P'_A)$ will insure that P is a Popper function.

This also indicates that when $\mathscr{E}$ is an algebra with structure we care about there will be many Popper functions on $\mathscr{E}$ that ignore and even clash with the structural properties of $\mathscr{E}$. This is an obvious result of the fact that the Popper function is not based on the structure of $\mathscr{E}$ but induces whatever structure it needs onto $\mathscr{E}$. Because we cared about the structure of the proposition space we made belief functions responsive to it by defining $K(P_A)$ on the proposition field itself. This does not mean, however, that Popper's important theorem about Popper function induced structure is not useful for representation of belief. In fact the equivalence classes induced by P play a very important role.

3. *Conceptual Frameworks*

Suppose that P is an extended belief function on a field $\mathscr{E}$ of subsets of T. The algebra of equivalence classes induced by P on $\mathscr{E}$ corresponds to an important conceptual structure for a P-agent. The P-equivalence relation,

$$A \underset{\widetilde{p}}{} B \quad \text{iff} \quad P(A/C) = P(B/C) \quad \text{for all} \quad C,$$

holds between just those pairs of propositions that, for a P-agent, are not distinguishable by means of any possible evidence. No assumption whatever counts as more evidence for one than the other.

The maximum element of $\mathscr{E}//p$ is the set of all A such that $P(A/\bar{A}) = 1$. By remark 1.3i, $P(A/\bar{A}) = 1$ if and only if $P(A/C) = 1$ for all C. Thus, A is P-valid just in case a P-agent would accept A relative to any assumption C. If A is P-valid then a P-agent will count nothing as evidence against A. The P-valid propositions can be regarded as postulates of the agent's conceptual framework.

The basic constraints on extended belief functions insure that P-validity must conform to semantical possibility relative to $\mathscr{E}$. Let

$$\Delta^*(P) = \{A : P(A/\bar{A}) = 1\}$$

and

$$K^*(P) = \cap \{A : P(A/\bar{A}) = 1\}.$$

From the axioms on Popper functions we have

$$\Delta^*(P) \subseteq \Delta(P_A) \quad \text{for all} \quad A$$

and, thus

$$K(P_A) \subseteq K^*(P) \quad \text{for all} \quad A.$$

From this it follows that

$$K^*(P) \neq \emptyset$$

by constraint 2ii on belief functions. If $K^*(P) = \emptyset$ then $K(P_A) = \emptyset$ for all A which violates 2ii.

The other natural assumption about $K^*(P)$ is that

$$K^*(P) = T$$

so that $P(A/\bar{A}) = 1$ only if A is true in every possible world of $\mathscr{E}$. This does not follow from the constraints on extended conditional belief functions, and it should not be added.[17] One of the beauties of the Popper function representation is that part of the conceptual framework can be read off from the belief function. Nothing is lost by letting the structure of $\mathscr{E}$ be less specific than that given by $\mathscr{E}//P$. By letting $K^*(P)$ be less than all of T we can impose whatever specific meaning postulates that may be peculiar to the agent's conceptual framework. Finally, when $K^*(P)$ is allowed to be less than all of T there is room for the kind of conceptual change would correspond to giving up some meaning postulate.

4. Expected Utility and Measurement

The basic assumption here is that if the agent can conceive of evidence that would support A then he ought to be able to make hypothetical choices relative to the assumption that A holds. Thus, I assume that whenever $P(\bar{A}/A) \neq 1$ the agent has expected utilities defined for $K(P_A)$.

One of the standard puzzles about counterfactual assumptions is that there are often incompatible alternative ways to alter one's background knowledge to accommodate the assumption. In the framework of partial belief this problem is not crucial. As we have been expounding the heuristics for it, $\Delta(P_A)$ will include only those propositions that the agent is sure he ought to hold if he were to accept A. The agent will not choose between alternatives unless he is sure of one of them. This in no way affects the construction $K(P_A)$. There is no need to choose one of them since the agent can utilize finer partitions of $K(P_A)$ to define his expected utilities for it.

No general method for constructing exactly what $K(P_A)$ ought to be, given what $K(P_T)$ and A are, has been provided. What has been done is to provide constraints 1–4 on what $K(P_A)$ constructions are permissible. These constraints have been shown to be sufficient to have P a Popper function.

Though I do not present the details here, the outline of a treatment of expected utility appropriate to the Popper function representation of belief is not difficult. Expected utility is relativized to assumptions. Where $B_0 \ldots B_{n-1}$ is a partition of $K(P_A)$ and $a_0 \ldots a_{k-1}$ are hypothetical acts to be decided upon and for all a_j and B_i $P(B_i/A)$ is independent of a_j; then

$$E_A(a_j) = \sum_{i<n} U(a_j B_i) \cdot P(B_i/A).$$

More generally when $B_0 \ldots B_{n-1}$ is a partition of $K^*(P)$

$$E_A(a_j) = \sum_{i<n} U(a_j B_i \cap K(P_A)) \cdot P(B_i/A).$$

Appropriate modifications for any of the present axiomatic treatments of expected utility should be fairly routine. The main change is that $K(P_A)$ rather than A is the appropriate proposition to relativize to when accessing $E_A(a_j)$.

Measurement of counterfactual conditional probabilities is made by relativizing the Ramsey measuring choice to $K(P_A)$.

Assume that it were that A, then relative to this assumption choose between

(a)   receive prize x if B and nothing if not
(b)   receive prize x if any one of m of the n random outcomes of gambling device comes up, nothing otherwise.

This choice involves making a hypothetical assumption that may conflict with what the agent accepts, but the ordinary Ramsey measurement may do so as well. For many propositions I am quite sure that there is no way of making sure that receipt of the prize attaches to the truth of B. I doubt whether the counterfactual choices my measurements would require need be any worse off than some of those Savage's framework would require.

Once again, the most important idea supporting the extension of expected utility reasoning to cover assumptions that conflict with what the agent accepts is this: To the extent that the agent can conceive of possible evidence that would lead him to accept something that conflicts with what he now accepts he ought to be able to make hypothetical judgments about what actions would be appropriate if it were the case. In a nut shell, to the extent the agent seriously allows for the possibility that some-thing he accepts may be mistaken he should be able to plan for such contingencies.

III. COUNTERFACTUALS AND ITERATED CONDITIONALIZATION

1. Iterative Probability Models

One motivation for extending belief functions from classical probabilities to Popper functions is to be able to revise previously accepted evidence. When belief functions are represented only as classical probabilities previously accepted evidence cannot be revised by conditionalization on later inputs.

If P is only a classical probability and $P(A)=1$ then $P(A/C)=1$ for every C such that $P(C)>0$ and $P(A/C)$ is not defined when $P(C)=0$. If P is a Popper function $P(A/C)$ can be non-trivially defined even when $P(C)=P(C/T)=0$. This goes part of the way toward a solution.

It is not hard to see, however, that extension to Popper functions cannot be the whole solution. Even though P_C may be well defined when $P(C)=0$ so that the agent can shift his absolute belief function from P_T to P_C many of the new conditional belief assignments are not specified. For any B such that $P_C(B)=0$, no values for $P_C(H/B)$ are specified. Failure to specify these conditional beliefs can result in later failure to specify absolute degrees of belief. If after having found reason to accept C one were to later find reason to accept some further proposition B such that $P_C(B)=0$, then his appropriate new absolute belief function would not be specified. Introducing the Popper function representation does allow corrigibility; but, without some further apparatus, this only extends as far as one correction.

The problem with iterated shifts would be solved if one were able to specify not only the new absolute belief function P_C but also the rest of the values for an appropriate new Popper function $P_{\langle C \rangle}$. If such a $P_{\langle C \rangle}$ is defined for each extended belief function P and proposition C such that $P(\bar{C}/C) \neq 1$, then the appropriate shifts in belief as one accepts new evidence can be iterated even when there are iterated clashes of evidence. The shift from P to $P_{\langle C \rangle}$ will give all the P_C values so that $P_C(B)= =P_{\langle C \rangle}(B/T)$, and it will define $P_{\langle C \rangle}(H/B)$ when $P_{\langle C \rangle}(B/T)=0$. Upon being confronted with B as a new input, a new shift from $P_{\langle C \rangle}$ to $P_{\langle C \rangle \langle B \rangle}$ is made. In order to achieve a representation with this kind of richness additional apparatus is needed.

In order to see how to go about doing this, it will be helpful to reconsider Carnap's confirmation and credence functions. If K expresses everything a rational agent accepts, then, according to Carnap, his credence function $\mathscr{C}_{(K)}$ should conform to the confirmation function $\mathscr{C}$ so that

$$\mathscr{C}_{(K)}(B/A)=\mathscr{C}(B/K \cap A).$$

Where $K \cap A$ is empty, $\mathscr{C}_{(K)}$ is undefined.

Let us investigate what happens when we represent rational credence by means of Popper functions. Since we shall want to speak of $\Delta(P_{\langle C \rangle}, A)$ we shift our notation from $\Delta(P_A)$ to $\Delta(P, A)$. We understand $\Delta(P, A)=$

$= \{B: P(B/A)=1\}$ to be the set of propositions a P-agent would accept were he to minimally revise his beliefs to accept A. If the set $K(P, A)=$ $=\cap \varDelta(P, A)$ is a member of $\mathscr{E}$, then it is a single proposition expressing the total evidence accepted after the shift to accept A. This suggests that the appropriate relationship between the Popper function representation of rational credence and its corresponding confirmation function ought to be

$$P(B/A)=\mathscr{C}(B/K(P, A))$$

provided $K(P, A)$ is non-empty. When $K(P, A)$ is empty, $P(\bar{A}/A)=1$ and A is regarded as absurd. An absurd A is not a possible candidate for being accepted as a new input. Thus, there is no loss from having $\mathscr{C}$ undefined for the empty proposition $\emptyset$.

Our constraints on $\mathscr{E}$ do not ensure that $K(P, A)$ will be a member of $\mathscr{E}$. If $K(P, A)$ were denumerable and $\mathscr{E}$ were not closed under denumerable intersections, then we could have it that $K(P, A)$ is not in $\mathscr{E}$. We shall assume that $\mathscr{E}$ is closed under the formation of $K(P, A)$ for every A, and that $\mathscr{C}$ is defined on $\mathscr{E}$.

The connection between P and $\mathscr{E}$ allows $P_{\langle c \rangle}(B/A)$ to be specified in terms of $K(P_{\langle c \rangle}, A)$ so that

$$P_{\langle c \rangle}(B/A)=\mathscr{C}(B/K(P_{\langle c \rangle}, A)).$$

We are assuming that $\mathscr{C}(B/K(P_{\langle c \rangle}, A))$ represents the degree of belief in B appropriate to a rational agent with $K(P_{\langle c \rangle}, A)$ as his total evidence, and that $K(P_{\langle c \rangle}, A)$ is the total evidence accepted by a $P_{\langle c \rangle}$ agent who minimally revises his beliefs to accept A.

This goes part of the way toward solving the shifting problem because $K(P_{\langle c \rangle}, A)$ can be defined in terms of acceptance alone. No values of $P_{\langle c \rangle}$ except those where $P_{\langle c \rangle}(B/A)=1$ need be considered. If $\mathscr{E}$ were closed under a binary propositional function f such that

$$P_{\langle c \rangle}(f(AB))=1 \quad \text{iff} \quad P_{\langle c \rangle}(B/A)=1$$

then $K(P_{\langle c \rangle}, A)$ could be defined in terms of P. We have $P_{\langle c \rangle}(A)=P(A/C)$ so that

$$\varDelta(P_{\langle c \rangle}, A)=\{B: P(f(AB)/C)=1\},$$
and
$$K(P_{\langle c \rangle}, A)=\cap\{B: P(f(AB)/C)=1\}.$$

Given all this, $P_{\langle c \rangle}$ could be defined in terms of $\mathscr{C}$ and P.

One paradigm for counterfactual conditionals is characterized by acceptability conditions that meet the requirements we want for f. I call this the Ramsey test paradigm because it is characterized by a version of Ramsey's test for acceptability of hypotheticals (Ramsey [47] p. 24. Robert Stalnaker is responsible for the specific version of the test. He uses it to characterize the use of conditionals he intends his theory to explicate (Stalnaker [53]. The test is summed up in the following slogan. Accept $A \square \!\!\to B$ (the conditional with antecedent A and consequent B) if and only if the minimal revision of your system of beliefs needed to accept A also requires accepting B. On the Popper function representation of belief, this comes to:

$$\text{Accept } A \square \!\!\to B \quad \text{iff} \quad K(P, A) \subseteq B.$$

Any conditional with these acceptability conditions for rational belief functions will satisfy the constraints on f.

Let us augment $\mathscr{E}$ to include closure under $A \square \!\!\to B$. We assume that both P and $\mathscr{C}$ are defined on the augmented $\mathscr{E}$. For each C such that $P(\bar{C}/C) \neq 1$ we have, for all A and B

I. (i) $\quad \Delta(P_{\langle C \rangle}, A) = \{B : P(A \square \!\!\to B/C) = 1\}$

(ii) $\quad K(P_{\langle C \rangle}, A) = \cap \Delta(P_{\langle C \rangle}, A)$

(iii) $\quad P_{\langle C \rangle}(B/A) = \mathscr{C}(B/K(P_{\langle C \rangle}, A)), \quad \text{if} \quad K(P_{\langle C \rangle}, A) \neq \emptyset$
$\quad$ and $\quad P_{\langle C \rangle}(B/A) = 1 \quad \text{if} \quad K(P_{\langle C \rangle}, A) = \emptyset.$

We are interested in iterated shifts. Write '$P_{\langle C \rangle_n}$' for '$P_{\langle C_0 \rangle \cdots \langle C_{n-1} \rangle}$' and let '$P_{\langle C \rangle_0}$' denote P. A Popper function adequate for iterated shifting must have $P_{\langle C \rangle_{n+1}}$ adequate for shifting whenever $P_{\langle C_n \rangle}(\bar{C}_n/C_n) \neq 1$. To this end we give the following definition of an iterative probability model relative to $\mathscr{C}$.

II. $\quad$ P is $I_p(\mathscr{CE})$ (P is an iterative probability model for $\mathscr{E}$ relative to $\mathscr{C}$).

iff P is a Popper function on $\mathscr{E}$, $\mathscr{C}$ is a confirmation function that agrees with P, and for any $n+1$ length sequence C of propositions such that $P(\bar{C}_n/C_n) \neq 1$

(a) $\quad P_{\langle C \rangle_n}(A \square \!\!\to B/T) = 1 \quad \text{iff} \quad P_{\langle C \rangle_n}(B/A) = 1$

(b) $\quad P_{\langle C \rangle_n}$ is a Popper function on $\mathscr{E}$

(c) $\quad P_{\langle C \rangle_{n+1}}(B/T) = P_{\langle C \rangle_n}(B/C_n).$

One least elementary remark.

(d)   If P is $I_p(\mathscr{E})$ and $P_{\langle C \rangle_n}(\bar{C}_n/C_n) \neq 1$ then $P_{\langle C \rangle_{n+1}}$ is $I_p(\mathscr{E})$.

Each non-trivial $P_{\langle C \rangle_n}$ is itself an iterative probability model.

In the presence of (a) the requirement that $P_{\langle C \rangle_n}$ be a Popper function is equivalent to three constraints on $P_{\langle C \rangle_n}$ assignments to conditionals.

R.1.1. $P_{\langle C \rangle_n}$ is a Popper function iff

(1) $P_{\langle C \rangle_n}(A \,\square\!\!\rightarrow A) = 1$
(2) $P_{\langle C \rangle_n}(A \,\square\!\!\rightarrow B) \neq 1$ for some A and B, and
(3)   If $P_{\langle C \rangle_n}(A \,\square\!\!\rightarrow \bar{B}) \neq 1$ then
           $P_{\langle C \rangle}(AB \,\square\!\!\rightarrow D) = 1$ iff $P_{\langle C \rangle_n}(A \,\square\!\!\rightarrow (B \supset D)) = 1$.

Proof. The most important step is that (3) is equivalent to

(4) $K(P_{\langle C \rangle_n}, AB) = K(P_{\langle C \rangle_n}, A) \cap B$,
           provided $K(P_{\langle C \rangle_n}, A) \cap B \neq \emptyset$.

That (3) and (4) are, thus, equivalent follows quite straight-forwardly from definition I and (a). To show that (1)–(3) are sufficient to have $P_{\langle C \rangle_n}$ be a Popper function we need only note that (1) and (2) together with definition I and (a) yield that $P_{\langle C \rangle_n}$ satisfies constraints c1–c3 on extended conditional belief functions. Since (4) is just constraint c4 applied to $P_{\langle C \rangle_n}$ theorem II 2.1 yields the desired result. To show that if $P_{\langle C \rangle_n}$ is a Popper function then (1)–(3) are satisfied note that by (a), (1) and (2) follow trivially from the basic constraints on Popper functions. Since c4, also, holds for Popper functions we have (3) as well. ∎

2. *Ip-validity and Conditional Logic*

Let us investigate validity relative to iterative probability models. Assume that $\mathscr{E}$ is a countable field of propositions closed under conditionals, so that $A \,\square\!\!\rightarrow B$ is in $\mathscr{E}$ for every A and B in $\mathscr{E}$. We define iterative probability validity for A in $\mathscr{E}$.

IV. A is Ip-valid iff for every iterative probability model P on $\mathscr{E}$
           $P(A/C) = 1$ for every C in $\mathscr{E}$.

For a given probability model P we already had a notion of P-validity in that A is P-valid just in case $P(A/C) = 1$ for all C. Our stronger notion of Ip-validity is the obvious one that A be P-valid relative to every iterative probability model P.

The following axiomatization characterizes validity for David Lewis' basic logic **VC** for counter-factual conditionals.[18] Where A, B, C and D are any propositions in $\mathscr{E}$ the following are the rules and axioms:

Rules (1) Modus Ponens
      (2) Deduction within conditionals: for any $n \geqslant 1$,

$$\frac{\vdash (\bigcap_{i<n} A_i) \supset B}{\vdash (\bigcap_{i<n} (C \square\!\!\!\to A_i)) \supset (C \square\!\!\!\to B)}$$

Axioms (1) All truth functional tautologies

    (2) $A \square\!\!\!\to A$
    (3) $(\bar{A} \square\!\!\!\to A) \supset (B \square\!\!\!\to A)$
    (4) $\overline{(A \square\!\!\!\to \bar{B})} \supset (((AB \square\!\!\!\to D) \equiv (A \square\!\!\!\to (B \supset D)))$
    (5) $(A \square\!\!\!\to B) \supset (A \supset B)$
    (6) $AB \supset (A \square\!\!\!\to B)$.

The class $V(T)$ of **VC**-valid propositions of $\mathscr{E}$ is the smallest subset of $\mathscr{E}$ containing every instance of each axiom and closed under the rules. The axiomatization also characterizes valid consequence for **VC**-logic. Suppose $\varDelta$ is a subset of $\mathscr{E}$. The class $V(\varDelta)$ of **VC**-valid consequences of $\varDelta$ is the smallest subset of $\mathscr{E}$ which includes $\varDelta \cup V(T)$ and is closed under Modus Ponens. **VC**-consistency is defined in the usual way, i.e. $\varDelta$ is **VC**-consistent just in case there is no B such that B and $\bar{B}$ are both in $V(\varDelta)$.

Theorem 2.1 A is Ip-valid iff A is **VC**-valid.

Proof. From right to left. We show that the **VC**-axioms are Ip-valid and that the **VC**-rules preserve Ip-validity. Thus, we show the soundness of the **VC**-axiomatization for Ip-models. This is facilitated by Remark 1.3 on Popper functions.

1.3 (i) $P(A/\bar{A}) = 1$ iff $P(A/C) = 1$ for all C
    (ii) $P(A/\bar{A}) \neq 1$ iff $P(A/C) = 1$ for some C.

If we cannot consistently assume that $P(A/C) = 0$ for some Ip-model P and proposition C then A is Ip-valid. By remark d (this section) we have that $P_{\langle C \rangle}$ is an Ip-model if P is and $P(\bar{C}/C) \neq 1$. Therefore, to show A is Ip-valid it suffices to show that there is no Ip-model P such that $P(A) = P(A/T) = 0$.

That modus-ponens preserves Ip-validity follows trivially from the fact that when $P(\bar{A} \cup B) = 1$ then $P(A) \leqslant P(B)$. Consider Rule 2: Assume $(\bigcap_{i<n} A_i) \supset B$ is Ip-valid and that P is an Ip-model such that

$$P((\bigcap_{i<n} (C \square\!\!\rightarrow A_i)) \supset (C \square\!\!\rightarrow B)) = 0.$$

We have $P(A_i/C) = 1$ for all A_i and, thus, that $P(\bigcap_{i<n} A_i/C) = 1$. We also have $P(B/C) = 0$. But, by the Ip-validity of $(\bigcap_{i<n} A_i) \supset B$ we also have $P(B/C) = 1$ which is impossible. The Ip-validity of the axioms is established similarly. This completes showing the soundness of the **VC**-axiomatization for Ip-validity.

It is worth pointing out, however, that some of Lewis' axioms correspond directly to salient features of Ip-models. Axiom 3, $(\bar{A} \square\!\!\rightarrow A) \supset \supset (B \square\!\!\rightarrow A)$, corresponds to R 1.3 (i) (above) and axiom 1, $A \square\!\!\rightarrow A$, corresponds to the Popper function constraint that $P(A/A) = 1$. The most striking case is axiom 4,

$$\overline{(A \square\!\!\rightarrow \bar{B})} \supset (((AB \square\!\!\rightarrow D) \equiv (A \rightarrow (B \supset D)))),$$

which corresponds to constraint (3) on acceptability of conditionals,

If $P(A \square\!\!\rightarrow \bar{B}) \neq 1$, then
$P(AB \square\!\!\rightarrow D) = 1$ iff $P(A \square\!\!\rightarrow (B \supset C)) = 1$,

and, thus, also to the main $K(P_A)$ condition,

$$K(P, AB) = K(P, A) \cap B, \quad \text{provided} \quad K(P, A) \cap B \neq \emptyset,$$

on extended conditional belief functions. Lewis apologizes for having to use such a long and unintuitive axiom. There is some interest in seeing that in the IP-framework this axiom corresponds directly to a very natural constraint on acceptance.

Let us turn now to the other half of the theorem and show that each Ip-valid A is also **VC**-valid. What we show here is that the **VC**-axiomatization is complete with respect to Ip-validity. If A is Ip-valid then $P(\bar{A}) = 0$ for every Ip-model P. Therefore, if for each **VC**-consistent proposition A there is an Ip-model P such that $P(A) = 1$ then every Ip-valid proposition is a theorem of the **VC**-axiomatization. Thus, the standard Henkin-Lindenbaum procedure for showing completeness is applicable.

Suppose that A is **VC**-consistent (i.e. $\{A\}$ is **VC**-consistent). Lindenbaum's lemma holds for **VC**-consistency (Lewis [37] p. 125). Therefore, there is a maximal **VC**-consistent subset $\varDelta$ of $\mathscr{E}$ such that $A \in \varDelta$. We use $\varDelta$ to define that part of P where $P(B/A)=1$ so that for every A, B in $\mathscr{E}$

$$P(B/A)=1 \quad \text{iff} \quad A\,\square\!\!\rightarrow B \in \varDelta.$$

One **VC**-theorem is $(T\,\square\!\!\rightarrow A)\equiv A$. Since $A \in \varDelta$ so is $T\,\square\!\!\rightarrow A$. Thus, $P(A)= = P(A/T)=1$. For any sequence $C \in \mathscr{E}^n$ and $A \in \mathscr{E}$ we have

$$K(P_{\langle C \rangle_n}, A)=\cap\{B:(C_0\,\square\!\!\rightarrow ...(C_{n-1}\,\square\!\!\rightarrow B))...)\in\varDelta\}$$

and

$$K(P_{\langle C \rangle_n}, A)=\emptyset \quad \text{iff} \quad P_{\langle C \rangle_n}(B/A)=1 \quad \text{for all } B.$$

These follow straight forwardly from the basic properties of **VC**-consistency and definition I.

Let $\mathscr{E}^*$ be the σ-field generated by $\mathscr{E}$. Since $\mathscr{E}^*$ is countable there exist classical measures M on $\mathscr{E}^*$ that assign $M(A)>0$ to every A in $\mathscr{E}^*$ such that $A \neq \emptyset$. Any such M generates a regular confirmation function $\mathscr{C}$ on $\mathscr{E}^* \times \mathscr{E}^* - \{\emptyset\}$ where

$$\mathscr{C}(H/E)=M(H \cap E)/M(E).$$

Let $\mathscr{C}$ be any such confirmation function on $\mathscr{E}^*$.

The rest of the P-values can be defined relative to $\mathscr{C}$ in the manner of definition I. For all A, B in $\mathscr{E}$ let

$$P(B/A)=\mathscr{C}(B/K(P, A))$$

provided $K(P, A)\neq\emptyset$. What remains is to show that P is an Ip-model for $\mathscr{E}$ relative to $\mathscr{C}$. The basic constraint that $P(A\,\square\!\!\rightarrow B/T)=1$ iff $P(B/A)=1$ results from the **VC**-validity of

(T1) $\quad (T\,\square\!\!\rightarrow B))\equiv(A\,\square\!\!\rightarrow B).$

One of the **VC**-axioms (VC2) is $A\,\square\!\!\rightarrow A$. Therefore $A\,\square\!\!\rightarrow A \in \varDelta$, for every A in $\mathscr{E}$ and

(1) $\quad P(A/A)=1 \quad \text{for all } A.$

The maximal consistent set $\varDelta$ cannot have $T\,\square\!\!\rightarrow \bar{T}$ in it or it would not be **VC**-consistent. Therefore,

(2) $P(B/A) \neq 1$ for some A and B.

Finally Lewis' axiom

(VC4) $\overline{(A \,\square\!\!\!\rightarrow\, \bar{B})} \supset (((AB \,\square\!\!\!\rightarrow\, D) \equiv (A \,\square\!\!\!\rightarrow\, (B \supset D))))$

insures that

(3)   If $P(A \,\square\!\!\!\rightarrow\, \bar{B}) \neq 1$ then
           $P(AB \,\square\!\!\!\rightarrow\, C) = 1$ iff $P(A \,\square\!\!\!\rightarrow\, (B \supset D)) = 1$.

Therefore since the basic constraint holds for P remark R.1.1 yields that P is a Popper function.

Assume C is a sequence of propositions of length $n+1$, and that $P_{\langle C \rangle_n}(\bar{C}_n / C_n) \neq 1$. The following derived inference rules hold for any sequence C.

(DR1)   If $A \in V(T)$ then
         $(C_0 \,\square\!\!\!\rightarrow\, \ldots (C_n \,\square\!\!\!\rightarrow\, A \ldots) \in V(T)$.

(DR2)   If $(A \equiv B) \in V(T)$ then
         $(C_0 \,\square\!\!\!\rightarrow\, \ldots (C_{n-1} \,\square\!\!\!\rightarrow\, A) \ldots) \equiv (C_0 \,\square\!\!\!\rightarrow\, \ldots (C_{n-1} \,\square\!\!\!\rightarrow\, B) \ldots$
         $\ldots) \in V(T)$.

When (DR2) is applied to T1) we have

(a) $P_{\langle C \rangle_{n+1}}(A \,\square\!\!\!\rightarrow\, B / T) = 1$ iff $P_{\langle C \rangle_{n+1}}(B/A) = 1$.

When (DR2) is applied to

(T2) $(C_n \,\square\!\!\!\rightarrow\, (T \,\square\!\!\!\rightarrow\, B)) \equiv (T \,\square\!\!\!\rightarrow\, (C_n \,\square\!\!\!\rightarrow\, B))$

we have

           $K(P_{\langle C \rangle_{n+1}}, T) = K(P_{\langle C \rangle_n}, C_n)$

which yields clause

(c) $P_{\langle C \rangle_{n+1}}(B/T) = P_{\langle C \rangle_n}(B/C_n)$.

By (c) and the assumption we have $P_{\langle C \rangle_{n+1}}(\bar{C}_n / T) \neq 1$ and thus that (2) holds for $P_{\langle C \rangle_{n+1}}$. That (1) holds for $P_{\langle C \rangle_{n+1}}$ follows from applying (DR1) to axiom VC-2. Similarly, that (3) holds follows from applying (DR1) to axiom (VC-4). Thus, we have shown that P is an Ip-model for $\mathscr{E}$ relative to $\mathscr{C}$. ∎

3. *Getting Rid of the Confirmation Function*

In the definition of Ip-model confirmation functions were appealed to. The idea was that there should be some probability function $\mathscr{C}$ such that the agents beliefs were as though they conformed to $\mathscr{C}$ by

(1) $\qquad P(B/A) = \mathscr{C}(B/K(P, A))$.

Suppose that $\mathscr{E}$ is closed under formation of $K(P_{\langle C \rangle_n}, A)$, and consider the following condition.

(2) $\qquad K(P_{\langle C \rangle_n}, K(P_{\langle C \rangle_{n+1}}, A)) = K(P_{\langle C \rangle_{n+1}}, A)$.

If this holds for every $P_{\langle C \rangle_n}$ then each $P_{\langle C \rangle_{n+1}}$ is definable in terms of $P_{\langle C \rangle_n}$ by

(3) $\qquad P_{\langle C \rangle_{n+1}}(B/A) = P_{\langle C \rangle_n}(B/K(P_{\langle C \rangle_{n+1}}, A))$.

There is no need to use any confirmation function, since the P values with $K(P, A)$ second arguments do the same job. In Part I it was suggested that credence and confirmation correspond to two different heuristics for conditional probability. Where $P(B/A)$ is the degree of belief in B appropriate to minimally revising to *add* A and nothing further to what one already has, $\mathscr{C}(B/A)$ is the degree of belief appropriate to a minimal revision to *reduce* everything one accepts to just A.

The idea behind (2) is that for $K(P_{\langle C \rangle_n}, A)$ these two procedures should come to the same thing. If $K(P, A)$ is the minimal revision to add A then the minimal revision to add $K(P, A)$ should just be $K(P, A)$ itself. Thus, $\mathscr{C}(B/K(P, A)) = P(B/K(P, A))$.

4. *Probability of Conditionals*

Except for the basic constraints on acceptance of conditionals Ip-models leave open what $P(A \,\square\!\!\!\rightarrow B)$ should be. The inspiration for the present treatment was a system of Stalnaker's (Stalnaker [52] pp. 74–79) based on the hypothesis

(SH) $\qquad P(A \,\square\!\!\!\rightarrow B) = P(B/A)$ all A and B.

David Lewis has shown that Stalnaker's system trivialized in that it can take at most only four values. (Lewis [39] pp. 4–7). Lewis' result applies to any system with (SH) together with

(Ip) $P(A \rightarrow B/C) = P(B/AC)$ all A, B and C

provided $P(AC) > 0$. Since (Ip) holds for Ip-models Lewis' result insures that all significantly non-trivial Ip-models do not satisfy the Stalnaker hypothesis.[19]

5. *Conceptual Change*

With the addition of the conditional operator the agent can use certain propositions to represent that other propositions are postulates of his conceptual framework. By the basic constraint on probability assignments to conditionals

(1) $P(\tilde{A} \square \rightarrow A) = 1$ iff $P(A/\bar{A}) = 1$.

Let '$\square A$' abbreviate $\bar{A} \square \rightarrow A$'. The following rule and axioms for this defined necessity operator are Ip-valid.

(K1) $\dfrac{\vdash A}{\vdash \square A}$

(m1) $\square A \supset A$

(m2) $\square(A \supset B) \supset (\square A \supset \square B)$.

These together with truth functional tautologies characterize necessity in modal system M, the weakest of Kripke's standard modal systems (Kripke [30]). The S4-axiom

(S4) $\square A \supset \square \square A$

is not Ip-valid, because one can have both

$P(A/\bar{A}) = 1$

and

$P(\bar{A} \square \rightarrow A / \overline{\bar{A} \square \rightarrow A}) \neq 1$.

Thus, system M is the modal logic that corresponds to P-validity in Ip-models.

Having Ip-models where the S4 axiom fails allows for the representation of conceptual change. We represent the minimal revision to add A as a new postulate as

$P_{\langle \square A \rangle}$.

We represent the minimal revision to remove postulate A from P-valid status as

$$P_{\langle \overline{\Box A} \rangle}$$

One interesting property of this is that

$$P_{\langle \overline{\Box A} \rangle}(A)=1$$

though of course $P_{\langle \overline{\Box A} \rangle}(\Box A)=0$. This is as it should be.

I think that being able to represent conceptual change in this way is a strong argument in favour of using M-necessity rather than S4 as the necessity induced by the conditional.[20]

6. *Stalnaker vs. Lewis*

Stalnaker's logic of counterfactuals **VC-S** is equivalent to the result of adding a single axiom

(7) $\overline{(A \Box\!\!\rightarrow B)} \supset (A \Box\!\!\rightarrow \bar{B})$

to Lewis' **VC** (Stalnaker [53], Stalnaker and Thomason [54], Lewis [37], [40]). There is now considerable controversy over the merits of Stalnaker's axiom (e.g. Lewis [37], [38], [39], van Fraassen [57], [60], Pollock [42]). If one adds the constraint

(7*) $P(A \Box\!\!\rightarrow B)=0$ iff $P(B/A)=0$

to those imposed on Ip-models then (7) would be valid and Stalnaker's logic would capture Ip-validity.

Even without this, Stalnaker's axiom is valid for 2-valued Ip-models. Since $P(B \cup \bar{B}/A)=1$ we cannot have both $P(B/A)=0$ and $P(\bar{B}/A)=0$. Therefore, either $P(B/A)=1$ or $P(\bar{B}/A)=1$ and in neither case can (7) be assigned zero.

7. *Ellis on the Logic of Subjective Belief*

In a very interesting investigation of the logic of subjective belief Brian Ellis argues that the correct logic of truth for a system of propositions ought to correspond to what would hold in every admissible two-valued probability system for those propositions (Ellis [11] p. 127). He gives the following reasons:

For if we are certain of the premises of a valid argument, we ought to be certain of its conclusion, and if we are not certain of the conclusion of a valid argument, then we ought not to be certain of all its premises....

Consequently, if there is any divergence between our logics of truth and certainty, then either something is wrong with our probability theory or with the way we have applied it to the analysis of arguments or something is wrong with our logic of truth.

If Ellis' principle were right Stalnaker's logic would be the correct logic of subjective belief. The argument he gives however, only supports the weaker principle that the truth logic should capture just those arguments where for any admissible probability system where all the premises receive probability 1.0 the conclusion must also receive probability 1.0. This weaker principle is just Ip argument-validity.

Ellis opens his concluding remarks as follows:

We have as yet no adequate logic of subjective probability. The classical probability calculus is not an adequate logic of subjective probability because

(a)   it is not capable of handling subjective probability claims concerning subjunctive conditions, and
(b)   it is not strong enough to deal with compound conditionals.

The present system of Ip-models has been constructed to answer just these needs.

8. *Concluding Remarks*

This is a good place to make it clear that Ip models and Ip-validity are not intended as a theory of rational belief change. The Ip-constraints characterize coherence and coherent shifting given an input. What they do not specify is what inputs are rational under what circumstances. Clearly, a full theory of rational belief change would have to include a theory of inputs.

I think that some discussions between Bayesians and classical testing theorists are confused by the fact that the testing theorist is talking about rational inputs while the Bayesian is talking about coherence. The orthodox Bayesian may say that no theory of inputs is needed because the only appropriate inputs are observations and they are so obvious as to requires no theory. Taking seriously the idea that observations are fallible, which can be done in the Ip-framework, indicates that some theory of inputs is needed. Working out the details of one might help bring together some of the good points in the two traditions.

Finally, I should like to point out that Ip-models are very much ideal-izations. No actual agent can be expected to have his belief function defined for all the propositions in a Lewis algebra, nor is any actual agent expected to attain the semantical omniscience built into the characteriza-tion of belief functions. There are two comments to be made on this. First the fragment for which an actual agent's belief function is defined can be expected to include conditionals and counterfactual conditions. Second the Ip-model is a constraint on actual rational degrees of belief in much the way that ordinary logic is a constraint on acceptance. To the extent that the belief function is defined and fails to meet the Ip-constraints it is incoherent.

University of Western Ontario

NOTES

[1] Ramsey [46], De Finetti [12] and Savage [49]. This orthodox tradition is a salient subclass of I. J. Good's 46656 (Good [17]) variations on the Bayesian theme. It is charac-terized by the representation of belief functions as point probability functions and by its emphasis on the role of belief in guiding decisions.

[2] One of the most ingenious alternative approaches is that of Cox and Good where certain modest assumptions about belief require that there exists a probability func-tion representing the beliefs. See Cox [10], Good [15]. For the most explicit treatment of the mathematical details see Aczel [1] pp. 319–24. See Shimony [51] for a recent ap-plication of this argument.

[3] Often the field of propositions can be restricted to what Savage calls a small world situation, Savage [49] pp. 87–90. Where each possible world can be regarded as no more than one of the alternative specifications of those factors that would be relevant to the decision problem.

[4] Everything I shall say about expected utility in this paper can be relativized to such situations. In fact I do not believe that the current treatments for cases where $P(A_i)$ depends on a_j are entirely adequate.

[5] The result would be more interesting if it applied to ordinary coherence as well as strict coherence. As we shall see strict coherence is a bit odd anyway.

[6] Originally Carnap's goal was to find constraints that would make $\mathscr{C}$ completely deter-mined by the semantical properties of $\mathscr{E}$. Thus, $\mathscr{C}$ would represent the logical probability function generated by the field of propositions $\mathscr{E}$. This goal of a single logical $\mathscr{C}$-function now seems unattainable and has been largely given up. For a discussion of the changes in Carnap's program from a basically Popperian point of view see Lakatos [32]. The best statement of the new more modest goals of the Carnapian program is Jeffrey [25].

[7] Carnap defines regularity so that certain propositions are exempt. This is a mistake on his part for the following two reasons. First, the strict coherence argument he ex-plicitly claims as the justification for regularity allows no such exceptions. Secondly, violations of full regularity clash with the basic heuristic that guides the program (see text).

[8] Thus, Carnap's objections to acceptance of hypotheses (Carnap [8] pp. 28–31) do not rest

108   WILLIAM L. HARPER

on strict coherence. Indeed the discussion of credence suggests strongly that Carnap allows acceptance of evidence claims.

[9] See May and Harper [41] for discussing the minimum change idea together with some metrics and optimization techniques.

[10] If we replace conditional bets by conditional expected utility and the assumption that the bets-beliefs postulate holds for both P_0 and P_1 by the assumption that utility assignments relative to propositions in $K(P_0) \cap K(P_1)$ remain unchanged, my version of the dutch book argument is transformed into a version of Savage's conditional expected utility argument for conditionalization.

[11] (Popper [34] pp. 318–358). Popper's published work on this subject is given in a series of papers starting in 1938. Most of these papers together with bibliographic material are included in the pages cited.

[12] A minimal algebra is simply a set closed under a unary and a binary relation.

[13] Popper's main axiomatization uses the weaker

a2′.   If $P(A/D) = P(B/D)$ for all D, then
$P(C/A) = P(C/B)$.

The present a2 is given as an alternative stronger version on p. 335 of Popper [1]. Popper uses the much weaker assumption

$$P(A/A) = P(B/B)$$

in place of a1. I use a1 only to make things more perspicuous.

[14] In the standard Kolmogorov treatment classical probabilities are defined in σ-fields and are σ-additive. If A is a denumerable sequence of pairwise disjoint sets then

$$P\left(\bigcup_{i<\omega} A_i\right) = \sum_{i<\omega} P(A_i).$$

We may, also, have σ-additive Popper functions. Nothing I shall say in the present paper will turn on the difference between finite additively and σ-additivity. In future work developing the measure theory for Popper functions σ-additivity will be important.

A very nice construction for Popper measures by combining even non-denumerably many classical probability measures has been developed by Bas C. van Fraassen [59].

[15] Proof of a1. 3 (i): We proceed by first showing a lemma. The proof of this is essentially that given in Popper [1] p. 352. In the present version a2 is not appealed to.

LEMMA. If $P(\bar{C}/C) \neq 1$ then $P(AB/C) + P(A\bar{B}/C) = P(A/C)$.
Assume $P(\bar{C}/C) = 1$ and note

(1) $P(B/AC) + P(\bar{B}/AC) = P(C/AC) + P(\bar{C}/AC)$.

holds in case $P(\overline{AC}/AC) = 1$ by constraint 2ii and in case $P(\overline{AC}/AC) \neq 1$ by a3. Multiply both sides of 1 by $P(A/C)$.

(2) $P(A/C) \cdot P(B/AC) + P(A/C) \cdot P(\bar{B}/AC) =$
$= P(A/C) \cdot P(C/AC) + P(A/C) \cdot P(\bar{C}/AC)$.

Using a4 on 2 we get:

(3) $P(AB/C) + P(A\bar{B}/C) = P(AC/C) + P(A\bar{C}/C)$.

Coherence (or a4, a5 and a4) yields $P(AC/C) = P(A/C)$, and $P(A\bar{C}/C) = 0$.
We turn now to the main result.

R1.3(i)   If $P(A/\bar{A}) = 1$ then $P(A/B) = 1$ all B.

Assume $P(A/\bar{A})=1$, and note that $P(A/B)=1$ trivially if $P(\bar{B}/B)=1$ (by constraint 2ii). Assume $P(\bar{B}/B)\neq 1$ and use the lemma.

(1) $P(AA/B)+P(A\bar{A}/B)=P(A/B)$

Use a4 on 1:

(2) $P(A/B)\cdot P(A/AB)+P(\bar{A}/B)\cdot P(A/\bar{A}B)=P(A/B)$.

By a1, a5 $P(A/AB)=1$. Since $P(A/\bar{A})=1$ we have $P(\bar{A}/\bar{A})=1$ and by 2ii $P(BA/\bar{A})=1$ which yields $P(A/\bar{A}B)=1$ by a4. Thus,

(3) $P(A/B)+P(\bar{A}/B)=P(A/B)$

which completes the proof.

[16] Assume $K(P_B)\sim A\neq\phi$. Thus, $P(\bar{B}/B)\neq 1$ and $P(A/B)\neq 0$. To see that $K(P_{AB})\subseteq$ $\subseteq K(P_B)\cap A$ assume $P(C/B)=1$. $P(A\bar{C}/B)=0$, therefore $P(A/B)\cdot P(\bar{C}/AB)=0$ and $P(C/AB)=1$. To see that $K(P_B)\cap A\subseteq K(P_{AB})$ assume $P(C/AB)=1$ and note that $P(\bar{A}\cup$ $\cup C/B)=1$.

[17] Hughes Leblanc attempted to show that Popper functions on sentences are equivalent to a natural extension of Carnap's treatment of confirmation functions for sentences (Leblanc [33]). The following definitions of a deducibility relation $\vdash$ and extended confirmation function $\mathscr{C}$ are equivalent to ones Leblanc gives (Leblanc [33]).

c0. $\vdash a$ iff a is a tautology

c1. $0\leqslant\mathscr{C}(b/a)\leqslant\mathscr{C}(a/a)=1$

c2.   If $\vdash a\equiv b$ and $\vdash c\equiv d$ then $\mathscr{C}(a/c)=\mathscr{C}(b/d)$

c3.   If not $\vdash\bar{a}$ then $\mathscr{C}(\bar{b}/a)=1=\mathscr{C}(b/a)$

c4. $\mathscr{C}(ab/c)=\mathscr{C}(a/c)\cdot\mathscr{C}(b/ac)$

Leblanc gives the following axioms for Popper functions

a1. $0\leqslant P(b/a)\leqslant P(a/a)=1$

a2.   If $a\sim b$ then $P(c/a)=P(c/b)$

a3.   If $P(c/a)\neq 1$ then $P(\bar{b}/a)=1=P(b/a)$

a4. $P(ab/c)=P(a/c)\cdot P(b/ac)$

a5. $P(ab/c)=P(ba/c)$,

to which he adds

a6. $\vdash_p a$ iff $P(a/\bar{a})=1$.

He then shows

(1)   Any $\vdash$ and $\mathscr{C}$ satisfying c0–c4 also satisfy a1–a6.

He also claims to show

(2)   Any P and $\vdash_P$ that satisfy a1–a6 also satisfy c0–c4.

The second claim is false, because $\vdash_P$ need not capture only tautologies. This was first pointed out by Stalnaker (Stalnaker [52] footnote p. 70). I include these remarks, because Stalnaker did not indicate how Leblanc went wrong, nor what was actually proved. Indeed one can have a Popper function where $P(a/\bar{a})=1$ or $P(\bar{a}/a)=1$ for all a. Leblanc misleads himself by using the axiomatization that is supposed to recursively define $\vdash$ as simple constraints on $\vdash$.

What Leblanc actually succeeds in proving is that any Popper function P satisfies c1–c4 relative to $\vdash_P$ and that $\vdash_P$ captures at least all tautologies. Furthermore, his proof of the converse is actually a proof of the stronger claim that any $\vdash$ which captures at least all tautologies and $\mathscr{C}$ which satisfies c1–c4 relative to $\vdash$ also satisfy a1–a6.

[18] Lewis [37] p. 132.
[19] After Lewis' trivialization [39], Stalnaker has given up (SH). (Comment delivered by Stalnaker at CPA 1972). Bas van Fraassen, however, has been developing ingenious attempts to circumvent Lewis' results (van Fraassen [57], [58]). These attempts all reject (Ip). Since (Ip) corresponds to iterated conditionalization I think that the price van Fraassen pays to keep (SH) is too high.
[20] This point was the motivation for developing the method of handling universal instantiation for M-versions of conditional logic given in Harper [19].
[21] See next section.

BIBLIOGRAPHY

[1] Aczel, J., *Lectures on Functional Equations and Their Applications*, Academic Press, New York, 1966.
[2] Adams, E., 'The Logic of Conditionals', *Inquiry* **8** (1965) 166–197.
[3] Adams, E., 'Probability and the Logic of Conditionals', in *Aspects of Inductive Logic* (J. Hintikka and P. Suppes, eds.), North Holland, Amsterdam, 1966, pp. 265–316.
[4] Adams, E., *The Logic of Conditionals: An Application of Probability to Deductive Logic*, D. Reidel, Dordrecht, 1975.
[5] Adams, E., 'Prior Probabilities and Counterfactual Conditionals', this volume, p. 1.
[6] Carnap, R., 'Inductive Logic and Rational Decisions', in Carnap and Jeffrey [8], pp. 7–31.
[7] Carnap, R., 'A Basic System of Inductive Logic, Part I', in Carnap and Jeffrey [8], pp. 35–165.
[8] Carnap, R. and Jeffrey, R. L. (eds.), *Studies in Inductive Logic and Probability*, Univ. of California Press, Los Angeles, 1971.
[9] Chisholm, R., 'The Contrary-to-Fact Conditional', in *Readings in Philosophical Analysis* (H. Feigl, and W. Sellars eds.), New York, 1949, pp. 482–497.
[10] Cox, R. T., *The Algebra of Probable Inference*, Johns Hopkins Press, Baltimore, 1960.
[11] Ellis, B., 'The Logic of Subjective Probability', *British Journal for the Philosophy of Science* **24** (1973) 125–152.
[12] de Finetti, B., 'Foresight: Its Logical Laws, Its Subjective Sources', in Kyburg and Smokler [31], pp. 93–158.
[13] de Finetti, B., *Probability, Induction and Statistics*, John Wiley and Sons, York, 1972.
[14] de Finetti, B., 'Initial Probabilities: A Prerequisite for any Valid Induction' *Synthese* **20** (1969) 2–16.
[15] Good, I. J., *Probability and the Weighing of Evidence*, Charles Griffin and Company Limited, London 1950.
[16] Good, I. J., 'The Bayesian Influence or How to Sweep Subjectivism Under the Carpet', Volume II of these proceedings, p. 125.
[17] Good, I. J., 'Explicativity, Corroboration, and the Relative Odds of Hypotheses', *Synthese* **30** (1975) 39–73.
[18] Goodman, N., *Fact, Fiction and Forecast* (paper back ed.) Bobbs Merrill Co., New York, 1965.
[19] Harper, W. L., 'A Note on Universal Instantiation in the Stalnaker-Thomason Conditional Logic', Forthcoming in *The Journal of Philosophical Logic*.
[20] Harper, W. L. and Kyburg, H. E., 'The Jones Case', *British Journal for the Philosophy of Science* **19** (1968) 247–258.

[21] Hintikka, J., *Knowledge and Belief*, Cornell Univ. Press, Ithaca, 1962.
[22] Hintikka, J., *Models for Modalities*, Reidel Publishing Co., Dordrecht, 1969.
[23] Jeffrey, R. C., *The Logic of Decision*, McGraw-Hill, New York, 1965.
[24] Jeffrey, R. C., 'Probable Knowledge', in Lakatos [1], pp. 166–180.
[25] Jeffrey, R. C., 'Carnap's Inductive Logic', *Synthese* **25** (1973) 299–306.
[26] Kemeny, J., 'Fair Bets and Inductive Probabilities', *Journal of Symbolic Logic* **20** (1955) 263–273.
[27] Kolmogovov, A. N., *Foundations of the Theory of Probability* 2nd English ed., Chelsea, New York, 1956.
[28] Krantz, D. H., Luce, R. D., Suppes, P. and Tversky, A., *Foundations of Measurement*, Academic Press, New York, 1971.
[29] Kripke, S., 'Semantical Considerations on Modal Logic', in *Modal and Many-Valued Logics, Acta Philosophica Fennica* **16** (1963) 83–94.
[30] Kripke, S., 'Semantical Analysis of Modal Logics, I', *Zeitschrift für mathematische Logik und Grundlagen der Mathematik* **9** (1963) 67–96.
[31] Kyburg, H. and Smokler, H. (eds), *Studies in Subjective Probability*, Wiley, New York, 1964.
[32] Lakatos, I., 'Changes in the Problem of Inductive Logic', in Lakatos (ed.): *The Problem of Inductive Logic*, North-Holland, Amsterdam, 1968, pp. 315–417.
[33] Leblanc, H., 'On Requirements for Conditional Probability Functions', *Journal of Symbolic Logic* **25** (1960) 238–242.
[34] Levi, I., 'Probability Kinematics', *British Journal for the Philosophy of Science* **18** (1967) 197–207.
[35] Levi, I., 'If Jones Only Knew More', *British Journal for the Philosophy of Science* **20** (1969), 153–159.
[36] Levi, I., 'On Indeterminate Probabilities', in *Journal of Philosophy* **71** (1974) 391–418.
[37] Lewis, D. K., *Counterfactuals*, Blackwell, London, 1973.
[38] Lewis, D. K., 'Counterfactuals and Comparative Possibility', in Hockney, Harper, Freed (eds.), *Contemporary Research in Philosophical Logic and Linguistic Semantics*, U.W.O. series, D. Reidel Publ. Co., Dordrecht and Boston.
[39] Lewis, D. K., 'Probabilities of Conditionals and Conditional Probabilities' (dittograph).
[40] Lewis, D. K., 'Completeness and Decidability of Three Logics of Counterfactual Conditionals', *Theoria* **37** (1971) 74–85.
[41] May, S. and Harper, W. L., 'Toward an Optimization Theory for Applying Minimum Change Principles in Probability Kinematics', this volume, p. 137.
[42] Pollock, J., *Counterfactuals, Dispositions and Causes* (Manuscript).
[43] Popper, K. R., *The Logic of Scientific Discovery*, Harper Torchbook edition, New York, 1959.
[44] Popper, K. R., *Conjectures and Refutations*, Harper Torchbook edition, New York, 1965.
[45] Popper, K. R., *Objective Knowledge*, Oxford, 1972.
[46] Ramsey, F. P., 'Truth and Probability', in Ramsey [47], pp. 156–211.
[47] Ramsey, F. P., *The Foundations of Mathematics*, Routledge & Kegan Paul Ltd., London, 193.
[48] Rescher, N., *Hypothetical Reasoning*, North Holland, Amsterdam, 1964.
[49] Savage, L. J., *The Foundations of Statistics*, John Wiley & Sons, Inc., 1954.
[50] Shimony, A., 'Coherence and the Axioms of Confirmation', *Journal of Symbolic Logic* **20** (1955) 1–28.

[51] Shimony, A., 'Scientific Inference', in *The Nature and Function of Scientific Theories* (ed. by R. Colodny), Univ. of Pittsburgh Press, Pittsburgh, 1970.
[52] Stalnaker, R., 'Probability and Conditionals', *Philosophy of Science* **37** (1970) 64–80.
[53] Stalnaker, R., 'A Theory of Conditionals', *Studies in Logical Theory* (*American Philosophical Quarterly*, Supplementary Monograph Series), Oxford, 1968.
[54] Stalnaker, R. and Thomason, R., 'A Semantical Analysis of Conditional Logic', *Theoria* **36** (1970) 23–42.
[55] Teller, P., 'Conditionalization and Observation', *Synthese* **26** (1973) 218–258.
[56] Teller, P., 'Conditionalization, Observation and Change of Preference', this volume, p. 205.
[57] van Fraassen, B. C., 'Probabilities of Conditionals', this volume, p. 261.
[58] van Fraassen, B. C., 'Notes on Probabilities of Conditionals' (dittograph).
[59] van Fraassen, B. C., 'Construction of Popper Probability Functions' (dittograph).
[60] van Fraassen, B. C., 'Hidden Variables in Conditional Logic', forthcoming in *Theoria*.
[61] van Fraassen, B. C., 'Theories and Counterfactuals', forthcoming in a festschrift for W. Sellars (ed. by H.-N. Castaneda).
[62] Vickers, J. M., 'Probability and Non-Standard Logics', in *Philosophical Problems in Logic* (Karel Lambert, ed.) Reidel Publ. Co., Dordrecht, 1970, pp. 102–120.
[63] von Neuman, J. and Morgenstern, O., *Theory of Games and Economic Behavior*, Princeton Univ. Press, Princeton, 1953.

Dear Bill:

Here is my argument to the effect that Harper's constraint leads to the consequence that the conditional collapses into the material conditional.

I need three assumptions: non-triviality, Harper's constraint, and an assumption about subfunctions. *Non-trivial* means what it meant for Lewis: at least 3 disjoint propositions with non-zero probability. Harper's constraint is the assumption that $P(A \,\square\!\!\rightarrow B) = 1$ iff $P(B/A) = 1$. The subfunction constraint requires a little explanation.

Define a *classical subfunction*, P_c, of a Popper function P as follows: $P_c(A) =_{df} P(A/C)$ for all A. Obviously, if $P(\bar{C}/C) \neq 1$, P_c will be a classical probability function. Now if Harper's constraint is to do the job which is its motivation – to solve the problem of iterated shifts – it is necessary that there be some way of extending the classical subfunctions to Popper subfunctions which also satisfy Harper's constraint. You have a specific way of doing this, using a Carnapian confirmation function. I shall assume only that there is *some* way of doing it. That is, I assume that for every proposition C such that $P(\bar{C}/C) \neq 1$, there exists a Popper function P_c such that (a) If $P(B/C) \neq 0$, $P_c(A/B) = P(AB/C)/P(B/C)$, and (b) P_c satisfies Harper's constraint. Call this: the *subfunction constraint*.

You don't make this last assumption explicit, but I believe it is implied by what you do say, and essential to the strategy you pursue for solving the iteration problem.

The meat of the proof is in the following lemma:

> I. *In any non-trivial Popper function P satisfying Harper's constraint and the subfunction constraint, if $P(B/A) = 0$, then $P_A(\bar{B}/B) = 1$ for all A, B.*

Proof. Let X, Y and Z be three propositions with non-zero probability which partition the sample space. That is X, Y and Z are such that

$$P(X) > 0, \qquad P(Y) > 0, \qquad P(Z) > 0$$
$$XY = XZ = YZ = \Lambda$$
$$P(X \vee Y \vee Z) = 1.$$

113

By non-triviality, there must be such an X, Y and Z. Let A and B be any propositions such that $P(B/A)=0$. We may assume, without loss of generality, that $P(A)\neq 0$.

1. $P_{Y\vee AX}(Y/Y\vee Z\vee B)=1$
2. $P_{Y\vee AX}((Y\vee Z\vee B)\square\!\!\rightarrow Y)=1$ by Harper's constraint, applied to subfunction.
3. $P_{Y\vee AX}((Y\vee Z\vee B)\square\!\!\rightarrow Y/X)=1$ (assuming $P(AX)\neq 0$)
4. $P_{AX}((Y\vee Z\vee B)\square\!\!\rightarrow Y)=1$ (assuming $P(AX)\neq 0$)
5. $P_{AX}(Y/Y\vee Z\vee B)=1$ (assuming $P(AX)\neq 0$) by Harper's constraint.

Now repeat steps 1–5, interchanging Y with Z, yielding

6. $P_{AX}(Z/Y\vee Z\vee B)=1$ (assuming $P(AX)\neq 0$)
7. $P_{AX}(YZ/Y\vee Z\vee B)=1$ (assuming $P(AX)\neq 0$)
    Combining steps 5 and 6.

But since $YZ\Vdash\bar B$, and $B\Vdash Y\vee Z\vee B$, it follows from step 7 that

8. $P_{AX}(\bar B/B)=1$ (assuming $P(AX)\neq 0$)
9. $P_{AX}(B\square\!\!\rightarrow\bar B)=1$ (assuming $P(AX)\neq 0$) by Harper's constraint.

Now repeat steps 1–9 twice more, interchanging first Y with X, and then Z with X, yielding

10. $P_{AY}(B\square\!\!\rightarrow\bar B)=1$ (assuming $P(AY)\neq 0$)
11. $P_{AZ}(B\square\!\!\rightarrow\bar B)=1$ (assuming $P(AZ)\neq 0$).

Now since the members of the set of propositions $\{AX, AY, AZ\}$ which have non-zero probability form a partition of A, it follows from 9, 10 and 11 that

12. $P_A(B\square\!\!\rightarrow\bar B)=1$, and so
13. $P_A(\bar B/B)=1$ by Harper's constraint.

The second lemma follows easily from the first.

> II. *If P is a non-trivial probability function satisfying Harper's constraint and the subfunction constraint, then for all A, B, C, $P_c(B/A)=P(B/AC)$.*

Proof. This is obvious if $P(A/C) \neq 0$. If $P(A/C) = 0$, then by I, $P_c(B/A) = 1$ for all B, and since $P(AC) = 0$, it also follows from I that $P(B/AC) = 1$ for all B. So $P(B/AC) = P_C(B/A)$ for all A, B, C.

Now the main theorem

III. *If P is a non-trivial probability function satisfying Harper's constraint and the subfunction constraint, then* $P(A \Box\!\!\!\rightarrow B) = $
$= P(A \supset B)$ *for all A, B.*

Proof.

1. $\quad P(A \Box\!\!\!\rightarrow B) = P(AB) \times P_{AB}(A \Box\!\!\!\rightarrow B) + P(A\bar{B} \wedge (A \Box\!\!\!\rightarrow B)) + $
$\quad + P(\bar{A}) \times P_{\bar{A}}(A \Box\!\!\!\rightarrow B)$

2. $\quad P_{AB}(A \Box\!\!\!\rightarrow B) = 1$ by Harper's constraint, since $P_{AB}(B/A) = 1$

3. $\quad P_{\bar{A}}(B/A) = P(B/A\bar{A})$ by II, and since $P(B/A\bar{A}) = 1$,

4. $\quad P_{\bar{A}}(A \Box\!\!\!\rightarrow B) = 1$ by Harper's constraint.

5. $\quad P(A \Box\!\!\!\rightarrow B/(A \Box\!\!\!\rightarrow B) \wedge A\bar{B}) = 1$, so

6. $\quad P_{(A \Box\!\!\!\rightarrow B) \wedge A\bar{B}}(B/A) = 1$ by Harper's constraint. So

7. $\quad P(B/(A \Box\!\!\!\rightarrow B) \wedge A\bar{B}) = 1$, from which it follows that

8. $\quad P((A \Box\!\!\!\rightarrow B) \wedge A\bar{B}) = 0$.

Now substituting 2, 4 and 8 in 1,

9. $\quad P(A \Box\!\!\!\rightarrow B) = P(AB) \times 1 + 0 + P(\bar{A}) \times 1 = P(A \supset B)$.

The problem is caused by what Bas called "metaphysical realism" (although I think this assumption has nothing to do with metaphysics): "metaphysical realism" says that there is a single interpretation of the conditional which does not vary with a charge in the epistemic situation. The reason this conflicts with Harper's constraint is that *that* assumption relates the interpretation of conditional to what is accepted – to a feature of the epistemic situation. As long as one begins with a probability function which is complex enough so that it can change in various different ways, one can bring these two assumptions into conflict.

Harper's constraint without metaphysical realism – so without the subfunction constraint is, I think, okay. But it won't help with the iteration problem.

Yours

Bob

WILLIAM L. HARPER

RAMSEY TEST CONDITIONALS AND ITERATED
BELIEF CHANGE*
(A RESPONSE TO STALNAKER)

I. A NEW DEFINITION OF ITERATIVE PROBABILITY MODEL

1. *Introduction*

In 'Rational Belief Change, Popper Functions and Counterfactuals' (Harper [1]), I used Popper's treatment of conditional probability and an account of conditionals motivated by Ramsey's test for acceptability of hypotheticals to construct iterative probability models.[†] An iterative probability model is to be an extension of the Bayesian representation of rational belief that allows for iterated shifting by conditionalization on new inputs even when this involves iterated revisions of previously accepted evidence. Stalnaker has raised serious difficulties with my construction. The source of these difficulties is my account of Ramsey test conditionals. In this paper I give an account of Ramsey test conditionals adequate to construct iterative probability models free of all such difficulties.

The source of the difficulties Stalnaker raises is the following result which holds for any iterative probability model P given by my construction in 'Rational Belief Change' (Harper [1], page 97). If $P(A/T)>0$, $P(B/A)=0$, and there is a partition $\{XYZ\}$ of T such that $P(X/T)>0$, $P(Y/T)>0$, and $P(Z/T)>0$, then $P_{\langle A\rangle}(\bar{B}/B)=1$ (Stalnaker [2]). This result defeats the purpose of using such probability models to solve the problem of iterated shifts that require iterated revisions of previously accepted evidence. It has the effect that $P_{\langle A\rangle\langle B\rangle}$ is the absurd function assigning 1 to every proposition whenever $P(A/T)>0$ and $P(B/A)=0$, for any P rich enough to assign non-zero probability to three disjoint propositions.

* This research was partially supported by Canada Council Grant 1302-C70-300.
[†] See Harper [1] for background on Popper's treatment of probability and Ramsey's test for hypotheticals. The basic framework and notation for this paper will be that of Harper [1].

2. *Ramsey Test Conditionals*

A Ramsey test conditional is one which satisfies the following accept-ability conditions. Accept $A\Box\!\!\rightarrow B$ if and only if the minimal revision of what you now accept needed to accept A requires also accepting B. Where K is what the agent now accepts and $K_{\langle A\rangle}$ is the minimal revision of K required to accept A this comes to

$$K\subseteq A\Box\!\!\rightarrow B \quad \text{iff} \quad K_{\langle A\rangle}\subseteq B.$$

On the Popper function representation of rational belief where $K=$ $=K(P, T)^*$ and $K_{\langle A\rangle}=K(P, A)$ the Ramsey test constraint is just Harper's condition

$$P(A\Box\!\!\rightarrow B)=1 \quad \text{iff} \quad P(B/A)=1.$$

Stalnaker's result shows that the treatment of Ramsey test conditionals I gave is inadequate. It does not show that there cannot be an adequate treatment of such conditionals.

The problem, as Stalnaker suggests (Stalnaker [2], p. 115) is that my treatment of Ramsey test conditionals used a single binary connective $A\Box\!\!\rightarrow B$ that could not vary with changes in the relevant acceptance con-text. I shall now argue that any adequate nearest worlds treatment of Ramsey test conditionals ought to allow for different conditional prop-ositions $A\Box\!\!\rightarrow_{K_1} B$ and $A\Box\!\!\rightarrow_{K_2} B$ for different acceptance contexts K_1 and K_2. This will provide independent motivation for a new definition of iterative probability model based on conditionals that are relativized to acceptance contexts.

For purposes of the present argument Lewis' most general truth con-ditions for conditionals based on comparative similarity are enough. The conditional $A\Box\!\!\rightarrow B$ holds in world w just in case some AB-world is closer to w than any $A\bar{B}$-world. (Lewis [1], p. 49).

We want acceptability of $A\Box\!\!\rightarrow_{K_1} B$ to satisfy Ramsey's test relative to K_1 and acceptability of $A\Box\!\!\rightarrow_{K_2} B$ to satisfy Ramsey's test relative to K_2 so that

$$K_1\subseteq A\Box\!\!\rightarrow_{K_1} B \quad \text{iff} \quad K_{1\langle A\rangle}\subseteq B$$
$$K_2\subseteq A\Box\!\!\rightarrow_{K_2} B \quad \text{iff} \quad K_{2\langle A\rangle}\subseteq B$$

---

* $K(P, A)=\bigcap\{B:P(B/A)=1\}$ (see Harper [1]) where P is the agents extended conditional belief function $K(P, T)$ is his present acceptance context and $K(P, A)$ is the minimal re-vision of $K(P, T)$ needed to accept A.

where $K_{1\langle A\rangle}$ and $K_{2\langle A\rangle}$ are respectively the minimal revisions of K_1 and K_2 required to have A accepted. For any acceptance context K we want

(c) $\qquad K_{\langle A\rangle} \subseteq K \cap A \quad$ if $\quad K \cap A \neq \emptyset.$

This follows from the principle that when minimally revising K to have A accepted one should not give up any propositions already accepted unless required to do so. In order to meet this motivation *A-worlds in K should be selected as nearer to worlds in K than any A-worlds outside of K*. If w is a $\bar{A}$-world in $K_1 \cap K_2$, both $K_1 \cap A$ and $K_2 \cap A$ are nonempty, $K_1 \cap A \subseteq B$, but $K_2 \cap A \subseteq \bar{B}$, then $A \square\!\!\rightarrow_{K_1} B$ is true in w while $A \square\!\!\rightarrow_{K_2} B$ is false in w. Thus $A \square\!\!\rightarrow_{K_1} B$ is not the same proposition as $A \square\!\!\rightarrow_{K_2} B$.

If one allows for a variety of conditional operators $\square\!\!\rightarrow_{K_1}$, $\square\!\!\rightarrow_{K_2}$, etc., then a formulation of Ramsey's test requires specification that the conditional be appropriate for the acceptance context from which it is evaluated. The proper formulation is

$$K \subseteq A \square\!\!\rightarrow_K B \quad \text{iff} \quad K_{\langle A\rangle} \subseteq B,$$

where K is an acceptance context, $A \square\!\!\rightarrow_K B$ is a Ramsey test conditional appropriate to this context, and $K_{\langle A\rangle}$ is the minimal revision of K to have A accepted. This leads to the following relativization of Harper's constraint:

$$P(A \square\!\!\rightarrow_{K(P,T)} B/T) = 1 \quad \text{iff} \quad P(B/A) = 1.$$

Here we specify that it is acceptability of the $K(P, T)^*$ conditional that corresponds to conditional probability 1.

This relativization of Harper's constraint does not lead to Stalnaker's result. Let $K_0 = K(P_{Y \vee AX}, T)$ and $K_1 = K(P_{AX}, T)$. With these abbreviations the key steps in Stalnaker's proof are as follows:

1. $P_{Y \vee AX}(Y/Y \vee Z \vee B) = 1$
2. $P_{Y \vee AX}((Y \vee Z \vee B) \square\!\!\rightarrow_{K_0} Y) = 1$
3. $P_{Y \vee AX}((Y \vee Z \vee B) \square\!\!\rightarrow_{K_0} Y/X) = 1$
4. $P_{AX}((Y \vee Z \vee B) \square\!\!\rightarrow_{K_0} Y) = 1$
5. $P_{AX}(Y/Y \vee Z \vee B) = 1.$

---

* $K(P, T)$ is what the P agent now accepts. (See note, p. 118.)

The steps from 1 to 2 and from 4 to 5 are to be authorized by Harper's constraint. On the relativized version the step from 1 to 2 is appropriate if the conditional is a K_0 conditional, but the step from 4 to 5 requires a K_1 conditional. Thus, the proof does not go through.

Ramsey's test gives independent motivation for having conditionals made relative to acceptance contexts, and with relativized conditionals Harper's constraint does not lead to Stalnaker's result. The next step is to give a definition of iterative probability model that uses relativized conditionals. We will then solve the shifting problem by developing relativized conditionals appropriate to construct such models.

3. *Rényi's Confirmation Functions*

It will be useful to modify the treatment of confirmation functions to allow $\mathscr{C}$ to be restricted to a set $\mathscr{K}$ of admissible acceptance contexts. The following definition is a modification of one given by Alfred Rényi in his elegant treatment of conditional probability (Rényi [1], p. 38). Suppose $\mathscr{K}$ is a subset of the non-empty sets in field $\mathscr{E}$ on T and $\mathscr{C}$ maps $\mathscr{E} \times \mathscr{K}$ into the reals.

$\mathscr{C}$ is a confirmation function for $\mathscr{E} \times \mathscr{K}$ iff

(1) $\mathscr{C}(K/K)=1$ and $\mathscr{C}_K$ is a classical probability on $\mathscr{E}$, for every K in $\mathscr{K}$.

(2)   If $K_0, K_1 \in \mathscr{K}$ and $K_1 \subseteq K_0$, then
        (a) $\mathscr{C}(K_1/K_0)>0$
        (b) $\mathscr{C}(A/K_1)=\dfrac{\mathscr{C}(AK_1/K_0)}{\mathscr{C}(K_1/K_0)}$,   for all A in $\mathscr{E}$.

Constraint (1) insures that $\mathscr{C}_K$ is a classical probability on $\mathscr{E}$ and $\mathscr{C}_K(A)=1$ for every A such that $K \subseteq A$. Constraint (2a) insures that when $K_1 \subseteq K_0$ the classical conditional probabilities $\mathscr{C}_{K_0}(A/K_1)$ exist, and (2b) adds the requirement that $\mathscr{C}(A/K_1)$ be the same as $\mathscr{C}_{K_0}(A/K_1)$. A confirmation function $\mathscr{C}$ for $\mathscr{E} \times \mathscr{K}$ is regular just in case

(3) $\mathscr{C}(A/K)=1$ only if $K \subseteq A$

for all $A \in \mathscr{E}$ and $K \in \mathscr{K}$.

Rényi requires that $\mathscr{K}$ satisfy conditions for what he calls a *bunch* (Rényi, p. 38). A subset $\mathscr{K}$ of $\mathscr{E}$ is a *bunch* just in case.

(1) $K_1, K_1 \in \mathcal{K}$ only if $K_0 \cup K_1 \in \mathcal{K}$

(2) $\underset{i \in \omega}{\cup} K_i = T$ for some sequence $K \in \mathcal{K}^\omega$

(3) $\emptyset \notin \mathcal{K}$.

He also requires that each $\mathcal{C}_K$ be σ-additive (Rényi, p. 39). The result of these restrictions is that every Rényi function is a conditional probability generated by a σ-finite* measure μ on ε where for every K in $\mathcal{K}$

(a) $0 < \mu(K) < \infty$

(b) $\mathcal{C}(A/K) = \dfrac{\mu(AK)}{\mu(K)}$ (Rényi, p. 40).

This measure μ is determined up to a positive constant factor (Rényi, p. 40).

The beauty of Rényi's treatment, and the main reason for wanting to be able to have $\mathcal{C}$ restricted to a subset $\mathcal{K}$ of admissible conditions, is that T does not have to be one of the admissible conditions. This allows the use of unbounded measures μ where $\mu(T) = \infty$ to be used to generate $\mathcal{C}$. On this treatment, for example, there is no problem having $\mu(\{x\})$ equal and positive for every element x of a denumerable T.

We have only imposed Rényi's condition 3 that the empty set is not an admissible acceptance context. More conditions on sets of acceptance contexts suitable for iterated conditionalization will emerge, but we will want to allow for sets $\mathcal{K}$ that violate Rényi's conditions 1 and 2. On dealing with conceptual change (in a forthcoming paper) it will be convenient to have disjoint acceptance contexts $K_{\langle C \rangle_n}$, $K_{\langle D \rangle_k}$ where $P_{\langle C \rangle_n}$ and $P_{\langle D \rangle_k}$ generate different algebras of equivalence classes without allowing $K_{\langle C \rangle_n} \cup K_{\langle D \rangle_k}$ to be an admissible acceptance context. One advantage of giving up 2 is that one can break up an uncountable T into an uncountable set of pairwise disjoint countable acceptance contexts. Each of these disjoint K's would be a maximum in a family of subcontexts. For each such family a distinct measure μ_K (where K is the maximum of the family) could generate $\mathcal{C}_K$. This would allow $\mathcal{C}$ to be σ-additive and *regular* even though T is uncountable. For every element x of T, $\mu(\{x\})$ could be positive where μ is the measure generating $\mathcal{C}_K$ for acceptance contexts K where $x \in K$.

* μ is a σ-finite iff there is a denumerable partition $\{K_i\}$ $i < \omega$ of T such that $\mu(K_i) < \infty$ for all i.

We leave out the requirement of σ-additivity in order to allow for finitely additive probability models. De Finetti, for one, has argued that finite additively is all one should require of a belief function (De Finetti [1]). Nothing we shall do here requires ruling out finitely additive functions, so we allow for them.

4. Iterative Probability Models

Suppose $\mathcal{K}$ is a set of subsets of T designated as admissible acceptance contexts, $\mathcal{E}$ is a field of subsets of T closed under a binary connective $\square\!\!\rightarrow_K$ for every K in $\mathcal{K}$ and P is a Popper belief function on $\mathcal{E}$ such that $K(P, C) \in \mathcal{K}$ for every C in $\mathcal{E}$. For each K in $\mathcal{K}$ and C in $\mathcal{E}$ let

(I) $\qquad K_{\langle C \rangle} = \cap \{B : K \subseteq C \square\!\!\rightarrow_K B\}.$

Assume $K = K(P, T)$ and $\square\!\!\rightarrow_K$ is an appropriate Ramsey test conditional for P.

Remark 1. $K_{\langle C \rangle} = K(P, C)$ for all C.

Proof. The basic Ramsey test requirement gives us $P(C \square\!\!\rightarrow_K B/T) = 1$ iff $P(B/C) = 1$, and the general definition of $K(P, A)$ gives $P(B/C) = 1$ iff $K(P, C) \subseteq B$ and $P(C \square\!\!\rightarrow_K B/T) = 1$ iff $K \subseteq C \square\!\!\rightarrow_K B$. ∎

One can view this remark as the essence of the Ramsey test idea.

In addition to the assumptions about K and $\square\!\!\rightarrow_K$, assume that $\square\!\!\rightarrow_{K_{\langle C \rangle}}$ is an appropriate Ramsey test conditional for $P_{\langle C \rangle}$.

Remark 2. $K(P_{\langle C \rangle}, A) = K_{\langle C \rangle \langle A \rangle}$.

Proof. Simply apply remark 1 to

$$K(P_{\langle C \rangle}, T) = K(P, C) = K_{\langle C \rangle}. \quad \blacksquare$$

With conditionals satisfying these Ramsey test assumptions our goal of defining $P_{\langle C \rangle}$ in terms of P and $\mathcal{C}$ can be realized, provided $\mathcal{C}$ and $\mathcal{K}$ are appropriate.

(II) $\qquad P_{\langle C \rangle}(B/A) = \mathcal{C}(B/K_{\langle C \rangle \langle A \rangle}),$ if $K_{\langle C \rangle \langle A \rangle} \neq \emptyset$
$\qquad\qquad\qquad = 1,$ otherwise.

The requirements on $\mathcal{C}$ and $\mathcal{K}$ are that $\mathcal{C}$ is defined for $\mathcal{E} \times \mathcal{K}$ and that for all A, $K_{\langle C \rangle \langle A \rangle} \in \mathcal{K}$ if $K_{\langle C \rangle \langle A \rangle} \neq \emptyset$.

We are interested in iterated shifts. In general, we shall want acceptance contexts appropriate to iterated conditionalization to be generated by appropriate Ramsey test conditionals so that

$$K(P_{\langle C_0 \rangle \cdots \langle C_{n-1} \rangle} A) = K_{\langle C_0 \rangle \cdots \langle C_{n-1} \rangle \langle A \rangle}.$$

to this end we require that $\mathcal{K}$ be closed under the formation of non-empty $K_{\langle C \rangle}$.

Remark 3. Having $\mathcal{K}$ closed under non-empty $K_{\langle C \rangle}$ requires that $\mathscr{C}$ be regular if (c) ($K_{\langle A \rangle} = K \cap A$ if $K \cap A \neq \emptyset$) holds for all A.

Proof. Suppose $\mathscr{C}(A/K) = 1$ but $K \nsubseteq A$. Then $K \cap \bar{A} \neq \emptyset$. By (c) $K_{\langle \bar{A} \rangle} = K \cap \bar{A}$. Since $\mathcal{K}$ is closed under non-empty $K_{\langle C \rangle}$, $K \cap \bar{A} \in \mathcal{K}$. By clause 2 on $\mathscr{C}$ we have $\mathscr{C}(K \cap \bar{A}/K) > 0$ which contradicts our assumption. ∎

It will be convenient to write '$P_{\langle C \rangle_n}$' for '$P_{\langle C_0 \rangle \cdots \langle C_{n-1} \rangle}$', '$K_{\langle C \rangle_n}$' for '$K_{\langle C_0 \rangle \cdots \langle C_{n-1} \rangle}$', and let $P_{\langle C \rangle_0} = P$ and $K_{\langle C \rangle_0} = K$. We are now ready to define iterative probability models.

(III) P is $I_p(\mathscr{E}, \mathcal{K}, \mathscr{C})$ (P is an iterative probability model relative to $\mathscr{E}$, $\mathcal{K}$ and $\mathscr{C}$) iff

(1) $\mathcal{K}$ is a non-empty set of non-empty subsets of a set T, $\mathscr{E}$ is a field of subsets of T closed under a binary operator $\Box \mapsto_K$ for each K in $\mathcal{K}$, for every C in $\mathscr{E}$ and K in $\mathcal{K}$ if $K_{\langle C \rangle}$ is non-empty $K_{\langle C \rangle}$ is in $\mathcal{K}$, and $\mathscr{C}$ is a regular confirmation function with $\mathscr{E} \times \mathcal{K}$ included in its domain.

(2) For some K in $\mathcal{K}$ all finite sequences $C_0 \ldots C_{n-1}$ drawn from $\mathscr{E}$ and elements A, B of $\mathscr{E}$,

(a) $P_{\langle C \rangle_n}(B/A) = \mathscr{C}(B/K_{\langle C \rangle_n \langle A \rangle})$ if $K_{\langle C \rangle_n \langle A \rangle} \neq \emptyset$
         $= 1$,   otherwise
(b) $P_{\langle C \rangle_n \langle A \rangle}(B/T) = P_{\langle C \rangle_n}(B/A)$
(c) $P_{\langle C \rangle_n}$ is a Popper function on $\mathscr{E}$, if $K_{\langle C \rangle_n \langle A \rangle} \neq \emptyset$.

(IV) P is $I_p(\mathscr{E}, \mathcal{K})$ (P is an iterative probability model for $\mathscr{E}$ and $\mathcal{K}$) iff P is $I_p(\mathscr{E}, \mathcal{K}, \mathscr{C})$ for some $\mathscr{C}$.

Constraint (a), in clause (2) of definition III, is the requirement that $\Box \mapsto_{K_{\langle C \rangle_n}}$ be appropriate Ramsey test conditional for $K_{\langle C \rangle_n}$.

We have two remarks worth noting. Every non-trivial submodel of an I_p-model is also an I_p-model.

Remark 4. If P is $I_p(\mathscr{E}, \mathcal{K})$ and $P(\bar{C}/C) \neq 1$ then

$$P_{\langle C \rangle} \text{ is } I_p(\mathscr{E}, \mathcal{K}).$$

Proof. From the definition of I_p-model. ∎

We also have a correlation between conditionals that vary appropriately with nesting to the right and acceptance contexts for iterated conditionalization. Suppose K is in $\mathscr{K}$ and $\mathscr{E}$ and $\mathscr{K}$ satisfy clause (1) of definition III, and $C_0 \ldots C_{n-1}$ is a sequence of elements of $\mathscr{E}$.

Remark 5. $K_{\langle C \rangle_n \langle A \rangle} = \bigcap \{B : K \subseteq (C_0 \square\!\!\mapsto_K \ldots$
$$(C_{n-1} \square\!\!\mapsto_{K_{\langle C \rangle_{n-1}}} (A \square\!\!\mapsto_{K_{\langle C \rangle_n}} B))\ldots)\}.$$

Proof. By induction on n using definition I. ∎

The following theorem establishes the basic Ramsey test constraints on acceptance contexts required by I_p-models. Suppose $\mathscr{K}$, $\mathscr{E}$ and $\mathscr{C}$ satisfy clause (1) of definition III and $K \in \mathscr{K}$ such that for all sequences $C_0 \ldots C_{n-1}$ and A, B from $\mathscr{E}$

$$P_{\langle C \rangle_n}(B/A) = \mathscr{C}(B/K_{\langle C \rangle_n \langle A \rangle}) \quad \text{if} \quad K_{\langle C \rangle_n \langle A \rangle} \neq \emptyset$$
$$= 1 \quad \text{otherwise.}$$

THEOREM 1. P is $I_p(\mathscr{E}, \mathscr{K}, \mathscr{C})$ iff for all sequences $C_0 \ldots C_{n-1}$ and elements, A, B of $\mathscr{E}$

(1) $K_{\langle C \rangle_n \langle A \rangle} \subseteq A$
(2) $K_{\langle C \rangle_n \langle T \rangle} = K_{\langle C \rangle_n}$
(3)   If $K_{\langle C \rangle_n \langle A \rangle} \cap B \neq \emptyset$ then $K_{\langle C \rangle_n \langle AB \rangle} = K_{\langle C \rangle_n \langle A \rangle} \cap B$.

Proof. Having clause 2a of definition III in the hypothesis, together with the regularity of $\mathscr{C}$, gives

(i) $K(P_{\langle C \rangle_n}, A) = K_{\langle C \rangle_n \langle A \rangle}$

for all $C_0 \ldots C_{n-1}$, A in $\mathscr{E}$. This, also, gives the equivalence of (2) with clause 2b of definition III. From (i) it follows that (1) and (3) are respectively equivalent to conditions 1 and 4 on extended conditional belief functions (ebf's). From left to right: regularity of $\mathscr{C}$ makes $P_{\langle C \rangle_n}$ compact. Thus, by Theorem 2.2 of Harper [1], part II $P_{\langle C \rangle_n}$ is an ebf. From right to left all that remains to be shown is clause 2c of definition III. The hypothesis gives us that $P_{\langle C \rangle_n}$ satisfies conditions 2 and 3 on ebf's, if $K_{\langle C \rangle_n} \neq \emptyset$, and (1) and (3) directly gave conditions 1 and 4. Therefore, by Theorem 2.1 of Harper [1], Part II, we have $P_{\langle C \rangle_n}$ is a Popper function if $K_{\langle C \rangle_n} \neq \emptyset$. ∎

II. RELATIVISED CONDITIONALS AND CONSTRUCTION OF NON-TRIVIAL ITERATIVE PROBABILITY MODELS

1. *Preliminaries*

In order to construct non-trivial Ip-models we must develop a conditional that is appropriately relativized to acceptance contexts. We have already given independent motivation for such relativiation by applying the Ramsey test to conditionals based on comparative similarity of possible worlds. Let us use this as our starting point.

Our basic treatment of nearest worlds conditionals is a version of Lewis' with weak centering and the limit assumption (Lewis [1], pp. 48, 120). To each world w assign a set $R(w)$ to be regarded as the worlds accessible from w and a binary relation $\leqslant w$ to be regarded as comparative similarity to w. Accessibility is to satisfy

(0) $w \in R(w)$ (Reflexivity)

so that each world is accessible from itself. Comparative similarity is to satisfy four constraints

(1) $w \leqslant^w v$ (weak centering)
(2) $u \leqslant^w v$ and $v \leqslant^w z$ only if $u \leqslant^w z$ (Transitivity)
(3) $u \leqslant^w v$ or $v \leqslant^w u$ (Connectedness).

Weak centering is the constraint that each world w is as much like itself as any other world is like it. Both weak centering and transitivity must hold if $\leqslant w$ is to be any reasonable relation of overall similarity to w. The comparative similarity relations we will consider all satisfy connectedness as well. We define the set of nearest accessible A-worlds to w.

(i) $f_A(w) = \{u \in A \cap R(w) : u \leqslant^w v \quad \text{for all } v \text{ in } A \cap R(w)\}$.

We can now state the limit assumption economically.

(4) $A \cap R(w) \neq \emptyset$ only if $f_A(w) \neq \emptyset$ (Limit Assumption).

The limit assumption is the constraint that there are some *nearest* accessible A-worlds from w if there are any A-worlds accessible from w at all. Having the limit assumption allows the following simplification of truth conditions for $A \,\square\!\!\rightarrow B$.

(ii) $w \in A \,\square\!\!\rightarrow B$ iff $f_A(w) \subseteq B$.

We have $A \Box\!\!\rightarrow B$ is true at w just in case all nearest accessible A-worlds are B-worlds. The logic of such conditionals is VW, the weakest logic Lewis considers to be a candidate for counterfactuals. (Lewis [1], p. 130).

One feature of nearest worlds conditionals that will be useful is given in the following remark. Suppose $K_{\langle A \rangle} = \cap \{B : K \subseteq A \Box\!\!\rightarrow B\}$ and $\Box\!\!\rightarrow$ is a basic nearest worlds conditional with f as selection function for worlds in T and propositions in $\mathscr{E}$.

Remark

(a) $\quad K \subseteq A \Box\!\!\rightarrow B \quad$ iff $\quad \bigcup_{w \in k} f_A(w) \subseteq B$

(b) $\quad K_{\langle A \rangle} \subseteq B \quad$ iff $\quad \bigcup_{w \in k} f_A(w) \subseteq B$

(c) $\quad \bigcup_{w \in k} f_A(w) \subseteq K_{\langle A \rangle}$

(d) $\quad \bigcup_{w \in K} f_A(w) = K_{\langle A \rangle}, \quad$ if $\quad \bigcup_{w \in k} f_A(w) \in \mathscr{E}.$

Proof. These all follow easily from definition ii. ∎

Let us delay consideration of the problem of relativizing accessibility and deal first with the problem of relativizing our basic conditional under the assumption that all worlds are mutually accessible. Relativized conditionals restricted in this way will correspond to Ip-models where for all sequences $\langle C \rangle_n$, $P_{\langle C \rangle_n}(A/\bar{A}) = 1$ only if $A = T$. It will also be convenient to restrict attention to $\mathscr{E}$ and $\mathscr{K}$ where $\mathscr{K}$ is included in $\mathscr{E}$ and where $\mathscr{E}$ is the set of all subsets of T.

2. Relativized Conditionals (Step One)

The first relativized conditional we consider is based on the principle that *A-worlds in K are to be selected as nearer to worlds in K than any A-worlds outside of K.* The principle can be formulated as a constraint on a relativized selection function f_A^K picking out nearest A-worlds relative to acceptance context K.

(PRINCIPLE 1)

$$f_A^K(w) \subseteq K, \quad \text{if} \quad K \cap A \neq \emptyset \quad \text{and} \quad w \in K.$$

A selection function conforming to this principle can be grafted onto the semantical apparatus for our basic nearest worlds conditional without relativizing the comparative similarity relation. Given our assumptions about closure and accessibility f_A^K can be defined as follows:

(DEFINITION 1)

$$f_A^K(w) = f_{A \cap K}(w), \quad \text{if} \quad w \in K \text{ and } K \cap A \neq \emptyset =$$
$$= f_A(w), \quad \text{otherwise.}^2$$

Definition 1 modifies the basic selection apparatus just enough to incorporate principle 1.

Let us examine the properties of $K_{\langle A \rangle}$ generated by a Ramsey test conditional based on definition 1. Suppose $\mathscr{E}$ is the set of all subsets of a non-empty set T, closed under a binary connective $\Box\!\!\rightarrow_K$ for each K in a subset $\mathscr{K}$ of $\mathscr{E} - \{\emptyset\}$. For each world w in T let $\leqslant w$ and $R(w)$ satisfy all our basic constraints on comparative similarity and accessibility, and let $R(w) = T$ for all w. For each A in $\mathscr{E}$ let f_A be defined according to (i) and for each $K \in \mathscr{K}$ and A in $\mathscr{E}$ let f_A^K be given by definition 1. Let $w \in A \Box\!\!\rightarrow_K B$ iff $f_A^K(w) \subseteq B$ and $K_{\langle A \rangle} = \bigcap \{B : K \subseteq A \Box\!\!\rightarrow_K B\}$ for all $w \in T$, $K \in \mathscr{K}$, A, $B \in \mathscr{E}$.

(*Remark* 1a) If these assumptions are met, then for all A, $B \in \mathscr{E}$ and $K \in \mathscr{K}$

(1) $K_{\langle A \rangle} \subseteq A$
(2) $K_{\langle T \rangle} = K$
(c) $K_{\langle A \rangle} = K \cap A$, if $K \cap A \neq \emptyset$
(d) $K_{\langle A \rangle} \subseteq B$ and $K_{\langle B \rangle} \subseteq A$ only if $K_{\langle A \rangle} = K_{\langle B \rangle}$.

Proof. (1) and (2) are given by the properties of the basic nearest worlds conditional. (c) is a consequence of definition 1 together with the weak centering constraint on $\leqslant w$ and having all worlds in K mutually accessible. (d) follows from definition 1, the limit assumptions and transitivity of $\leqslant w$. ∎

Constraint (c) is what insures that minimal revision of K to accept an A compatible with K generates classical conditionalization (see Harper [1], pp. 82–83). Constraint (d) corresponds to the axiom

$$P(B/A) = 1 = P(A/B) \quad \text{only if} \quad P(C/A) = P(C/B)$$

on Popper functions, and to the theorem

$$((A \Box\!\!\rightarrow B) \wedge (B \Box\!\!\rightarrow A)) \supset ((A \Box\!\!\rightarrow C) \equiv (B \Box\!\!\rightarrow C))$$

for basic nearest worlds conditionals. Both (c) and (d) follow from

(3) $K_{\langle A \rangle} \cap B \neq \emptyset$ only if $K_{\langle AB \rangle} = K_{\langle A \rangle} \cap B$

in the Ip-model framework. By theorem I.1 we have that all of (1)–(3) are needed if the relativized conditional is to generate $K_{\langle A \rangle}$ suitable for constructing iterative probability models. By Remark 1a we have seen that both (1) and (2) hold when $\square\!\!\rightarrow_K$ is given by definition 1. We now show that the definition 1 conditional is not sufficient to guarantee (3). (*Remark* 1b) There exist T, $\mathscr{E}$, $\mathscr{K}$, K, A and B meeting the hypothesis of Remark 1a where (3) fails.

Proof. Consider the following example where T is the set of eight worlds corresponding to all the possible combinations of three atomic propositions P, Q and R.[3] Let $\mathscr{E}$ be the set of all subsets of T and $\mathscr{K}$ be $\mathscr{E} - \{\emptyset\}$. Let K be the set of QR worlds A be the $\bar{Q} \cap (\bar{R} \cup P)$ worlds, and B be the $\bar{Q} \cap (\bar{R} \cup \bar{P})$ worlds. Suppose that nearness is measured by counting atomic propositions on which worlds differ. We let all worlds be mutually accessible and note that all of the basic constraints on comparative similarity and accessibility are met. For each K let $\square\!\!\rightarrow_K$ be defined by definitions (i) and (1). $\mathscr{E}$ is closed under all the $\square\!\!\rightarrow_K$ since each proposition $C \square\!\!\rightarrow_K D$ is just the set of worlds in T where $f_C^K(w) \subseteq D$. Finally we let $K_{\langle C \rangle} = \cap \{D : K \subseteq C \square\!\!\rightarrow_K D\}$ for all $K \in \mathscr{K}$, C, $D \in \mathscr{E}$. Given this

$$K_{\langle A \rangle} = \bar{Q} \cap (P \equiv R)$$
$$K_{\langle A \rangle} \cap B = \bar{Q} \bar{P} \bar{R}$$
$$K_{\langle AB \rangle} = \bar{Q} \bar{R}$$

which violates (3). ■

If $K_{\langle A \rangle}$ is to give a Ramsey test construction of $K(P, A)$ for an extended belief function P, then constraint (3) is just condition

(4) $K(P, AB) = K(P, A) \cap B$,   provided $K(P, A) \cap B \neq \emptyset$

on acceptable extended belief functions, Harper [1]. If $K(P, A)$ is constructed by a definition 1 conditional where (3) fails the resulting function is not acceptable. Therefore definition 1 conditionals are not adequate to construct Ip-models.

The remarks used to motivate condition (4) in 'Rational Belief Change', Part II only argue for the weaker condition

(4−) $K(P_{\langle A \rangle}, B) = K(P, A) \cap B$,   if $K(P, A) \cap B \neq \emptyset$

which, given Ramsey test assumptions, corresponds to

$$(3-) \qquad K_{\langle A \rangle \langle B \rangle} = K_{\langle A \rangle} \cap B, \quad \text{if} \quad K_{\langle A \rangle} \cap B \neq \emptyset$$

and is just an application of constraint (c) to $K_{\langle A \rangle}$. Since definition 1 conditionals do give (c) and therefore $(3-)$ one is tempted to entertain a weaker notion of acceptable extended belief function. This would allow iterative extended belief functions to be generated relative to confirmation function using definition 1 conditionals. The temptation is strengthened by the fact that definition 1 conditionals yield (d) which corresponds to Popper's substitution axiom (a2). Indeed, iterative extended belief functions defined in this way would satisfy all of the Popper axioms and other constraints on Ip-models accept for Popper axiom

$$(a4) \qquad P(XB/A) = P(B/A) \cdot P(X/AB).$$

This probability axiom of Popper's corresponds to condition 4 on acceptable extended belief functions and would be violated by a belief function generated by a Ramsey test construction where (3) fails. In place if (a4) condition $(4-)$ would give

$$(a4-) \qquad P(XB/A) = P(B/A) \cdot P_{\langle A \rangle}(X/B).$$

This is weaker than a4 because it allows $P_{\langle A \rangle}(X/B)$ to differ from $P(X/AB)$ even when $P(B/A) > 0$.

The temptation to weaken constraint (4) ought to be resisted. Consider the following condition which is equivalent to (4):

$$(4') \qquad \text{If } B \subseteq A \quad \text{and} \quad K(P, A) \cap B \neq \emptyset \quad \text{then}$$
$$K(P, B) = K(P, A) \cap B.$$

That $(4')$ follows from (4) is obvious. To see that (4) follows from $(4')$ substitute $B \cap A$ for B in $(4')$.[4]

Suppose $B \subseteq A$ and $K(P, A) \cap B \neq \emptyset$. Given that $B \subseteq A$ we have that any revision of K sufficient to accept B is also sufficient to accept A. This requires that $K(P, B) \subseteq K(P, A)$ unless the minimal revision needed to accept B requires rejecting something accepted relative to $K(P, A)$. The assumption that B is compatible with $K(P, A)$ rules out this case. Therefore, $K(P, B) \subseteq K(P, A) \cap B$. The other inclusion follows because in order for $K(P, B)$ to be smaller than $K(P, A) \cap B$ one would have to accept more than is needed to in order to accept B.

3. Relativized Conditionals (Step Two)

Principle 1 is not enough to generate an adequate Ramsey test conditional. A suggestion about what else is needed can be obtained by examining the proof of Remark 1b. The comparative similarity relation used in this proof does not correspond to a reasonable way of making minimal changes in an acceptance context. In the example K was QR, A was $\bar{Q} \cap (R \supset P)$ and $K_{\langle A \rangle}$ was $\bar{Q} \cap (R \equiv P)$, where P, Q and R were atomic propositions. To go from QR to $\bar{Q} \cap (R \equiv P)$ upon minimally revising to accept $\bar{Q} \cap (R \supset P)$, when no special connections among Q, R, and P are assumed, hardly seems appropriate. The problem is that the comparative similarity relation makes differences in P count even though P is not decided by K. If we restrict our measure of nearness so that only propositions decided by K count then $K_{\langle A \rangle}$ becomes $\bar{Q} \cap (RP)$. In the absence of special knowledge of connections among P, Q and R this seems to be a very reasonable way of minimally revising QR to accept $\bar{Q} \cap \cap (R \supset P)$.

The implausibility of the minimal revision given in the proof of 1b suggests that one ought to make comparative similarity itself depend on K. The remedy given in our discussions suggests the principle that *only propositions decided by K should count in determining comparative similarity relative to K.* We shall take this principle as our guide in developing an adequate Ramsey test conditional.

If one reflects on the role of Ramsey test conditionals the new principle is very plausible. As an acceptance context the total content of K is given by the propositions it decides, therefore it is just these propositions that should form the basis of judgments of comparative similarity relative to K. For subsets S of $\mathscr{E}$, acceptance contexts K, and worlds x and u let $S_u^x K$ be the set of K-decided propositions in S on which x and u differ.

(iii) $S_u^x K = \{A \in S : (K \subseteq A \text{ or } K \subseteq \bar{A}) \text{ and}$
          $((x \in A \text{ and } u \notin A) \text{ or } (x \notin A \text{ and } u \in A))\}.$

We can now give a precise formulation of our new principle.

(PRINCIPLE 2)

$\quad$ If $\mathscr{E}_u^x K = \mathscr{E}_u^y K$ and $\mathscr{E}_v^x K = \mathscr{E}_v^y K$, then
$\quad u \leqslant_K{}^x v \quad$ only if $\quad u \leqslant_K{}^y v.$

If u and v both differ from x on exactly the same K-decided propositions on which they differ from y, then their comparative similarity to x relative to K must agree with their comparative similarity to y relative to K.

Suppose E is the set of all subsets of a non-empty set T, and $\mathscr{K}$ is a subset of $\mathscr{E} - \{\emptyset\}$. For each world w in T and $K \in \mathscr{K}$ let $\leqslant_K w$ be a relativized comparative similarity relation satisfying principle 2, and let $R(w) = T$ and have both $R(w)$ and $\leqslant_K w$ satisfy all of our basic constraints on comparative similarity and accessibility. For each A in $\mathscr{E}$ let f_A^K be defined in terms of $\leqslant_K$ according to (i), $A \square\!\!\mapsto_K B$ be defined according to (ii) and $K_{\langle A \rangle} = \cap \{B : K \in A \square\!\!\mapsto_K B\}$.

(*Remark 2*) If these assumptions are met, then

(a) $f_A^K(x) = f_A^K(y)$, if $x, y \in K$
(b) $f_A^K(w) \subseteq K$ if $K \cap A \neq \emptyset$ and $w \in K$ (Principle 1)
(1) $K_{\langle A \rangle} \subseteq A$
(2) $K_{\langle T \rangle} = K$
(3) $K_{\langle A \rangle} \cap B \neq \emptyset$ only if $K_{\langle AB \rangle} = K_{\langle A \rangle} \cap B$.

Proof. (a) follows directly from Principle (2) and the definition of $f_A^K(w)$. If x and y are both in K then comparative similarity to x must agree with comparative similarity to y since all worlds in K agree on all propositions decided by K. (b) is just principle 1 and it follows because, by (a), the A-worlds in K are as close to w as any worlds can be. (1) and (2) follow from (b) according to Remark 1a. What remains is to show (3). Suppose $K_{\langle A \rangle} \cap b \neq \emptyset$. That $K_{\langle A \rangle} \cap B \subseteq K_{\langle AB \rangle}$ is obvious, become $f_A^K(w) \cap B \subseteq f_{AB}^K(w)$ for all w in K. To show that $K_{\langle AB \rangle} \subseteq K_{\langle A \rangle} \cap B$ assume $x \in K_{\langle AB \rangle}$. For some z in $Kx \leqslant_K^z v$ for all v in AB. Since $K_{\langle A \rangle} \cap B \neq \emptyset$ there is a y in $A \cap B$ such that for some w in $Ky \leqslant_K^w v$ all v in A. By (a) we have $x \leqslant_K^w y$ therefore by transitivity of $\leqslant_K w$ we have $x \in K_{\langle A \rangle} \cap B$. ∎ Remark 2 shows us that a Principle 2 relativized conditional satisfies Principle 1 and all of conditions (1)–(3) on formation of $K_{\langle A \rangle}$. Thus, by Theorem 1, Principle 2 conditionals are adequate for generating iterative probability models.

4. A Relational measure of nearness

Let S be a set of propositions designated as important for nearness. In the example from the proof of 1b used to motivate Principle 1 S was the set of atomic propositions. An obvious way to define comparative similarity relative to K and such a set S of designated propositions is

$$u \leqslant_K {}^w v \quad \text{iff} \quad |S_u^w K| \leqslant |S_v^w K|.$$

Relative to K, u is as similar to w as v is just in case the cardinality of the K-decided designated propositions on which u differs from w is less than or equal to the cardinality of the K-decided designated proposition on which v differs from w.

One example of an intuitive relational measure of nearness based on this idea is as follows. Suppose $\mathscr{E}$ is a field of propositions generated by a set S_0 of atomic propositions. Let $\mathscr{B}$ be the set of atomic propositions together with their complements. Let S_1 be the set of propositions in $\mathscr{E}$ that are unions of intersections of pairs drawn from B, S_2 be the set of unions of intersections of triples. For positive finite i, S_i is the set of unions of $i+1$ tuples drawn from $\mathscr{B}$. *To decide nearness first use S_0, then break ties with S_1, and so on for all the S_i's.* This has the effect that

(iv) (a) $u <_K {}^w v$ iff for some i, $|S_{iu}^w K| < |S_{iv}^w K|$
            and $|S_{ju}^w K| = |S_{jv}^w K|$ all $j < i$.
      (b) $u \overset{w}{=} v$ iff $|S_{iu}^w K| = |S_{iv}^w K|$ all i
      (c) $u \leqslant_K {}^w v$ iff $u <_K {}^w v$ or $u \overset{w}{=} v$.

On this measure u is strictly nearer to w than v judged from K just in case u is judged nearer to w than v at the first stage where there is no tie. Only if there is a tie at every stage are u and v judged equally far from w.

This method for measuring nearness can be extended to a transfinite series of stages in order to accommodate fields closed under infinite intersections. If $\mathscr{E}$ is an $\aleph\alpha$-complete field and S_0 is of cardinality $\aleph\alpha$ or greater then S_{w+a} is the set of unions of intersections of subsets of $\mathscr{B}$ of cardinality $\aleph\alpha$.

Suppose that $\mathscr{E}$ is the set of all subsets of a non-empty set T, S_0 is a set of generators designated as atomic proposition in $\mathscr{E}$, $\mathscr{K}$ is a set of non-empty subset of T and for each $K \in \mathscr{K}$ and $w \in T$ let $\leqslant_K w$ be given by *iv*. Suppose $R(w) = T$ for all $w \in T$, so that all worlds are mutually accessible. For all A, B, K, W let $f_A^K(w)$ be defined in terms of $\leqslant_K w$ according to (i) and $A \square\!\!\rightarrow_K B$ be defined according to (ii).

Remark

(1) $\quad w \leqslant_K{}^w v$

(2) $\quad u \leqslant_K{}^w v$ and $v \leqslant_K{}^w z$ only if $u \leqslant_K{}^w z$

(3) $\quad u \leqslant_K{}^w v$ or $v \leqslant_K{}^w u$

(4) $\quad f_A^K(w) \neq \emptyset$, if $R(w) \cap A \neq \emptyset$.

PRINCIPLE 2.

(If $\mathscr{E}_u^x K = \mathscr{E}_u^y K$ and $\mathscr{E}_v^x K = \mathscr{E}_v^y K$, then

$u \leqslant_K{}^x v$ only if $u \leqslant_K{}^y v$.

Proof. (1)–(3) and Principle 2 are obvious. The limit assumption (4) follows because less-than is well founded in the class of cardinal numbers. ∎

Our relativized nearness measure satisfies Principle 2 and all the basic constraints on comparative similarity. By Remark 2 it also satisfies Principle 1 and constraints (1)–(3) on the formation to acceptance context $K_{\langle A \rangle}$.

5. *A non-trivial Ip-Model*

Suppose worlds in T are identical when they determine the same atomic propositions, and $\mathscr{E}$ is closed under intersections of cardinality up to and including the cardinality of S_0. Under these conditions $\mathscr{E}$ includes all subsets of T (but there need not be worlds in T corresponding to each combination of atomic propositions). Such an $\mathscr{E}$ is already closed under the formation of $\square\!\!\rightarrow_K$, since each proposition $A \square\!\!\rightarrow_K B$ is just the set of worlds w in T where $f_A^K(w) \subseteq B$. If we let $\mathscr{K}$ be the set of all non-empty subsets of T, then $\mathscr{K}$ is closed under the formation of non-empty $K_{\langle A \rangle}$ for all A in $\mathscr{E}$. In this case $\mathscr{K} \subseteq \mathscr{E}$ and each $K_{\langle A \rangle} = \bigcup_{w \in K} f_A^K(w)$. Suppose further that T is countable. Under the assumptions we have made when T is denumerably infinite then $\mathscr{E}$ and $\mathscr{K}$ have the cardinality of the continuum.

Suppose that μ is a measure on $\mathscr{E}$ such that for all K in $\mathscr{K}$

$$0 < \mu(K) < \infty$$

and let $\mathscr{C}$ map $\mathscr{E} \times \mathscr{K}$ into the reals so that

$$\mathscr{C}(A/K) = \frac{\mu(AK)}{\mu(K)}.$$

Suppose K is in $\mathcal{K}$ and for all sequences $C_0 \ldots C_{n-1}$ drawn from $\mathcal{E}$ and A, B in $\mathcal{E}$

$$P_{\langle C \rangle_n}(B/A) = \mathcal{C}(B/K_{\langle C \rangle_n \langle A \rangle}) \quad \text{if} \quad K_{\langle C \rangle_n \langle A \rangle} \neq \emptyset =$$
$$= 1, \quad \text{otherwise.}$$

Remark. P is $\text{Ip}(\mathcal{E}\mathcal{K}\mathcal{C})$.

Proof. Trivially $\mathcal{E}$ and $\mathcal{K}$ satisfy the Ip-closure conditions. By remarks 1, 2 and 3 we have the constraints (1)–(3) on formation of $K_{\langle C \rangle_n \langle A \rangle}$ for all $C_0 \ldots, C_{n-1}$. Therefore by Theorem 1, P is an Ip-model for $\mathcal{E}$ and $\mathcal{K}$ with respect to C, if $\mathcal{C}$ is a regular confirmation function with $\mathcal{E} \times \mathcal{K}$ included in its domain. Since $\mathcal{K}$ includes all non-empty elements in $\mathcal{E}$ and each has positive measure $\mathcal{C}$ is regular. ∎

This remark allows for the construction of a wide range of non-trivial iterative probability models. Any $\mathcal{E}$ and $\mathcal{K}$ where for some K in $\mathcal{K}$ and A in $\mathcal{K}$, $K \cap A \neq \emptyset$, $K_{\langle A \rangle} \cap B = \emptyset$, there are X, Y, Z partitioning T such that $K_{\langle X \rangle}$, $K_{\langle Y \rangle}$ and $K_{\langle Z \rangle}$ are all non-empty and $K_{\langle A \rangle \langle B \rangle}$ is also non-empty generates an Ip-model where Stalnaker's trivialization does not hold. A simple example is the eight world space generated by atomic propositions P, Q and R where $K = QR$, $A = R\bar{P}$, $B = \bar{Q}\bar{P}$, $X = R$, $Y = \bar{R}Q$ and $Z = \bar{R}\bar{Q}$. We let $\mathcal{E}$ be the set of all subsets of T, $\mathcal{K}$ be $\mathcal{E} - \{\emptyset\}$. For each $w \in T$ let $R(w) = T$ and let $\leqslant_K w$ be given by iv for K in $\mathcal{K}$. In this example all of $K_{\langle X \rangle}$, $K_{\langle Y \rangle}$ and $K_{\langle Z \rangle}$ are non-empty, $K_{\langle A \rangle} = Q\bar{P}R$ which has an empty intersection with B, and $K_{\langle A \rangle \langle B \rangle} = \bar{G}R\bar{P}$ which is not empty.

Having a relativized conditional based on a comparative similarity relation satisfying principle 2 allows us to get around Stalnaker's objection and construct Ip-models of considerable richness.[5]

NOTES

[1] Van Fraassen has used the idea that conditional propositions should be allowed to vary with context in order to circumvent Lewis' trivialization of Stalnaker's original probability semantics. (Van Fraassen [1]). In their exchange of letters (this volume, pp. 302–308) Stalnaker and van Fraassen discuss the role of the assumption that $A \square\!\!\rightarrow B$ is independent of context in Lewis's trivialization. The assumption plays a similar role in Stalnaker's trivialization of the non-relational Ip-models.

The relativized conditionals to be developed here are different from those previously proposed in that an explicit way of defining $A \square\!\!\rightarrow_K B$ relative to K is provided.

[2] This particular formulation was suggested to me by Nuel Belnap.

[3] Counter examples of this kind were first suggested to me by Allan Gibbard.

[4] This equivalence was suggested to me by reading van Fraassen [3] where the corre-

sponding equivalence between Popper's a4 and Rényi's constraint 2b (see this paper, p. 120) is reported. A fairly cumbersome probability proof of the equivalence of a4 and 2b is given in Rényi [1].

[5] The main limitation on construction of Ip-models is that the confirmation function must be regular. Our reason for having T countable was that we could have regular confirmation functions where $\mathscr{K}$ was all non-empty subsets of T. This made construction of $\mathscr{K}$ particularly simple.

In a forthcoming paper the properties of the Rényi's confirmation function will be exploited is construct iterative probability models where the set T is uncountable. This paper will also use relativized accessibility in order to handle conceptual change.

BIBLIOGRAPHY

de Finetti, B. [1] *Probability, Induction and Statistics,* John Wiley and Sons, New York, 1972.

Harper, W. L. [1] 'Rational Belief Change, Popper Functions and Counterfactuals', this volume, p. 73.

Lewis, D. K. [1] *Counterfactuals,* Basil Blackwell, London, 1973.

Lewis, D. K. [2] 'Probabilities of Conditionals and Conditional Probabilities' (Manuscript).

Rényi, A. [1] 'On a New Axiomatic Theory of Probability' *Acta Math. Acad. Sci. Hung.* **6** (1955) 285–335.

Rényi, A. [2] *Foundations of Probability*, Holden Day, San Francisco, 1970.

Stalnaker, R. [1] Letter to Van Fraassen, this volume, p. 302.

Stalnaker, R. [2] Letter to Harper, this volume, p. 113.

Stalnaker, R. [3] 'Probability of Conditionals', *Philosophy of Science* **37** (1970).

Stalnaker, R. [4] 'A Theory of Conditionals', *Studies in Logical Theory. (American Philosophical Quarterly*, Supplementary Monograph Series), Oxford, 1968.

Van Fraassen, B. [1] 'Probabilities of Conditionals', this volume, p. 261.

Van Fraassen, B. [2] Letter to Stalnaker, this volume, p. 307.

Van Fraassen, B. [3] Representation of Popper Conditional Probabilities (Manuscript).

SHERRY MAY AND WILLIAM HARPER

TOWARD AN OPTIMIZATION PROCEDURE FOR
APPLYING MINIMUM CHANGE PRINCIPLES
IN PROBABILITY KINEMATICS

I. PROBABILITY KINEMATICS

"Probability kinematics" is Richard Jeffrey's term for the study of how a rational agent ought to revise his beliefs in response to inputs from experience.[1] Typically, we have P_0 representing a rational agent's belief state just before making an observation. The input from the observation is a change in his assignment to some preposition E from $P_0(E)$ to some new value γ.[2] The problem is to specify what the agent's new belief state P_1 ought to be, given what P_0 was and that $P_1(E) = \gamma$.

On this characterization probability kinematics does not include study of what inputs are appropriate for various situations. The input, changing $P_0(E)$ to $P_1(E)$, is treated as one of the conditions on the problem. The new function P_1 is to be such that it would be a shift from P_0 that would be justified by this input.

Probability kinematics is only a part of the larger study of rational belief change; nevertheless, it is an interesting part. Good, Kyburg, Shafer, Levi, and others have proposed interval valued probability functions as representations of rational belief.[3] On such representations the usual method of conditionalization is inadequate as a device for probability kinematics. Moreover, Dempster's rule of combination (see Shafer), Kyburg's interval version of conditionalization, and Levi's recommendations for handling belief change seem to constitute quite different, and perhaps mutually incompatable, approaches to probability kinematics.[4]

Gudder's generalized probability measures are point functions, but they do not allow for the usual definition of conditional probability.[5] At least part of Gudder's motivation was to provide a notion of probability general enough to do justice to the vagaries of actual measurements.[6] If this motivation is to be realized some way of defining conditional probability is desirable.

Even when belief states are represented by point probability functions on Boolean algebras, there are grounds for investigating probability

Harper and Hooker (eds.), Foundations of Probability Theory, Statistical Inference, and Statistical Theories of Science,
Vol. I, 137–166. All Rights Reserved. Copyright © 1976 by D. Reidel Publishing Company, Dordrecht-Holland.

kinematics. The classical Bayesian solution to probability kinematical problems is conditionalization. Where $P_0(E) > 0$ and $P_1(E) = 1$ conditionalization sets the appropriate $P_1(H)$ for each H to be $P_0(H \wedge E)/P_0(E)$. Both the restrictions that $P_0(E) > 0$ and that $P_1(E) = 1$ are necessary if conditionalization is to apply, and there are reasons for wanting to avoid each of them.

Richard Jeffrey has proposed examples of learning from experience where it seems appropriate to represent the input from observation as a change from $P_0(E)$ to some $P_1(E)$ that is strictly less than 1. Suppose that $P_0(E) = 0.6$, where P_0 is the belief function of a rational tailor just before observing a piece of cloth and E is the proposition that the cloth is green. Upon observing the cloth in poor (perhaps yellowish) light the tailor changes $P_0(E)$ to $P_1(E) = 0.9$. Jeffrey argues that in many such cases there will be no particular proposition E^* such that $P_1(E^*) = 1$, $P_0(E/E^*) = 0.9$, and the appropriate way to represent the input is as accepting E^* as evidence.[7]

In order to handle cases like this Jeffrey provides the following recommendation. When the change from $P_0(E)$ to $P_1(E)$ is the input from experience the conditional probabilities $P_0(H/E)$ and $P_0(H/\bar{E})$ should not be changed. This recommendation yields a generalization of conditionalization, where

$$P_1(H) = P_0(H/E) \cdot P_1(E) + P_0(H/\bar{E}) \cdot P_1(\bar{E}).^8$$

Given Jeffrey's recommendation his rule follows, but the recommendation has been challenged by Isaac Levi.[9] The rule and the recommendation are given an interesting defense by Paul Teller.[10] The point here is that the arguments over Jeffrey's rule are arguments over how to do probability kinematics.

One crucial requirement for a theory of rational belief change is that earlier inputs should be correctable on the basis of later ones. If observation inputs are to be fallibilistic, as seems appropriate to do justice to human limitations, then provision must be made for revision of previous inputs. Conditionalization, by itself, cannot be a method of probability kinematics that meets this demand. If the input is to assign $P_1(E) = 1$ then any E' that conflicts with E has $P_1(E') = 0$ and cannot itself be used as a later input.

If one never assigns $P_1(E) = 1$ for inputs and represents all belief change by Jeffrey's rule then the requirement of corrigibility is met. Future inputs can result in new changes to the assignment to E. On such an approach, conditionalization would be seen as an approximation used to describe cases where $P_1(E)$ is quite close to 1. This is exactly the picture of rational belief change appropriate to having strict coherence as an additional constraint on rational belief functions.

If one wants a fallibilistic account of rational belief change where inputs are accepted so that $P_1(E) = 1$, then conditionalization must be augmented in ways that go beyond Jeffrey's rule. William Harper has provided an extension of Savage's representation of rational belief which allows $P_1(H/E)$ to be non-trivially defined even when $P_0(E) = 0$.[11] With some additional apparatus this representation allows for a fallibilistic account of probability kinematics where one can have both genuine acceptance in the sense that $P_1(E) = 1$ and corrigibility in that this assignment to E can be changed on the basis of later inputs. The basic idea used to develop this representation is a special case of the general principle for probability kinematics that we discuss next.

II. A LEAST CHANGE PRINCIPLE

We propose a principle that ought to apply quite generally to probability kinematical problems. In such a problem we have as given the initial rational belief function P_0, the input change of $P_0(E)$ to γ, and the general rationality constraints that P_1 must meet. We want P_1 to be a rational belief function which results from a shift from P_0 that is justified by the input change of $P_0(E)$ to $P_1(E) = \gamma$. The principle is this:

(LCP) P_1 is to be as near a belief function to P_0
        as is possible given that
        (1) $P_1(E) = \gamma$, and
        (2) P_1 meets the rationality constraints.

If the rationality constraints include specification of the requirements for point probability functions on a Boolean domain then (LCP) will require that P_1 be as near a probability function to P_0 as possible where $P_1(E) = \gamma$. If the rationality constraints allow interval values γ may be an interval and (LCP) will require that P_1 be a function satisfying the interval constraints which agrees with P_0 as closely as possible while having $P_1(E) = \gamma$.

One basic argument for the principle is simple. The change from P_0 to P_1 is to be justified by the input change of $P_0(E)$ to $P_1(E)=\gamma$. Any differences between P_0 and P_1 that go beyond what is minimally required in order to have $P_1(E)=\gamma$ would be gratuitous. Our idea is that a proper specification of coherence constraints and allowable inputs ought to be such that a change from P_0 to P_1 is justified by an input just to the extent that the change is forced by the new input together with the coherence constraints.

Further motivation for (LCP) comes from the fact that a special case of it has been used to reconstruct Savage's argument for conditionalization in a way that meets objections to it raised by Paul Teller.[12]

Some of the present work is motivated by the following conjecture relative to a measure of nearness.

(c)   Suppose P_0 is a probability function on the Lindenbaum algebra $L/\equiv$ of a predicate calculus L and $0<P_0(E)<1$ for some $E\in L/\equiv$. Then
(i) the conditional probability function, $P_{0_E}(H)=P_0(H/E)$ for all H, is the unique nearest element to P_0 in the set K of probability functions P on $L/\equiv$ such that $P(E)=1$.
(ii) more generally, if K is the set of P where $P(E)=\gamma\neq 1$ then the Jeffrey rule solution is the unique nearest element of K to P_0.

Should conjectures of this sort work out for a suitably plausible measure of nearness, then (LCP) could be used to give a very general definition of conditional probability that could be applied in probability kinematical problems where ordinary conditionalization fails.

Our ultimate goal is to provide intuitively plausible measures of nearness and optimization procedures that will allow (LCP) to be applied to solve probability kinematical problems in a wide variety of contexts. Our present research is restricted to the problem of developing optimization techniques adequate to investigate measures of nearness for point probability functions on a predicate calculus. Even this more limited problem is so much more abstract than ordinary optimization problems as to require considerable extension of existing techniques.

We expect much of our optimization work to carry over to investigations of more exotic representations of rational belief.

III. A MEASURE OF NEARNESS

Functionals and metrics defined on various classes of probability functions will be discussed as possible measures of nearness for subjective probability functions. One such measure is chosen for investigation. In discussing the various possibilities it is not intended to dismiss each of the measures not chosen as being totally inappropriate. Rather it is intended to point out the conceptual or technical difficulties involved in applying those measures to subjective probability functions.[13]

The Kullback-Leibler entropy (or information) functional is defined as follows [10]. Let $(X, \mathscr{E}, \mu_i, i = 1, 2$, be a probability space. If μ_1 is absolutely continuous with respect to μ_2, then by the Radon-Nikodym theorem there exists a non-negative function $f(x)$ on X such that

$$\mu_1(E) = \int_E f(x) \, d\mu_2(x)$$

for every E in $\mathscr{E}$, and $f(x)$ is unique up to a set of μ_2-measure zero. In this case it is possible to define a functional on the set of measures μ_1 on $\mathscr{E}$ which are absolutely continuous with respect to μ_2 by

$$I(1:2) = \int_X f(x) \log[f(x)] \, d\mu_2(x).$$

The functional $I(1:2)$ is called the information of μ_1 with respect to μ_2 [13, page 5].

It should be noted that other information measures have been defined. However they are related to Kullback's definition.[14] Since Kullback's definition is given in the greatest generality, it was chosen as a representative of the class of information measures and studied as a possible measure of nearness for the probability kinematics problem.

In his book Kullback discussed the role of $I(1:2)$ in statistics. An information statistic (that is, an estimate of the minimum discrimination information between two sets of density functions describing the same population) is used to test the hypothesis that an estimator is distributed according to one of a class of specified density functions. A description of the testing procedure follows. An estimator is a function T defined on the sample space. It is used to estimate a given parameter θ (characteristic) of the population from which the samples come. Let the null hypothesis H_2 be that the distribution of T belongs to the set $\{\mu_k\}_{k \in K}$, and the alternative hypothesis H_1 be that the distribution of T belongs to the set $\{\mu_j\}_{j \in J}$ (J and K are index sets). Let $I(*:k)$ be the minimum value over $j \in J$ of $I(j:k)$, where the minimization is subject to the constraint

$$\int_X T(x) f_j(x) \, d\mu_k = \theta \, ;$$

that is, subject to the constraint that T is an unbiased estimator. The value of $I(*:k)$ is known as a function of θ, so given a sample x in X, an estimate of $I(*:k)$, $\hat{I}(*:k;x)$, is made using the value of the estimator of θ, $T(x)$. The minimum of $\hat{I}(*:k;x)$ over K is defined to be $\hat{I}(*:H_1)$. The estimate $\hat{I}(*:H_2)$ of the minimum discrimination information of H_1 with respect to H_2 is similarly defined. Since the distribution P of the information statistic is known, the null hypothesis is rejected if

$$P((\hat{I}(*:H_2) - \hat{I}(*:H_1) \geqslant c) \mid H_2) \leqslant \alpha$$

where c and α determine the probabilities of type I and type II errors occurring.

It is not immediately evident what conceptual relationship exists between the statistics problem just described and the problem of probability kinematics. However finding the solution to each problem involves a minimization procedure, and $I(1:2)$ does provide a measure of nearness for probability functions.

If X is finite and p_1 and p_2 are probability functions defined on X, then $I(1:2)$ becomes

$$\sum_X p_1(x) \log[p_1(x)/p_2(x)].$$

Since X is finite this sum exists. It has been shown [10] that for X finite, conceiving the problem of defining conditional probability given the expected values μ_i of certain random variables V_i, as one of renormalization yields a solution equivalent to minimizing $I(1:2)$ subject to the constraints

$$\sum_X V_i(x)\, p_1(x) = \mu_i.$$

Remark 1. The sum $I(1:2)$ will not in general exist for X infinite.

Remark 2. Consider the problem of conditionalizing on an event y which has zero probability. The value of $p_1(y)$ would be one, and the value of $p_2(y)$ would be zero. In this case the term $p_1(y) \log[p_1(y)/p_2(y)]$ in the sum $I(1:2)$ is undefined. The difficulty is that for problems of this sort, p_1 is not absolutely continuous with respect to p_2; $p_2(y)=0$ but $p_1(y)=1$.[15] Therefore the information measure could not be used for defining conditional probabilities on propositions which have zero probability.

This is not a surprising result. Minimizing $I(1:2)$ in order to define conditional probabilities is equivalent to conceiving the conditionalization problem as one of renormalization (for X finite), and renormalization on events with zero probability has no meaning.

Remark 3. Jamison [10] defined the conditional probability of A given the expected values μ_i of certain random variables V_i, as a limit of probabilities. He then asked the question: can the probability defined by this limit be interpreted as a classical conditional probability? If X is finite, his answer is yes (he appeals to an argument involving Markov chains and Martin boundaries). However he pointed out that if X is infinite but discrete there are technical difficulties involved with a similar argument and for X uncountable there are conceptual difficulties as well. Since the limit definition is equivalent to minimizing $I(1:2)$ for X finite, it might be the case that related technical and conceptual difficulties would arise in any attempt to define conditional probabilities in those cases, by minimizing an information measure.

Remark 4. Although the Kullback-Liebler information functional is not defined in terms of conditional probabilities, the justification for calling it an information functional draws heavily upon an argument involving an application of Bayes' theorem and a subsequent interpretation of a conditional probability defined in the traditional sense $(P(A \mid B) = (A \cap B)/P(B))$ as a posterior probability. Therefore the use of an information functional as a measure of nearness for probability functions in order to define conditional probabilities, would appear to be circular.

The minimization problem for probability kinematics was first proposed for the case of a sentential calculus with finitely many atomic sentences. For that case the Lindenbaum algebra is finite (that is, there are finitely many equivalence classes of formulas), and the obvious metric for probability functions p_1 and p_2 defined thereon is a Euclidean metric:

(i) $\qquad d_1(p_1, p_2) = \sum_{|\varphi| \in L/\equiv} |p_1(\varphi) - p_2(\varphi)|$

(ii) $\qquad d_2(p_1, p_2) = \sum_{|\varphi| \in L/\equiv} (p_1(\varphi) - p_2(\varphi))^2$

(iii) $\qquad d_3(p_1, p_2) = \sup_{|\varphi| \in L/\equiv} (p_1(\varphi) - p_2(\varphi)).$

The obvious difficulty with metrics (i) and (ii) is that for the case of $L/\equiv$ infinite, the sums in (i) and (ii) do not exist. Since the case for predicate calculus is fundamentally more interesting than the case for sentential calculus, being able to make the extension from sentential to predicate calculus is very important. The third metric is easily extended, and has strong mathematical appeal. Since a form of (iii) was ultimately chosen as the measure of nearness to begin our investigation of the probability kinematics problem, a discussion of its appeal will be postponed until the Prokhorov metric has been discussed.

In a paper entitled 'Convergence of Random Processes and Limit Theorems in Probability Theory' [21] Yu. V. Prokhorov defined a metric for probability measures such that convergence in the sense of this metric is equivalent to weak convergence of distributions of random processes.[16] The metric is defined as follows.

DEFINITION 1. Let $(\mathscr{E}, d)$ be a complete separable metric space and F be any closed set in $\mathscr{E}$. For any $x \in \mathscr{E}$ define $d(x, F)$ to be the infimum of $d(x, y)$ for $y \in F$, and for any $\varepsilon > 0$ define F^ε to be the set of $x \in \mathscr{E}$ such that $d(x, F) < \varepsilon$. If μ_1 and μ_2 are measures defined on $\mathscr{E}$, the 'distance' $L(\mu_1, \mu_2)$ between them is defined as the maximum of $\varepsilon_{1, 2}$ and $\varepsilon_{2, 1}$ where

$$\varepsilon_{i, j} = \inf\{\varepsilon : \mu_j(F) < \mu_i(F^\varepsilon) + \varepsilon \quad \text{for all } F\}.$$

The obvious difficulty with using Prokhorov's metric for subjective probability functions is the requirement that $\mathscr{E}$ be a metric space. To date there have been no attempts to define such metrics where the algebra is the Lindenbaum algebra of a language. Producing such a metric would itself be a task of some interest; however, we do not want to start our investigation of the probability kinematics problem with a measure of nearness that presupposes such a metric.

The measure of nearness we use to conduct our investigation of the optimization techniques results from the natural embedding of the set of probability functions into the Banach space of bounded additive functions defined on the Stone space of the Lindenbaum algebra. If $f_p(h(x))$ is defined to be $p(x)$ where $x \in \Sigma$ and $h(x)$ is the set of ultrafilters on Σ which contain x, then the function f_p is a bounded additive function defined on $h(\Sigma)$. If p is identified with f_p, the norm on the Banach space can be used to induce a metric on the set of probability functions as follows:

$$d(p_0, p_1) = \|f_{p_0} - f_{p_1}\| =$$
$$= \sup_{\substack{x \in \Sigma}} (\sup_{\substack{y \in \Sigma \\ y \leqslant x}} (p_0(y) - p_1(y)) - \inf_{\substack{y \in \Sigma \\ y \leqslant x}} (p_0(y) - p_1(y))).$$

Since Σ has a maximum element, $d(p_0, p_1)$ will equal $\sup_{y \in \Sigma}(p_0(y) - p_1(y)) - \inf_{y \in \Sigma}(p_0(y) - p_1(y))$. Also, since $\inf_{y \in \Sigma}(p_0(y) - p_1(y))$ will equal $[(1 - p_0(z)) - (1 - p_1(z))]$ if $\sup_{y \in \Sigma}(p_0(y) - p_1(y))$ equals $[p_0(z) - p_1(z)]$, then $d(p_0, p_1) = 2 \sup_{y \in \Sigma}(p_0(y) - p_1(y))$. (Note that $\Sigma = L/\equiv$).

One important feature of this metric is that Jeffrey's rule (and, thus, conditionalization as a special case) optimizes it.[17] Suppose that the input is to change $p_0(\varphi)$ to $p_1(\varphi) = \gamma$ where $0 < p_0(\varphi) < 1$ and $\gamma \neq 0$. Let p^* be computed according to Jeffrey's rule. Then $d(p_0, p^*) = 2|p_0(\varphi) - \gamma|$. The proof goes as follows. Let ψ be any element of the algebra. Then by the laws of probability calculus,

$$p_0(\psi) - p^*(\psi) = (p_0(\psi \wedge \varphi) + p\ (\psi \wedge \neg\varphi))$$
$$- (p^*(\psi \mid \varphi)\, p^*(\varphi) + p^*(\psi \mid \neg\varphi)\, p^*(\neg\varphi))$$
$$= (p_0(\psi \wedge \varphi) + p_0(\psi \wedge \neg\varphi))$$
$$- (p_0(\psi \mid \varphi)\, p^*(\varphi) + p_0(\psi \mid \neg\varphi)\, p^*(\neg\varphi))$$
$$= (p_0(\psi \wedge \varphi) + p_0(\psi \wedge \neg\varphi))$$
$$- ((p_0(\psi \wedge \varphi)/p_0(\varphi))\, p^*(\varphi)$$
$$+ (p_0(\psi \wedge \neg\varphi)/p_0(\neg\varphi))\, p^*(\neg\ \varphi))$$
$$= p_0(\psi \wedge \varphi)\, ((p_0(\varphi) - p^*(\varphi))/p_0(\varphi))$$
$$+ p_0(\psi \wedge \neg\varphi)\, ((p^*(\varphi) - p_0(\varphi))/(1 - p_0(\varphi))).$$

It must be the case that either $p_0(\varphi) - p^*(\varphi) < 0$ or $p^*(\varphi) - p_0(\varphi) < 0$. Assume that $p^*(\varphi) - p_0(\varphi) < 0$. Then

$$p_0(\psi) - p^*(\psi) \leqslant p_0(\psi \wedge \varphi)\left((p_0(\varphi) - p^*(\varphi))/p_0(\varphi)\right)$$
$$\leqslant p_0(\varphi)\left((p_0(\varphi) - p^*(\varphi))/p_0(\varphi)\right)$$
$$= p_0(\varphi) - \gamma.$$

Therefore $\sup_{\psi \in \Sigma}(p_0(\psi) - p^*(\psi)) = p_0(\varphi) - \gamma$ and $d(p_0, p^*) = 2(p_0(\varphi) - \gamma) = 2|p_0(\varphi) - \gamma|$. Similarly it can be shown that $d(p_0, p^*) = 2|p_0(\varphi) - \gamma|$ if $p_0(\varphi) - p^*(\varphi) < 0$.

Any p' in the set of functions p where $p'(\varphi) = \gamma$ must differ from p_0 by at least as much as $2|p_0(\varphi) - \gamma|$. Therefore this result makes it clear that Jeffrey's rule does optimize the metric.

Note that this metric would be unsuitable for defining conditional probabilities on propositions φ with zero probability. This is because for any probability function p which satisfied the constraint $p(\varphi) = 1$, it would be true that $d(p_0, p) = 2 \sup_{\psi \in \Sigma}(p_0(\psi) - p(\psi)) = 2(p_0(\neg\varphi) - p(\neg\varphi)) = 2$. Consequently it would be impossible to distinguish between admissible probability functions using this metric. The difficulty lies in the sensitivity of the metric. For non-extreme cases, that is, in cases where $p_0(\varphi) > 0$ and φ is the conditionalizing event, some discrimination of admissible probability functions is possible, and discrimination increases as $p_0(\varphi)$ approaches $\frac{1}{2}$. In fact, for cases in which $p_0(\varphi) > \frac{1}{2}$ this metric should yield very good results for the optimization problem.

Though this metric is not suitable for defining conditional probabilities on propositions with zero probability there is reason to believe that the metric could be suitably sensitized. It may be that the correspondence between charges on the power set of the set of ultrafilters of the algebra with this norm and the normed conjugate of the space of bounded functions defined on the set of ultrafilters, will provide a means for realizing adequate sensitivity. Failing that, there are at least two other possibilities. However, increasing the sensitivity of this metric is a matter for further research and will not be discussed here in any further detail.

The supremum metric was chosen for the following reasons. It can handle the infinite case. It is a mathematically natural measure of nearness, since it is induced by a norm on a Banach space in which the probability functions can be embedded. Since Jeffrey's rule does optimize it, there is hope that conjecture (c) could be realized using it.

In choosing this metric as our first to investigate we are not disregarding its difficulties for handling extreme input changes. Nor are we dismissing the other measures of nearness considered. In this paper all we intend to establish are the basic details of the optimization procedure.

IV. OPTIMIZATION THEORY

At first glance it would appear that the problem to be solved here fits into the broad class of optimization problems called mathematical programming. However, because the 'choice variables' are probability functions, a difficulty arises; a theory of mathematical programming more general than the standard is needed to solve this problem. Such a theory was found in a paper by the late Lucien Neustadt entitled 'A General Theory of Extremals' [19]. In this paper Neustadt presented a unified theory of necessary conditions for optimization problems. He stated and proved a general theorem called the abstract maximum principle, which contains as special cases necessary conditions in the theories of mathematical programming and optimal control. In order to do this, it was necessary for him to work in a very general framework; in fact, a framework sufficiently general to encompass the probability kinematics problem. This section contains a description of Neustadt's theory in its most general format, and also in the particular format amenable to the probability kinematics problem.

The starting point of a general theory of optimization must be a general concept of extremality. Neustadt calls his generalized optimal solution a (π, Φ, Z)-extremal. It is possible to define a (π, Φ, Z)-extremal under the following very general conditions. There must be given:

(1)   an arbitrary set $\mathscr{E}'$ and a subset $\mathscr{E}$ of $\mathscr{E}'$;
(2)   a real linear topological space $\mathscr{L}$;
(3)   an open convex cone Z in $\mathscr{L}$;[18]
(4)   a function Φ which maps $\mathscr{E}'$ into $\mathscr{L}$;
(5)   a positive integer m;
(6)   a function π which maps $\mathscr{E}'$ into R^m
        (Euclidean m-space).

DEFINITION 2. If $\mathscr{E}'$, $\mathscr{E}$, $\mathscr{L}$, Z, Φ, m and π have properties (1) to (6), then an element $e_* \in \mathscr{E}'$ will be called a (π, Φ, Z)-extremal if

(i) $\pi(e_*) = 0$

(ii) $\Phi(e_*) \in \bar{Z}$ (the closure of Z)

(iii) $\{e \in \mathscr{E} : \pi(e) = 0 \quad \text{and} \quad \Phi(e) \in Z\} = \emptyset$.

Remark 1. Note that a (π, Φ, Z)-extremal need not be unique.

Remark 2. Consider the following optimization problem:

> Let A be a set, and f and g be real-valued functions defined on A. Find an element $q \in A$ that achieves a minimum for f subject to the constraint $g(q) = 0$.

If q is a solution of this problem, and if the following definitions are made, then q is a (π, Φ, Z)-extremal:

$$\mathscr{E} = \mathscr{E}' = A$$
$$\mathscr{L} = R^1$$
$$Z = \{\gamma \in R^1 : \gamma < 0\}$$
$$m = 1$$
$$\pi = g$$
$$\Phi = f - f(q).$$

Clearly $\pi(q) = g(q) = 0$, and $\Phi(q) = f(q) - f(q) = 0$ where 0 is an element of $\bar{Z} = \{\gamma \in R^1 : \gamma \leqslant 0\}$. Also, if $a \in \mathscr{E}$ and $\pi(a) = 0$, then $f(a) \geqslant f(q)$ since q is a solution of the optimization problem. Therefore $\Phi(a)$ equals $f(a) - f(q) \geqslant 0$ and $\Phi(a) \notin Z$. Hence $\{a \in \mathscr{E} : \pi(a) = 0 \text{ and } \Phi(a) \in Z\} = \emptyset$ and q is a (π, Φ, Z)-extremal.

Remark 3. Observe that if A is the set of all subjective probability functions defined on the Lindenbaum algebra of predicate calculus, $g : A \to R^1$ is defined by $g(p) = p(\varphi) - \gamma$, and $f : A \to R^1$ is defined by $f(p) = d(p_0, p)$, then p_* is a (π, Φ, Z)-extremal where $\mathscr{E}, \mathscr{E}', \mathscr{L}, Z, m, \pi$ and Φ are defined as above, and p_* minimizes $d(p_0, p)$ subject to $p(\varphi) = \gamma$.

It is not surprising to discover that in order to obtain a maximum principle for (π, Φ, Z)-extremals, a certain condition must be satisfied by the set $\mathscr{E}$, the cone Z, the maps π and Φ, and the (π, Φ, Z)-extremal e_*. This condition is quite complex in its most general format. However if one considers how this condition is met in more standard optimization problems, then intuitively its imposition will seem reasonable. Furthermore, although the probability kinematics optimization problem is non-standard, imposing Condition 1 (below) on it will be quite natural.

CONDITION 1. There exists a non-empty convex set $\mathcal{M}$ in a real vector space $\mathcal{U}$, an affine function $h: \mathcal{M} \to R^m$, and a Z-convex function $\hat{h}: \mathcal{M} \to \mathcal{L}$ such that:

> For every triple $(\mathcal{C}, N, \eta_1)$ – where $\mathcal{C}$ is a k-simplex in $\mathcal{M}$, $k \leqslant m$, $0 \in h(\mathcal{C})$, N is a neighbourhood of 0 in $\mathcal{L}$, and η_1 is a positive real number – there exists $\varepsilon_0 \neq 0$, $\varepsilon_1 \in (0, \eta_1)$ and a map $\Theta: \mathcal{C} \to \mathcal{E}$ such that
> (1) $\| \{\pi(\Theta(y)) - \pi(e_*)\}/\varepsilon_0 - h(y)\|_m < \eta_1$ for all $y \in \mathcal{C}$[19]
> (2) $\{\Phi(\Theta(y)) - \Phi(e_*)\}/\varepsilon_1 \in \mathrm{co}\, \hat{h}(\mathcal{C}) + \bar{Z} + N$ for all $y \in \mathcal{C}$
> (3) $\pi \circ \Theta: \mathcal{C} \to R^m$ is continuous as a function of the barycentric coordinates of a point in $\mathcal{C}$.

The reader may find Figure 1 a convenient representation of Condition

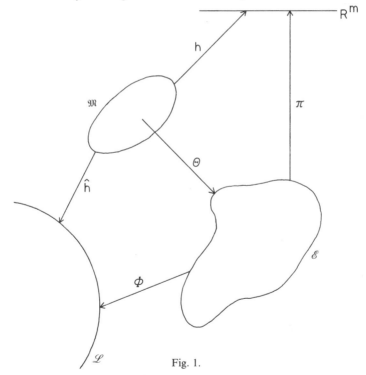

Fig. 1.

1 for the discussion which follows. Firstly, an affine function defined on a vector space is just a linear function plus a constant, and if $Z = \{\gamma \in R^1 : \gamma < 0\}$ then Z-convexity is equivalent to ordinary convexity. Secondly, for standard mathematical programming problems, one can derive necessary conditions for an optimal solution when the choice variables form a convex set, the constraint function is linear (or alternately, affine), and the objective functional is convex. Immediately one is led to suspect that Condition 1 is an attempt to 'appropriately approximate' the set of choice variables $\mathscr{E}$ by the convex set $\mathscr{M}$, the constraint function π by the affine function h, and the objective functional Φ by the convex function $\hat{h}$. Equation (1) in Condition 1 would guarantee the 'appropriateness' of h as an 'approximation' to π, equation (2) would guarantee the 'appropriateness' of $\hat{h}$ as an 'approximation' to Φ, and (3) would guarantee the 'appropriateness' of $\mathscr{M}$ as an 'approximation' to $\mathscr{E}$.[20] It is certainly not intuitively obvious from this discussion exactly what convex set, affine function and 'convex' function would constitute 'appropriate approximations' to $\mathscr{E}$, π and Φ respectively. However, Condition 1 will simplify considerably for the problem under investigation here.

THEOREM 1. [19] (Abstract Maximum Principle). Let e_* be a (π, Φ, Z)-extremal such that Condition 1 is satisfied. Then there exists a continuous linear functional $l \in \mathscr{L}^*$ ($\mathscr{L}^*$ is the dual of $\mathscr{L}$) and a vector $\alpha \in R^m$ such that

(1) $\alpha \cdot h(y) + l(\hat{h}(y)) \leq 0$ for all $y \in \mathscr{M}$

(2)   if $\alpha = 0$ then $l \neq 0$

(3) $l(z) \geq 0$ for all $z \in \bar{Z}$

(4) $l(\Phi(e_*)) = 0$.

The proof of this theorem is very long and involved, and consequently shall be omitted. Furthermore any discussion of exactly what information about e_* this theorem provides shall be postponed until section six when a detailed analysis will be made for a particular problem.

For the special case of e_* a (π, Φ, Z)-extremal where $\mathscr{L} = R^1$ and $Z = \{\gamma \in R^1 : \gamma < 0\}$, the conclusions of Theorem 1 can be somewhat sharpened if some additional conditions are imposed on the problem data. Recall that in this case, Z-convexity of a function is equivalent to ordinary convexity.

ADDITIONAL CONDITION 2. There exists a convex set H_1 in $\mathcal{M}$ such that

(a) $\mathcal{M} \subset H_1$

(b)   there exists a $\bar{y}_1 \in \mathcal{M}$ which is also an internal point of H_1

(c) $\hat{h}$ can be extended to H_1 such that it remains a convex function.

Note that Condition 2 is automatically satisfied if $\hat{h}$ can be extended to all of $\mathcal{U}$ in such a way that $\hat{h}$ remains a convex function. In fact for the probability kinematics problem, this is exactly the way Condition 2 will be met. The strengthened maximum principle which results is given in the following theorem.

THEOREM 2. Let e_* be a (π, Φ, Z)-extremal, where $\mathcal{L} = R^1$ and $Z = \{\gamma \in R^1 : \gamma < 0\}$, and suppose that Condition 1 is satisfied. Then there exist vectors $\alpha \in R^m$ and $\beta \in R^1$ such that

(5) $\alpha \cdot h(y) + \beta \cdot \hat{h}(y) \leqslant 0$ for all $y \in \mathcal{M}$

(6) $\|\alpha\|_m + |\beta| > 0$

(7) $\beta \leqslant 0$

(8) $\beta \cdot \Phi(e_*) = 0$.

If in addition, Condition 2 is satisfied, there exists an affine function $\Lambda : \mathcal{U} \to R^1$ such that

(9) $\alpha \cdot h(y) + \beta \cdot \Lambda(y) \leqslant 0$ for all $y \in \mathcal{M}$

(10) $\Lambda(y) \leqslant \hat{h}(y)$ for all $y \in H_1$

If also $0 \in \mathcal{M}$ and 0 is an internal point of H_1, if h is linear on $\mathcal{M}$, and if $\hat{h}(0) = 0$, then Λ may be taken to be linear. If $\mathcal{U}$ is a linear topological space, H_1 has a non-empty interior, $\hat{h}$ is continuous at some interior point $\hat{y}_1$ of H_1, and Λ is an affine functional on $\mathcal{M}$ which satisfies (10), then Λ is continuous on $\mathcal{M}$.

Once again the proof will be omitted because it is very long. Note that the only difference between the four equations of Theorem 1 and the first four equations of Theorem 2 is that β appears in Theorem 2 wherever the linear functional l appears in Theorem 1. This follows trivially from the fact that any linear functional defined on R^1 can be represented as the product of each member of R^1 with some real number β.

Now attention must turn from the general to the specific. How can the problem data for the probability kinematic problem be made to satisfy Conditions 1 and 2? Furthermore, what information about p_* will be derivable from Theorems 1 and 2? These questions will be dealt with in subsequent sections.

V. EMBEDDING FORMULATION

It will be shown that the probability kinematics problem belongs to the class of optimization problems to which the following lemma applies.

LEMMA 1. Suppose that e_* is a (π, Φ, Z)-extremal where $e_* \in \mathscr{E}$, $\mathscr{E}$ is a convex set in a vector space $\mathscr{U}$, $Z = \{\gamma \in R^1 : \gamma < 0\}$, and $\pi|_{\mathscr{E}}$ and $\Phi|_{\mathscr{E}}$[21] are respectively affine and convex. Then Condition 1 is satisfied with $\mathscr{M} = \mathscr{E}$, $h = \pi|_{\mathscr{E}}$ and $\hat{h} = \Phi|_{\mathscr{E}} - \Phi(e_*)$.

Proof. For every triple (β, N, η_1), if ε_0 and ε_1 are chosen to be $\eta_1/(1 + \eta_1)$, and $\Theta(y)$ is defined to be $e_* + \varepsilon_1 (y - e_*)$ for $y \in \beta$, then it is not difficult to establish the following:

(11) $[\pi(\Theta(y)) - \pi(e_*)]/\varepsilon_0 = h(y);$

(12) $[\Phi(\Theta(y)) - \Phi(e_*)]/\varepsilon_1 \leqslant \hat{h}(y).$

Therefore (1) and (2) of Condition 1 are satisfied trivially. Also, $\pi \circ \Theta$ is continuous on β as a function of the barycentric coordinates of a point in β since it is an affine function defined on a simplex.

It was established   that the optimal solution p_* to the probability kinematics problem is a (π, Φ, Z)-extremal where $\mathscr{E} = \mathscr{E}' = A$, $Z = \{\lambda \in R^1 : \lambda < 0\}$, $\pi(p) = p(\varphi) - \gamma$, and $\Phi(p) = d(p_0, p) - d(p_0, p_*)$. The following theorem combined with lemma 1, establishes that Condition 1 is satisfied for this problem with $\mathscr{M} = \mathscr{E}, h = \pi$ and $\hat{h} = \Phi - \Phi(p_*)$. (Note that $\Phi(p_*) = 0$.)

THEOREM 3. With A, π and Φ defined as above, A is a convex subset of the vector space $B(\Sigma)$ of bounded functions defined on Σ, π is an affine function, and Φ is a convex function.

Proof. It is a simple matter to show that A is convex and π is affine. In order to show that Φ is convex, let $p_1, p_2 \in A$ and $\alpha \in (0, 1)$. Then

$$\Phi(\alpha p_1 + (1 - \alpha) p_2) = d(p_0, \alpha p_1 + (1 - \alpha) p_2) - d(p_0, p_*)$$

$$= \|p_0 - \alpha p_1 - (1-\alpha) \, p_2\| - d(p_0, \, p_*)$$
$$= \|\alpha p_0 + (1-\alpha) \, p_0 - \alpha p_1 - (1-\alpha) \, p_2\| - d(p_0, \, p_*)$$
$$\leqslant \|\alpha p_0 - \alpha p_1\| + \|(1-\alpha) \, p_0 - (1-\alpha) \, p_2\|$$
$$\quad - \alpha d(p_0, \, p_*) - (1-\alpha) \, d(p_0, \, p_*)$$
$$= \alpha \|p_0 - p_1\| - \alpha d(p_0, \, p_*)$$
$$\quad + (1-\alpha) \|p_0 - p_2\| - (1-\alpha) \, d(p_0, \, p_*)$$
$$= \alpha (d(p_0, \, p_1) - d(p_0, \, p_*))$$
$$\quad + (1-\alpha) \, (d(p_0, \, p_2) - d(p_0, \, p_*))$$
$$= \alpha \Phi(p_1) + (1-\alpha) \, \Phi(p_2).$$

Therefore Φ is convex.

Remark 1. Note that the proof of the convexity of Φ depends upon the fact that the metric was induced by a norm.

Remark 2. Neustadt's theory of extremals is formulated in much greater generality than was required here. Consequently it allows us considerable flexibility in choosing a measure of nearness for probability functions.

Remark 3. Since A is convex, π is affine and $f(p) = d(p_0, p)$ is convex, by analogy with the Kuhn-Tucker condition for convex programming, necessary conditions for an optimal solution of the probability kinematics problem may coincide with sufficient conditions.

Remark 4. Also by analogy with convex programming, if f were strictly convex on A then it would probably be the case that the optimal solution p_* was unique. However, f is not strictly convex as the following lemma shows.

LEMMA 2. For any $p_1 \in A$ and $\alpha \in (0, 1)$, $\alpha p_1 + (1-\alpha) \, p_0 \in A$ and $f(\alpha p_1 + (1-\alpha) \, p_0) = \alpha f(p_1) + (1-\alpha) \, f(p_0)$.

Proof. Since A is convex, $\alpha p_1 + (1-\alpha) \, p_0 \in A$ for any $p_1 \in A$ and $\alpha \in (0, 1)$.

Also,
$$f(\alpha p_1 + (1-\alpha) \, p_0) = d(p_0, \alpha p_1 + (1-\alpha) \, p_0)$$
$$= \|p_0 - \alpha p_1 - (1-\alpha) \, p_0\|$$
$$= \|\alpha p_0 - \alpha p_1\|$$
$$= \alpha \|p_0 - p_1\| + (1-\alpha) \|p_0 - p_0\|$$
$$= \alpha d(p_0, p_1) + (1-\alpha) \|p_0 - p_0\|$$
$$= \alpha f(p_1) + (1-\alpha) \, f(p_0).$$

Since f is not strictly convex, nothing can be said as yet about the uniqueness of p_*.

It is also possible to impose condition 2 on the problem data for the probability kinematics problem. The function $\hat{h}$ can be extended to all of $B(\Sigma)$ as follows. For any $x \in B(\Sigma)$ define $\hat{h}(x)$ by

$$\hat{h}(x) \equiv \sup_{\psi \in \Sigma}(p_0(\psi) - x(\psi)) - \inf_{\psi \in \Sigma}(p_0(\psi) - x(\psi)) - d(p_0, p_*).$$

LEMMA 3. The function $\hat{h}$ so defined is convex on $B(\Sigma)$.

Proof. Let $x_1, x_2 \in B(\Sigma)$ and $\alpha \in (0, 1)$. Then

$$\hat{h}(\alpha x_1 + (1 - \alpha)x_2) = \sup_{\psi \in \Sigma}(p_0(\psi) - \alpha x_1(\psi) - (1 - \alpha) x_2(\psi))$$
$$- \inf_{\psi \in \Sigma}(p_0(\psi) - \alpha x_1(\psi) - (1 - \alpha) x_2(\psi))$$
$$- d(p_0, p_*).$$

But

$$\sup_{\psi \in \Sigma}(p_0(\psi) - \alpha x_1(\psi) - (1 - \alpha) x_2(\psi))$$
$$= \sup_{\psi \in \Sigma}(\alpha p_0(\psi) - \alpha x_1(\psi)$$
$$+ (1 - \alpha) p_0(\psi) - (1 - \alpha) x_2(\psi))$$
$$\leqslant \sup_{\psi \in \Sigma}(\alpha p_0(\psi) - \alpha x_1(\psi))$$
$$+ \sup_{\psi \in \Sigma}((1 - \alpha) p_0(\psi) - (1 - \alpha) x_2(\psi))$$
$$= \alpha \sup_{\psi \in \Sigma}(p_0(\psi) - x_1(\psi))$$
$$+ (1 - \alpha) \sup_{\psi \in \Sigma}(p_0(\psi) - x_2(\psi)).$$

Similarly,

$$- \inf_{\psi \in \Sigma}(p_0(\psi) - a x_1(\psi) - (1 - \alpha) x_2(\psi))$$
$$\leqslant - \alpha \inf_{\psi \in \Sigma}(p_0(\psi) - x_1(\psi))$$
$$- (1 - \alpha) \inf_{\psi \in \Sigma}(p_0(\psi) - x_2(\psi)).$$

Therefore

$$\hat{h}(a x_1 + (1 - \alpha) x_2) \leqslant a \sup_{\psi \in \Sigma}(p_0(\psi) - x_1(\psi))$$

$$-\alpha \inf_{\psi \in \Sigma} (p_0(\psi) - x_1(\psi))$$

$$+(1-\alpha) \sup_{\psi \in \Sigma} (p_0(\psi) - x_2(\psi))$$

$$-(1-\alpha) \inf_{\psi \in \Sigma} (p_0(\psi) - x_2(\psi))$$

$$-\alpha d(p_0, p_*) - (1-\alpha) d(p_0, p_*)$$

$$= \alpha \hat{h}(x_1) + (1-\alpha) \hat{h}(x_2),$$

and $\hat{h}$ is convex on $B(\Sigma)$.

Since $\hat{h}$ can be extended to all of $B(\Sigma)$ in such a way that $\hat{h}$ remains a convex function, condition 2 is automatically satisfied. Therefore Theorems 1 and 2 can be applied to the probability kinematics problem. Since Theorem 1 is contained in Theorem 2 only the latter will be discussed here. The last two sentences in the statement of Theorem 2 contain additional results for (π, Φ, Z)-extremals which satisfy supernumerary conditions. The first result applies to situations in which $0 \in \mathcal{M}$. Since the zero function is not a probability function, this result does not apply to p_*. The second reads as follows:

> If $\mathcal{U}$ is a linear topological space, H_1 has a non-empty interior, $\hat{h}$ is continuous at some interior point $\hat{y}_1$ of H_1, and Λ is an affine functional on $\mathcal{U}$ which satisfies (10), then Λ is continuous on $\mathcal{U}$.

LEMMA 4. The vector space $\mathcal{U} = B(\Sigma)$ is a linear topological space, $H_1 = \mathcal{U}$ has a non-empty interior, and $\hat{h}$ is continuous on $\mathcal{U}$.

Proof. A norm $\| \ \|_2$ can be defined on $B(\Sigma)$ as follows:

$$\text{for } x \in B(\Sigma), \|x\|_2 = \sup_{\psi \in \Sigma} |x(\psi)|.$$

Since $B(\Sigma)$ is a normed linear space the operations of vector addition and scalar multiplication are continuous. Hence $B(\Sigma)$ is a linear topological space where the topology on $B(\Sigma)$ is induced by the norm $\| \ \|_2$.

Since H_1 is the whole space $\mathcal{U}$, H_1 has a non-empty interior.

Let x be any member of $B(\Sigma)$. In order to show that $\hat{h}$ is continuous at x, for any $\varepsilon > 0$ there must exist a $\delta > 0$ such that $|\hat{h}(y) - \hat{h}(x)| < \varepsilon$ whenever $\|y - x\|_2 < \delta$. If v and w are any two members of $B(\Sigma)$, then

$$\hat{h}(v) - \hat{h}(w) = \sup_{\psi \in \Sigma} (p_0(\psi) - v(\psi)) - \inf_{\psi \in \Sigma} (p_0(\psi) - v(\psi))$$

$$-d(p_0, p_*) - [\sup_{\psi \in \Sigma}(p_0(\psi) - w(\psi))$$
$$- \inf_{\psi \in \Sigma}(p_0(\psi) - w(\psi)) - d(p_0, p_*)]$$
$$= [\sup_{\psi \in \Sigma}(p_0(\psi) - v(\psi)) + \sup_{\psi \in \Sigma}(w(\psi) - v(\psi))$$
$$- \inf_{\psi \in \Sigma}(p_0(\psi) - v(\psi)) - \inf_{\psi \in \Sigma}(w(\psi) - v(\psi))]$$
$$- [\sup_{\psi \in \Sigma}(p_0(\psi) - w(\psi)) + \sup_{\psi \in \Sigma}(w(\psi) - v(\psi))$$
$$- \inf_{\psi \in \Sigma}(p_0(\psi) - w(\psi)) - \inf_{\psi \in \Sigma}(w(\psi) - v(\psi))].$$

But,

$$- [\sup_{\psi \in \Sigma}(p_0(\psi) - w(\psi)) + \sup_{\psi \in \Sigma}(w(\psi) - v(\psi))]$$
$$\leqslant - \sup_{\psi \in \Sigma}(p_0(\psi) - w(\psi) + w(\psi) - v(\psi))$$
$$= - \sup_{\psi \in \Sigma}(p_0(\psi) - v(\psi)),$$

and

$$\inf_{\psi \in \Sigma}(p_0(\psi) - w(\psi)) + \inf_{\psi \in \Sigma}(w(\psi) - v(\psi))$$
$$\leqslant \inf_{\psi \in \Sigma}(p_0(\psi) - w(\psi) + w(\psi) - v(\psi))$$
$$= \inf_{\psi \in \Sigma}(p_0(\psi) - v(\psi)).$$

Therefore

$$(13) \quad \hat{h}(v) - \hat{h}(w) \leqslant [\sup_{\psi \in \Sigma}(p_0(\psi) - v(\psi)) + \sup_{\psi \in \Sigma}(w(\psi) - v(\psi))$$
$$- \inf_{\psi \in \Sigma}(p_0(\psi) - v(\psi)) - \inf_{\psi \in \Sigma}(w(\psi) - v(\psi))]$$
$$- \sup_{\psi \in \Sigma}(p_0(\psi) - v(\psi)) + \inf_{\psi \in \Sigma}(p_0(\psi) - v(\psi))$$
$$= \sup_{\psi \in \Sigma}(w(\psi) - v(\psi)) - \inf_{\psi \in \Sigma}(w(\psi) - v(\psi))$$
$$\leqslant 2 \sup_{\psi \in \Sigma}|w(\psi) - v(\psi)|$$
$$= 2\|w - v\|_2.$$

Since v and w are arbitrary members of $B(\Sigma)$, (13) gives the following inequalities:

(14) $\qquad \hat{h}(y) - \hat{h}(x) \leqslant 2 \|x - y\|_2$
$\qquad\qquad\qquad = 2 \|y - x\|_2;$

(15) $\qquad -(\hat{h}(y) - \hat{h}(x)) = \hat{h}(x) - \hat{h}(y)$
$\qquad\qquad\qquad \leqslant 2 \|y - x\|_2.$

Combining (14) and (15) gives

(16) $\qquad |\hat{h}(y) - \hat{h}(x)| \leqslant 2 \|y - x\|_2.$

Thus, if δ is chosen to be $\varepsilon/2$, then $\|y - x\|_2 < \delta$ implies

$$|\hat{h}(y) - \hat{h}(x)| \leqslant 2 \|y - x\|_2 < 2(\varepsilon/2) = \varepsilon.$$

Therefore $\hat{h}$ is continuous on $\mathcal{U}$.

VI. RESULTS OBTAINED FROM THE ABSTRACT MAXIMUM PRINCIPLE

Theorem 2 guarantees the existence of two real numbers α and β and an affine functional Λ defined on $B(\Sigma)$ such that the following conditions are satisfied:

(17) $\qquad \alpha(p(\varphi) - \gamma) + \beta(d(p_0, p) - d(p_0, p_*)) \leqslant 0 \quad$ for all $\quad p \in A$

(18) $\qquad |\alpha| + |\beta| > 0$

(19)[22] $\qquad \beta \leqslant 0$

(20) $\qquad \alpha(p(\varphi) - \gamma) + \beta\Lambda(p) \leqslant 0 \quad$ for all $\quad p \in A$

(21) $\qquad \Lambda(x) \leqslant \sup_{\psi \in \Sigma} (p_0(\psi) - x(\psi)) - \inf_{\psi \in \Sigma} (p_0(\psi) - x(\psi)) - d(p_0, p_*)$
$\qquad\qquad\qquad\qquad\qquad\qquad\qquad\qquad$ for all $\quad x \in B(\Sigma).$

Also, by lemma 4 and theorem 2, Λ is continuous on $B(\Sigma)$. The following lemmas can be derived from inequalities (17) to (21) and the continuity of Λ.

LEMMA 5. The constant α cannot be zero, and sign $(\alpha) = -\text{sign}$ $(p_0(\varphi) - \gamma)$. Furthermore, without loss of generality it may be assumed that α is positive.

Proof. Inequality (17) is true for all probability functions p. In particular it is true for p_0:

$$\alpha(p_0(\varphi)-\gamma)+\beta(d(p_0, p_0)-d(p_0, p_*))\leqslant 0;$$

that is.

$$\alpha(p_0(\varphi)-\gamma)-\beta d(p_0, p_*)\leqslant 0.$$

Since by (19) $\beta\leqslant 0$, and $d(p_0, p_*)\geqslant 0$,

$$(22)\qquad \alpha(p_0(\varphi)-\gamma)\leqslant\beta d(p_0, p_*)\leqslant 0.$$

This implies sign $(\alpha)=-$ sign $(p_0(\varphi)-\gamma)$.

If α were zero, then by (18) β could not be zero. Also, (22) would imply $\beta d(p_0, p_*)=0$. Since $\beta\neq 0$, $d(p_0, p_*)$ would have to be zero, and hence $p_0=p_*$. Since $p_0=p_*$ is the trivial case, it may be assumed that $\alpha\neq 0$.

If $p_0(\varphi)-\gamma<0$, then α is positive. A problem equivalent to adjusting the probability of φ from $p_0(\varphi)$ to γ is adjusting the probability of $\neg\varphi$ from $p_0(\neg\varphi)=1-p_0(\varphi)$ to $1-\gamma$. If $p_0(\varphi)-\gamma>0$, then

$$p_0(\neg\varphi)-(1-\gamma)=1-p_0(\varphi)-1+\gamma=-(p_0(\varphi)-\gamma)<0.$$

If $p_0(\varphi)-\gamma>0$, the equivalent problem of adjusting $p_0(\neg\varphi)$ will yield $\alpha>0$. Hence, without loss of generality it may be assumed that $\alpha>0$.

LEMMA 6. If $0<\gamma<1$, then $\beta<0$.

Proof. Suppose that $0<\gamma<1$ and $\beta=0$. Then by (17),

$$\alpha(p(\varphi)-\gamma)\leqslant 0$$

for all $p\in A$. Since $\alpha>0$, this implies $p(\varphi)\leqslant\gamma<1$ for all probability functions p. But this is absurd since it is always possible to find a probability function p_1 such that $\gamma<p_1(\varphi)\leqslant 1$. Therefore $\beta\neq 0$ and since $\beta\leqslant 0$, $\beta<0$.

Remark. It is clear that a similar argument for the case $\gamma=0$ or $\gamma=1$ would yield no information about β. Furthermore no manipulations of any of inequalities (17) through (21) will provide any more information about β than (19).

LEMMA 7. Since Λ is a continuous affine functional defined on $B(\Sigma)$, it admits an integral representation.

In order to simplify the proof of lemma 7, a theory of integration with respect to charges will be discussed before the proof is presented. Charges can be used to represent the normed conjugate of the space $B(T)$ where T is any non-empty set. If x' is a member of the normed conjugate of $B(T)$, (that is, if x' is a continuous linear functional defined on $B(T)$), then it will be possible to define a charge μ on the ring of all subsets of T, such that for any $x \in B(T)$,

$$(23) \qquad x'(x) = \int_{\substack{T \\ S}} x \, d\mu,$$

where the S-integral is an integral with respect to a charge and has yet to be defined.

Let T be any non-empty set, R be the ring of all subsets of T, μ be any charge on R, and x be any member of $B(T)$. A partition of T is a finite collection $E_1, \ldots, E_n$ of pairwise disjoint non-empty sets in R which cover T. If $\pi_1 = (E_1, \ldots, E_n)$ and $\pi_2 = (F_1, \ldots, F_m)$ are partitions of T where $m \geqslant n$, then it will be said that π_2 is a refinement of π_1 if each F_j is a subset of some E_k.

DEFINITION 3. The function $x \in B(T)$ is said to be S-integrable with integral I, if given $\varepsilon > 0$ there exists a partition π_ε of T, such that for any partition $\pi = (G_1, \ldots, G_q)$ which is a refinement of π_ε and any $t_i \in G_i$,

$$\left| \sum_{i=1}^{q} x(t_i) \mu(G_i) - I \right| < \varepsilon.$$

In this case I will be denoted by $\int_{\substack{T \\ S}} x \, d\mu$.

It can be shown that when (T, R, μ) is a totally charged space, any bounded measurable function on T is S-integrable [26, pp. 402–3]. Since R is the ring of all subsets of T, any member of $B(T)$ is measurable and hence integrable with respect to any charge on R.

THEOREM 4 [26, page 403]. The normed conjugate of $B(T)$ is congruent to the space of all charges on R, the correspondence between a continuous linear functional x' and its associated charge μ being indicated by the two formulas:

$$(24) \qquad x'(x) = \int_S^T x \, d\mu \quad x \in B(T)$$

$$\mu(E) = x'(\chi_E) \qquad E \in R$$

where χ_E is the characteristic function of E.

Proof of Lemma 7. Since Λ is an affine functional defined on the vector space $B(\Sigma)$, it can be written as follows:

$$(25) \qquad \Lambda(x) = b(x) + \Lambda(0),$$

where b is a linear functional defined on $B(\Sigma)$. Since Λ is continuous b is continuous, and by Theorem 4 there exists a charge μ on the set of all subsets of Σ such that

$$(26) \qquad b(x) = \int_S \int_\Sigma x \, d\mu$$

for all $x \in B(\Sigma)$. The charge μ is defined by (24). If (26) is substituted in (25), the following integral representation for Λ results:

$$(27) \qquad \Lambda(x) = \int_S \int_\Sigma x \, d\mu + \Lambda(0).$$

LEMMA 8. The following inequality must be satisfied by all $p \in A$, where $\beta_1 = \beta/\alpha$ and μ is defined by (24):

$$(28) \qquad p(\varphi) - \gamma + \beta_1 \int_S \int_\Sigma (p - p_*) \, d\mu \leq 0.$$

Proof. Equation (20) must be true for all $p \in A$; in particular it must be true for $p = p_*$. Therefore

$$(29) \qquad \beta \Lambda(p_*) \leq 0.$$

Also, by (21),

$$\Lambda(p_*) \leq [\sup(p_0(\psi) - p_*(\psi)) - \inf(p_0(\psi) - p_*(\psi))] - d(p_0, p_*).$$

Since the term in square brackets is by definition $d(p_0, p_*)$, $\Lambda(p_*) \leqslant 0$. Since $\beta \leqslant 0$, this implies $\beta \Lambda(p_*) \geqslant 0$. This, combined with (29), implies $\beta \Lambda(p_*) = 0$.

Using lemma 7, Equation (20) can be rewritten as follows:

$$(30) \qquad \alpha(p(\varphi) - \gamma) + \beta \int_{\substack{\Sigma \\ S}} p \, d\mu + \beta \Lambda(0) \leqslant 0.$$

But

$$\beta \left(\int_{\substack{\Sigma \\ S}} p \, d\mu + \Lambda(0) \right) = \beta \left(\int_{\substack{\Sigma \\ S}} (p - p_*) \, d\mu + \int_{\substack{\Sigma \\ S}} p_* \, d\mu + \Lambda(0) \right)$$

$$= \beta \int_{\substack{\Sigma \\ S}} (p - p_*) \, d\mu + \beta \Lambda(p_*),$$

and since $\beta \Lambda(p_*) = 0$, (30) becomes

$$\alpha(p(\varphi) - \gamma) + \beta \int_{\substack{\Sigma \\ S}} (p - p_*) \, d\mu \leqslant 0.$$

Since $\alpha > 0$, this implies

$$(31) \qquad p(\varphi) - \gamma + \beta_1 \int_{\substack{\Sigma \\ S}} (p - p_*) \, d\mu \leqslant 0$$

for all $p \in A$.

The following theorem actually puts constraints on the optimal probability function p_*.

THEOREM 5. The optimal solution p_* is the probability function which maximizes $F(p) \equiv p(\varphi) + \beta_1 \int_{\Sigma}^{S} p \, d\mu$ over A.

Proof. For any $p \in A$, Lemma 8 guarantees that

$$(32) \qquad p(\varphi) - \gamma + \beta_1 \int_{\substack{\Sigma \\ S}} (p - p_*) \, d\mu \leqslant 0.$$

For $p = p_*$, the left-hand side of (32) is zero. Therefore the function $G(p) \equiv p(\varphi) - \gamma + \beta_1 \int_{\Sigma} (p - p_*) \, d\mu$ attains its maximum over A at p_*. Since $F(p) = G(p) + a$ constant, F also attains its maximum at p_*.

Theorem 5 provides a means for generating information about p_*. Whatever can be said about the probability function which maximizes $F(p)$ can be said about p_*. Thus we have the following theorem.

THEOREM 6. The probability function p_* which minimizes the distance $d(p_0, p)$ between the original probability function p_0 and those probability functions p which satisfy $p(\varphi) = \gamma$, satisfies the following:

(i)   Where the charge μ, defined by (24), is positive p_* should be small. Furthermore the larger the value of μ, the more important it is that p_* be small.

(ii)   Where μ is negative p_* should be large, and the more negative the value of μ, the more important it is that p_* be large.

Proof. Since $p(\varphi)$ is fixed for $p = p_*$, the first term in $F(p)$ will not affect the maximization of F. Also, since $\beta_1 \leqslant 0$, $F(p)$ will attain its maximum value when $\int_{\Sigma} p \, d\mu$ attains its minimum value. The function p takes on values between zero and one. The charge μ is bounded, but may take on positive and negative values. Consequently, in order to minimize $\int_{\Sigma} p \, d\mu$ and hence maximize $F(p)$, p must be small whenever μ is positive, and large whenever μ is negative. Also, the larger the absolute value of μ, the more important it is that p meet these restrictions.

Note that Theorem 6 does not specify the exact relationship which exists between p_* and μ. In order to be precise, one would have to consider (a) the effect of the 'limit' operation on the finite sum $\sum_{i=1}^{q} p(t_i) \mu(G_i)$ (see definition 3), and (b) the coherency requirements for p_*. Despite its generality Theorem 6 does provide a way to generate p_*. The charge μ is known, and its sign on a subset of Σ determines the magnitude of p_* there.

Theorem 6 provides as much information about the optimal solution as one would expect from any theory of optimization. Thus it has been established that it is possible to treat the probability kinematics problem as a 'minimum change' problem. As was demonstrated in Sections IV and V, Neustadt's theory of extremals can be applied to problems even more general than the probability kinematics problem.

The purpose of the present work was to generate abstract optimization procedures for measures of nearness on probability functions. Theorem 6 shows that Neustadt's theory of extremals can be applied to problems of this sort. The further problem of analyzing the information provided by the optimization result is non-trivial and will be dealt with in future work.

APPENDIX

Mathematical Definitions

These definitions are listed here in the same order as the expressions to which they refer first appear in the text.

A subset Z of a linear topological space $\mathscr{L}$ is a *convex cone* if $Z \neq \emptyset$ and $\alpha Z + \beta Z \subset Z$ whenever $\alpha > 0$ and $\beta \geqslant 0$.

Let $\mathscr{M}$ be a convex set in $\mathscr{U}^{23}$ and suppose V is a vector space. Then a function h which maps $\mathscr{M}$ into V is *affine* if

$$h(\alpha y_1 + (1 - \alpha) y_2) = \alpha h(y_1) + (1 - \alpha) h(y_2)$$

for all $y_1, y_2 \in \mathscr{M}$ and $\alpha \in (0, 1)$.

Suppose $\mathscr{M}$ is a convex set in $\mathscr{U}$, $\mathscr{L}$ is a linear topological space, and Z is a convex cone in $\mathscr{L}$. A function $\hat{h}$ which maps $\mathscr{M}$ into $\mathscr{L}$ is *Z-convex* if

$$\hat{h}(\alpha y_1 + (1 - \alpha) y_2) - \alpha \hat{h}(y_1) - (1 - \alpha) \hat{h}(y_2) \in \bar{Z}$$

whenever $y_1, y_2 \in \mathscr{M}$ and $\alpha \in (0, 1)$.

A set S in $\mathscr{U}$ is a *k-simplex* if $S = co\{y_0, y_1, ..., y_k\}$ for some $k+1$ points $y_0, y_1, ..., y_k \in \mathscr{U}$ which are in general position. (The points $y_0, y_1, ..., y_k \in \mathscr{U}$ are in general position if $\sum_{i=0}^{k} \beta^i y_i = 0$ and $\sum_{i=0}^{k} \beta^i = 0$ imply $\beta^i = 0$ for all i, where β^i is a real number for any i.)

Let $S = co\{y_0, y_1, ..., y_k\}$ be a simplex in $\mathscr{U}$ where $y_0, y_1, ..., y_k$ are in general position. Then for each $y \in S$, $y = \sum_{i=0}^{k} \beta^i y_i$ for some $\beta^0, \beta^1, ..., \beta^k \in R^1$ with $\sum_{i=0}^{k} \beta^i = 1$.

These β^i are uniquely determined by y and are called the *barycentric coordinates* of y with respect to S.

Let S be a simplex in $\mathcal{U}$. A function f which maps S into R^m is *continuous as a function of the barycentric coordinates* of a point in S if, for any $\bar{y} \in S$ where $\bar{y}$ has barycentric coordinates $\bar{\beta}^0, \bar{\beta}^1, \ldots, \bar{\beta}^k$, given $\varepsilon > 0$ there exists $\delta > 0$ such that

$$\| f(\bar{y}) - f(y) \|_m < \varepsilon$$

whenever $\| (\bar{\beta}^0, \bar{\beta}^1, \ldots, \bar{\beta}^k) - (\beta^0, \beta^1, \ldots, \beta^k) \|_{k+1} < \delta$, where y has barycentric coordinates $\beta^0, \beta^1, \ldots, \beta^k$.

The point y is an *internal point* of a convex set $\mathcal{M}$ in $\mathcal{U}$ if $y \in \mathcal{M}$ and if, for any point $y_0 \in \mathcal{U}$, there exists a real number $\varepsilon_0 > 0$ such that $y + \varepsilon y_0 \in \mathcal{M}$ whenever $|\varepsilon| \leqslant \varepsilon_0$.

NOTES

[1] Richard Jeffrey [11]
[2] Where $E_0, \ldots, E_{n-1}$ is a partition, a simultaneous change of each $p_0(E_i)$ to γ_i can count as an input. In this generalized framework the input change of $p_0(E)$ to γ is equivalent to the two element partition input changing $p_0(E)$ to γ and $p_0(\bar{E})$ to $1 - \gamma$.
[3] See Good [6], Kyburg [14], [15], Shafer [24], and Levi [17].
[4] See especially, Shafer [24], Kyburg [14], and Levi [17].
[5] Gudder [7].
[6] Gudder [7].
[7] Jeffrey [11]; see also Teller [25].
[8] Where the input is a simultaneous change in the assignments to n-membered partition $E_0, \ldots, E_{n-1}$, Jeffrey's rule is

$$p_1(H) = \sum_{i < n} p_0(H/E_i) \cdot p_1(E_i).$$

[9] Levi [16].
[10] Teller [25].
[11] Harper [9].
[12] Harper [9].
[13] The reader should keep in mind that no attempt to define a measure of nearness for subjective probability functions has previously been made. Therefore most of the measures of nearness which are discussed have been defined in a context other than for subjective probability functions.
[14] Kullback discussed many of these other information measures (Fisher's, Chernoff's, Hartley's, Lindley's, Savage's, Shannon's and Wiener's) and their relationship to his measure in his book. For references to these information measures, see the index in the book by Kullback [13, page 391].
[15] Recall that absolute continuity was required in order to define $I(1:2)$.
[16] Let Σ be a complete separable metric space. Then the sequence $\{\mu_n\}$ of measures in Σ is called weakly convergent to the measure μ if, for any continuous and bounded functional $f(x)$,

$$\int_{\Sigma} f(x) \, d\mu_n \to \int_{\Sigma} f(x) \, d\mu.$$

[17] This result is due to Dr. S. Burris, Mathematics Faculty, University of Waterloo.

[18] For the definition of a convex cone, see the appendix of mathematical definitions. Any non-standard definitions will be listed there.

[19] The symbol $\| \ \|_m$ denotes the usual Euclidean norm in R^m.

[20] For a good discussion of this type of 'approximation' in optimization problems of a less general nature, see [2, Chapter 2].

[21] The functions $\pi|_{\mathscr{E}}$ and $\Phi|_{\mathscr{E}}$ are π restricted to $\mathscr{E}$ and Φ restricted to $\mathscr{E}$ respectively.

[22] Since $\Phi(p_*) = d(p_0, p_*) - d(p_0, p_*) = 0$, (8) in Theorem 2 is trivially satisfied and therefore omitted here.

[23] Throughout the appendix $\mathscr{U}$ will denote a real vector space.

BIBLIOGRAPHY

[1] Bell, J. L. and Slomson, A. B., *Models and Ultraproducts: An Introduction*, North-Holland Publishing Company, Amsterdam, 1971.

[2] Canon, M. D., Cullum, C. D., and Polak, E., *Theory of Optimal Control and Mathematical Programming*, McGraw-Hill, Inc., U.S.A., 1970.

[3] Dudley, R. M., 'Distances of Probability Measures and Random Variables', *Annals of Mathematical Statistics* **39** (1968) 1563–72.

[4] Dunford, N. and Schwartz, J. T., *Linear Operators Part I*, Interscience Publishers, Inc., New York, 1958.

[5] Finkbeiner, D. T., II, *Introduction to Matrices and Linear Transformations* (2nd ed.), W. H. Freeman and Company, San Francisco, 1966.

[6] Good, I. J., 'The Bayesian Influence or How to Sweep Subjectivism Under the Carpet', Volume II, p. 125.

[7] Gudder, S., 'A Generalized Measure and Probability Theory for the Physical Sciences', Volume III, p. 121.

[8] Harper, W. and Gibbard, A., 'Conditionals and Decision with Act-Dependent States', to appear in *Proceedings of the Fifth International Congress of Logic Methodology and Philosophy of Science* (ed. by Robert Butts and Jaakko Hintikka), D. Reidel, Dordrecht, Holland, 1976.

[9] Harper, W. L., 'Rational Belief Change, Popper Functions and Counterfactuals', this volume, p. 73.

[10] Jamison, B., Unpublished manuscript.

[11] Jeffrey, R. C., *The Logic of Decision*, McGraw-Hill Book Company, U.S.A., 1965.

[12] Kuhn, H. W. and Tucker, A. W., 'Nonlinear Programming', in Proceedings of the Second Berkeley *Symposium on Mathematical Statistics and Probability* (ed. by J. Neyman), Berkeley-Los Angeles, 1951, pp. 481–492.

[13] Kullback, S., *Information Theory and Statistics*, John Wiley and Sons, Inc., New York, 1959.

[14] Kyburg, H. E., *Foundations of Statistical Inference*, Reidel, Dordrecht, Holland, 1974.

[15] Kyburg, H. E., 'Statistical Knowledge and Statistical Inference', Volume II, p. 315.

[16] Levi, I., 'Probability Kinematics', *British Journal for Philosophy of Science* **18** (1967) 197–209.

[17] Levi, I., 'On Indeterminate Probabilities', *Journal of Philosophy* **LXXI** (1974) 381–418.

[18] May, S. J., 'On the Application of a Minimum Change Principle to Probability Kinematics', unpublished Ph.D. dissertation, University of Waterloo, Waterloo, Ontario, 1973.

[19] Neustadt, L. W., 'A General Theory of Extremals', *Journal of Computer and System Sciences* **3** (1969) 57–92.

[20] Popper, K. R., *The Logic of Scientific Discovery*, New York, 1961.

[21] Prokhorov, Yu. V., 'Convergence of Random Processes and Limit Theorems in Probability Theory', *Theory of Probability and its Applications* (1956) 157–214.

[22] Schay, G., 'Optimal Joint Distributions of Several Random Variables with Given Marginals', unpublished manuscript, University of Massachusetts at Boston, 1973.

[23] Schay, G., 'Nearest Random Variables with Given Distributions', to appear in *Annals of Probability*.

[24] Shafer, G., 'A Theory of Statistical Evidence', Volume II, p. 365.

[25] Teller, P., 'Conditionalization, Observation, and Change of Preference', this volume, p. 205.

[26] Taylor, A. E., *Introduction to Functional Analysis*, John Wiley and Sons, Inc., New York, 1961.

R. D. ROSENKRANTZ

SIMPLICITY

1. INTRODUCTION

Scientific inference is hypothetical-deductive: from given facts or experimental findings we infer laws or theories from which the facts follow or which account for the facts. This schema is somewhat oversimplified, for experimental findings are rarely if ever logical consequences of an explanatory theory, but merely 'agree' with or 'fit' the theory, a criterion of fit being presupposed. We may then consider the findings 'probabilistic consequences' of the theory. Bayes' theorem then enters as a more general schema of hypothetical deduction: from given experimental findings, to infer the most plausible theory that affords those findings highest probability.

Another constraint – simplicity – is thought to apply: among all those theories which fit the facts, we are enjoined to choose the simplest. The philosopher of science then faces the twofold task of characterizing simplicity and then justifying our preference for theories that are simplest in the indicated sense. At first blush the task seems formidable, for the simplest theory is often implausible. In estimating the number of distinct species or kinds in a population, for example, the simplest hypothesis, many would agree, is that which equates the total number of species in the population with the number found in the sample. If simplicity and plausibility are at odds, then it would appear impossible to rationalize preference for simplicity within a Bayesian framework. Perhaps such considerations have led others to think of simplicity as an epistemic utility. But without at all wishing to deny that simpler theories may have various practical advantages, our tact will be to show why simpler theories are indeed preferable from a strictly Bayesian point of view. In fact, while simpler theories may and may not be more plausible, they invariably have higher likelihood, and so, on a Bayesian analysis, they are better supported. In this paper, attention will be confined to the class of multinomial models – i.e., to categorized data – but much of what we show about this class holds quite generally.

Harper and Hooker (eds.), Foundations of Probability Theory, Statistical Inference, and Statistical Theories of Science, Vol. I, 167–203. All Rights Reserved. Copyright © 1976 by D. Reidel Publishing Company, Dordrecht-Holland.

Before developing our thesis that simplicity is an ingredient of likelihood or support, we must of course characterize simplicity in an intuitively acceptable way. By way of setting the stage for our explication, consider the following illustration.

Example 1. The A-B-O system of blood groups was initially thought to comprise four types: A, B, AB, and O (A is now known to have sub-types). The existence of four phenotypes demands a complication of the one-factor bi-allelic model considered by Mendel: either we must posit more than one factor or gene-pair, or we must assume more than two alleles. Consider a complication of the latter type first: the simplest, clearly, is to assume three (instead of two) alleles, call them A, B, O. These occur in the population with frequencies p, q, and r, with $p+q+r=1$. (Since r is determined from $r=1-p-q$, there are only two free parameters.) The O-allele of the gene functions as mere absence of A or B. Hence the genotypes AA or AO give type A, etc.. In a panmictic population, an individual inherits A-alleles from both parents with probability p^2, and inherits one A-allele and one O-allele with probability $2pr$. Hence type A individuals occur with frequency p^2+2pr. The other phenotypic probabilities are easily calculated and are given in Table I.

The simplest form of the polygenic model, on the other hand, posits two independent factors A and B. An individual has type A iff he receives at least one dominant A gene, and similarly for type B. If s, t are the proportions of dominant A and dominant B genes, resp., then an individual has type A blood with probability $s(1-t)$, since he receives a dominant A gene with probability s, and independently of this, a recessive allele of the B gene with probability $1-t$. The other phenotypic probabilities for this model are also listed in Table I.

TABLE I

	A	B	AB	O
one factor (H_1)	p^2+2pr	q^2+2qr	$2pq$	r^2
two factor (H_2)	$s(1-t)$	$(1-s)t$	st	$(1-s)(1-t)$

Which of these models, H_1 or H_2, is simpler? Since both have two free parameters, they cannot be compared just by counting parameters. As we will see farther on, there is a rather subtle difference between them that makes H_1 slightly simpler. But it is not obvious that H_1 is simpler, and, indeed, the example was chosen partly to highlight the need for the more powerful methods of effecting simplicity comparisons developed in this paper.

One important feature of the example already points the way to a new approach. H_1 and H_2 are models of the same experiment or class of experiments; both are intended to account for the observed frequencies with which the four blood groups occur in various genetically isolated human populations. Rather than ask which model is simpler, it might be more fruitful to ask which model is simpler relative to the contemplated class of experiments.

2. SAMPLE COVERAGE

By the *sample coverage* of a theory for an experiment I mean the proportion of possible experimental outcomes fitted by the theory. Standard statistical criteria of fit, like the chi square test, are suspect from a Bayesian point of view (cf. Section 5), and so we shall employ direct confidence criteria. A $100\,(1-\alpha)\%$ *direct confidence region* (or *DCR*) of a model H has probability $1-\alpha$ of containing the observed outcome of the experiment conditional on H. We call $1-\alpha$ the *confidence level* of the DCR. For the class of multinomial models, at least, DCR's are always *highest density regions*, or approximately so. That is, the probability (on H) of each included sample point exceeds that of any excluded sample point. This property makes DCR's more acceptable criteria of fit from a Bayesian standpoint. Their distribution theory will be outlined in Section 4.

We propose to measure the simplicity of a model relative to an experiment by its sample coverage: the smaller the sample coverage (at the chosen confidence level), the simpler the model. (This measure does not depend on the chosen confidence level.) A few words in defense of this explication are in order.

The paucity-of-parameters criterion[1] is perhaps one of the few widely shared intuitions about simplicity. Clearly, when we add a parameter to a model we increase its sample coverage, for the extended model will fit every outcome fitted by its special case. However, *it doesn't follow that a model with fewer parameters is simpler*. In Section 6 we will produce a one-parameter multinomial model that is simpler than a no-parameter multinomial model. The sound core of the paucity-of-parameters criterion is thus the rule that *special cases of a model are always simpler*. In any case, this rule is broad enough to cover most of the paradigms of simplicity that have been cited in the literature:

(1)   Polynomials of lower degree are simpler,
(2)   A regression surface is simpler the fewer the number of independent variables,
(3)   Differential equations of lower order are simpler, and among equations of the same order, those of lower degree are simpler,
(4)   'Simple' hypotheses in the statistician's sense (viz. point hypotheses) are simpler than composite hypotheses that include them,
(5)   Markov chains of lower order are simpler,
(6)   'Strong generalizations' which specify the number of non-empty kinds in a population are simpler than laws which specify those and additional kinds.

The reader will be able to supply additional examples. In each of these sequences, the complicated models include all their predecessors in the sequence (e.g. polynomials of lower degree) as special cases. And the one thing we always do when we pass from a special case to an extension thereof is to increase sample coverage. More evidence of the intuitive acceptability of our explication will be offered in the concluding section and in a forthcoming paper.[2]

In essence, our measure is an adaptation of Sir Karl Popper's notion of simplicity as logical strength. We have relativized the notion to an experiment (Popper himself has suggested relativizing it to a *problem*) and have dropped what seemed to us the unhelpful reference to logical strength in favor of sample coverage. (Sample coverage is not unrelated to Popper's notion of 'dimension'; *in an extended sense*, the dimension of a theory at a given confidence level is just the smallest sample size at which the sample coverage falls below 100%.) The reference to sample coverage seems unavoidable for, as we remarked earlier, experimental outcomes are not generally logical consequences of a theory but only fit the theory more or less well. (One cannot do better by appealing to what the theory logically excludes for the same reason; theories do not generally logically exclude anything, but only render certain outcomes extremely improbable.)[3] Still, Popper's contention that a model which yields sharper predictions is simpler is fully borne out by our analysis, for sharper predictions mean smaller sample coverage.

Several additional advantages flow from relativizing simplicity to an experiment. Most important of all, simplicity comparisons can always be effected. The question which of two models of an experiment has smaller sample coverage for that experiment always admits of a definite answer (once we fix the criterion of fit). Secondly, the so-called 'tacking paradox' to which Popper's formulation seems open, is averted by our formulation to at least this extent. By adjoining an extraneous hypothesis to a theory, we increase logical strength, and so, on Popper's formulation, simplify the theory – contrary to almost everybody's intuitions. But, on our formulation, the affect on sample coverage will be nil, provided that the conjoined hypothesis is truly extraneous. (A precise formulation of 'extraneous' might be that $P(x, H/K) = P(x, H)$ and $P(H/K) = P(H)$, whence $P(x/H, K) = P(x/H)$ for every outcome x, where K is the conjoined hypothesis.) Conjoining an hypothesis about atomic spectra to the blood group model H_1 should have no affect whatever on the latter's sample coverage for the experiment of counting the frequencies with which the four blood types occur in a human population (e.g., Latvians or Navaho Indians). By contrast, conjoining hypotheses always diminsihes prior probability (just by virtue of the probability calculus).

It should be emphasized, finally, the advantages that flow from relativization are purchased at no appreciable loss of generality. For we can just as easily relativize to a class of experiments, and so compare theories over the intended range of their application. Indeed, complicating a model vis-à-vis one experiment may result in an overall simplification when other experiments are considered. Adding a linkage parameter certainly complicates the Mendelian model applied to two-factor mating experiments, but the union of genetics and cytology this 'local' complication cemented added powerful new constraints that lead to greatly reduced sample coverage over the whole range of the theory's application. For example, the number of linkage groups must now coincide with the (directly observable) number of chromosome pairs, chromosomal aberrations lead to genetic predictions, and so on. Important as this feature is, we shall have no more to say about it in this paper.

3. SUPPORT

In this section we adduce some considerations on behalf of our thesis that simplicity is a component of support.

We are given a theory of small sample coverage. Suppose the theory false. Then it is rather improbable that the outcome of the contemplated experiment will fit the theory. As matters turn out, though, the outcome does fit. Then either a very improbable coincidence has occurred (how improbable depending on the sample coverage), or else the theory is at least an adequate model of the experiment in hand. From this perspective, the aim of experimentation is to make it virtually impossible for the data to fit one's theory unless that theory is true.[4]

The reasoning is reminiscent of that advanced by R. A. Fisher in connection with significance tests, except that, where I have accentuated the positive, Sir Ronald was wont to accentuate the negative. He argued from the improbability with which an outcome would fall outside the direct confidence region of a sharp null hypothesis to the falsehood of that hypothesis. Granted, inferences to the truth of an hypothesis are more hazardous. But I have expressed myself with some caution. Given improbably good agreement with a theory, I said, unless a rare chance has occurred, the theory is an adequate model of the experiment. The possibility always remains that other theories adequately represent the experiment. Save for those rarely encountered cases in which an exhaustive list of exclusive hypotheses is at hand, we should be properly circumspect in concluding that a theory in good agreement with the data is true, even if it is a very simple theory.

Whether we accentuate the positive or the negative seems of distinctly secondary importance, however, for Fisher's primary insight, surely, was to have seen the importance of the question: How probable is it that the data should agree this well with our hypothesis if, in fact, chance prevails? Fisher's own 'psycho-physical experiment' nicely illustrates the point. We must ask how probable it is that the Lady would classify that many cups of tea correctly were she merely guessing. Or, again, a man who claims he can detect water with a hazel prong fails to impress us merely by scoring a high percentage of hits. He must score an appreciably higher percentage than somebody digging at random in the chosen area. And so it is with theories. A theory fails to impress us merely by being accurate

– witness Ptolemaic astronomy – it must be *improbably* accurate.[5] These ideas can be given some precision as follows.

Let $1 - \alpha^*$ be the confidence level of the smallest DCR which contains the observed outcome of the experiment. The sample coverage of the theory at level $1 - \alpha^*$ is called the *observed sample coverage* (or *OSC*) of the theory. OSC, then, is just the proportion of possible outcomes which fit the theory as well as (or better than) the actually observed outcome. What I have loosely been calling 'improbably good agreement' then translates into low OSC.

Clearly, OSC decreases with accuracy or goodness-of-fit and with simplicity. Likelihood, by contrast, *increases* with simplicity and accuracy, as we will see. Hence we should expect OSC and likelihood to be inversely related, which suggests

THE SAMPLE COVERAGE RULE. The smaller the OSC of a theory, the greater its support.

Where comparisons are in question and a computable likelihood function exists, the rule asserts that the theory with the smaller OSC has the higher likelihood. (Some evidence for this, as a rule-of-thumb, will be marshalled in Section 6.)

The importance of the sample coverage rule lies in its wider applicability. It can be applied when no alternative hypotheses are in question to give an *absolute test* of conformity of the sort envisaged by Fisher. That is, we measure conformity, or better, support, not merely by accuracy, but by how improbably well the theory fits the data. We are then asking in effect: How probable would results this good be in a maximally disorderly universe? – viz., one in which all experimental outcomes were equiprobable. There is no pretense of giving an absolute cut-off that separates supporting from infirming outcomes. Yet a very reasonable *convention* might classify as supporting only outcomes at which the OSC is less than 50%, so that the theory would have less than an even chance of fitting the outcome as well in a disorderly universe.

The assumption of a maximally disorderly universe seems to play very much the role Fisher envisaged for null hypotheses in the evaluation of all experimental data. The reader may indeed be wondering whether there is any substantive difference between the two concepts. Certainly there is a difference in how the concepts are to be applied. Having rejected a sharp null hypothesis, the Fisherian procedure is to run an additional test to determine whether the departure from null is in the direction of the alternative hypothesis of interest – e.g., by an orthogonal partition of the chi square statistic. Our proposal is to assess support directly in terms of the OSC (in terms of how improbably well the data fit the theory). OSC serves, then, as a fixed pedal point of comparison for all theories, and thus allows us to compare the evidential support of not necessarily exclusive hypotheses, and to compare the severity of different experimental tests of the same theory. These points are considered further in Section 7.

The sample coverage rule implies, in particular, that a less accurate theory can be better supported, for what it loses in accuracy it may more than make up for in simplicity, so that its OSC is actually smaller. Though rigorous comparison would be difficult, one suspects that the simplified Copernican model, unencumbered by epicycles, had smaller OSC than various versions of the Ptolemaic model. At any rate, the sample coverage rule makes it clear why it is occasionally entirely rational to regard a less accurate upstart theory as more promising than a more accurate and well-entrenched rival. A theory may be inaccurate and yet be improbably accurate.

The point we wish to stress here, however, is that even where the sample coverage rule transcends Bayesian analysis (which always applies to a well-defined 'ultimate partition' of hypotheses), a roughly Bayesian justification of the rule can be provided.

It is by now a familiar point that Bayes' theorem implies, not only that hypotheses are confirmed (made more probable) by their consequences, but that they are confirmed the more strongly the more unexpected those consequences. For when x is a consequence of H, Bayes' theorem reduces to $P(H): P(x) = P(H/x)$, so that the posterior probability $P(H/x)$ of H varies inversely with the outcome probability $P(x)$. Of course, probability models of an experiment never imply particular outcomes, yet, for all practical purposes, they do imply that the outcome will fall within their 99.99% DCR. *This event*, call it E, *is the more improbable the smaller the sample coverage of the model.* That is, for comparing a pair of non-exclusive alternatives, we assume the universe maximally disorderly and on this basis compute $P(E)$ directly as sample coverage. Of course information is lost whenever we report merely whether or not the outcome falls within a DCR, and for this reason, the more elaborate Bayesian arguments of the next two sections are necessary.

4. MULTINOMIAL MODELS WITHOUT FREE PARAMETERS

Let the random vector $(X_1, ..., X_k)$ have the multinomial distribution with k categories and category probabilities $p_1, ..., p_k$:

(1) $$P(X_1 = x_1, ..., X_k = x_k) = \frac{n!}{x_1! ... x_{k-1}! x_k!} \prod_i p_i^{x_i} ... p_k^{x_k}$$

where x_i is the observed number in category i and $\sum x_i = n$, the sample size. Then the vector $(X_1, ..., X_{k-1})$ has, asymptotically, a non-singular normal distribution with density

(2) $$(2\pi)^{-(k-1)/2} \Lambda^{-1/2} e^{-Q/2}$$

where $Q = \sum_{i,j}^{k-1} (\Lambda_{ij}/\Lambda)(X_i - np_i)(X_j - np_j)$ is the quadratic form whose matrix of coefficients is inverse to the covariance matrix of the multinomial with diagonal entries $np_i(1-p_i)$ and off-diagonal entries $-np_i p_j$, $i \neq j$. The Λ_{ij} are thus the co-factors of the covariance matrix and Λ the determinant. Given the obvious analogy to the univariate case, Λ is called the *generalized variance* of the distribution and its square root the *generalized standard deviation*. It is well known that Q has a chi square distribution with $k-1$ degrees of freedom (abbreviated: df). Hence

(3) $$Q = \chi^2_{\alpha, k-1}$$

is the equation of a $100(1-\alpha)\%$ direct confidence ellipsoid for the vector $(X_1, ..., X_{k-1})$, writing $\chi^2_{\alpha, k-1}$ for the upper $100\alpha\%$ point of the chi square distribution with $k-1$ df. The volume of the ellipsoid $Q = c$ is given by:

(4) $$V = \frac{(\pi c)^{(k-1)/2}}{\Gamma\left(\dfrac{k-1}{2} + 1\right)} \Lambda^{1/2}$$

and further

(5) $$\Lambda = n^{k-1} P$$

where n is the sample size and $P = \prod_i p_i$ is the product of all category probabilities $p_1, ..., p_k$. Hence, sample coverage increases with Q, a measure of the deviation between observed and predicted values, and with P, the product of category probabilities. (Here we are using the obvious approximation to the number of sample points inside an ellipsoid by the volume of the ellipsoid.) It follows, in particular, that at a given confidence level and sample size, two multinomial models can be compared for simplicity by comparing the products of their category probabilities. The rougher the frequency counts predicted by the model, the smaller its sample coverage – as we would expect. It also follows from (4) and (5) that a model simpler than another at one sample size is simpler at every sample size.

Example 2. Consider the uniform trinomial $(k=3, p_1=p_2=p_3=\frac{1}{3})$ at sample size $n=36$. The covariance matrix is $\begin{pmatrix} 8 & -4 \\ -4 & 8 \end{pmatrix}$ with determinant $\Lambda=48$ (also given directly by (5)), whence $\Lambda_{ii}/\Lambda=\frac{1}{6}$ and $\Lambda_{ij}/\Lambda=\frac{1}{12}$. The upper 5% point of the chi square distribution with $k-1=2$ df is 5.99. The equation of the 95% confidence ellipse is thus $\frac{1}{6}(X-12)^2+\frac{1}{6}(X-12)$ $(Y-12)+\frac{1}{6}(Y-12)^2=5.99$. By setting $k=3$, $c=5.99$ and $\Lambda=48$ in (4), the area of the ellipse is found to be 130.3. The exact number of included sample points is 127. Since there are $\binom{n+k-1}{k-1}=\binom{38}{2}=703$ sample points altogether, the exact sample coverage for $n=36$ and $c=5.99$ is $\frac{127}{703}=0.1806$, or roughly 18%.

Suppose next that at $n=81$ we observe 32, 22, and 27 in categories 1, 2, 3. The covariance matrix is now $\begin{pmatrix} 18 & -9 \\ -9 & 18 \end{pmatrix}$ and $Q=\frac{2}{27}(32-27)^2+$ $+\frac{2}{27}(32-27)(22-27)+\frac{2}{27}(22-27)^2=1.85$. The number of sample points is now $\binom{81+3-1}{3-1}=3403$, the number of included sample points at $Q=1.85$ is approximately 90.6, using (4), and so the OSC is about 2.66%. I.e., if chance prevailed, one would expect agreement this good in only 2 or 3 cases in every 100 replications of the experiment.

Consider, finally, the trinomial model with category probabilities $\frac{1}{6}, \frac{1}{3}, \frac{1}{2}$. The ratio of the product of these to the product $(\frac{1}{3})^3=\frac{2}{27}$ for the uniform trinomial is 3:4. Hence, if the two models fit the data of any experiment equally well (i.e., if their quadratic forms assumed the same value at the observed outcome x), then the OSC of the model $\frac{1}{6}, \frac{1}{3}, \frac{1}{2}$ wpuld be three-fourths that of the uniform trinomial, and so it would be slightly better supported.

(Incidentally, one might reasonably argue a better definition of 'disorderly universe' would take the order in which the sample points fall into the different categories into account. The result would be to make fat models fatter and thin models thinner. However, the calculation of sample coverage would be considerably complicated and so I have chosen, for expository convenience, not to follow this course.)

Returning to the general case, let H', H'' be two no-parameter multinomial models with quadratic forms Q', Q'', and assume that at the outcome x of the experiment $Q'(x) = Q''(x)$, so that the two models fit the outcome equally well in a direct confidence sense. Then, by (2), the likelihood ratio quickly reduces to:

(6) $P(x/H''): P(x/H') = (\Lambda': \Lambda'')^{1/2} = (P': P'')^{1/2}$

where we continue to write P's for the products of category probabilities. Consequently, *the simpler model has higher likelihood.* In a quite literal sense, then, simplicity is an ingredient of support, for simplicity and likelihood both increase with decreasing generalized standard deviation. Notice, too, that if $P' = P''$, so that the models are of equal simplicity, the natural logarithm of the likelihood ratio reduces to $\frac{1}{2}(Q'(x) - Q''(x))$, so that the better fitting model then has higher likelihood.

5. MULTINOMIAL MODELS WITH FREE PARAMETERS

On a strict Bayesian analysis, support for a model with many parameters is measured by the integral of the likelihood function with respect to the probability distribution of the parameters. If some parameters can be independently measured or are partially known from earlier experimentation, that will of course effectively reduce sample coverage. The interesting case, for our purposes, is that in which the parameters are unknown, or, as in the blood group example, vary from experiment to experiment. In that event, we may suppose the prior probability mass assigned the model to be uniformly distributed over its parameter space. The posterior probability of the model is then obtained by multiplying the prior probability by the average likelihood.

In practice, the likelihood function seldom lends itself to analytic integration. Where it does not, we must numerically integrate it, e.g., by summing the likelihood over a grid of values and then dividing the result by the number of parameter values in the grid. We will assume for convenience that all parameters range from 0 to 1. Suppose the mesh of the grid is .01. Then, if there is one free parameter, the likelihood is summed over the 101 values 0, .01, .02, ..., .99, 1.00 and the result divided by 101 to yield an approximation to the average likelihood. If there are two parameters, we sum over 101^2 values and then divide by 101^2, and so on. Since most of the likelihood is contained in a small neighborhood of the vector of maximum likelihood estimates of the parameters, we should expect that when a model fits the outcome as well as an extension thereof, its average likelihood will be larger. Here is a proof.

Let K have free parameters $\theta_1, ..., \theta_t$, and let H be the special case of K obtained by setting $\theta_t = \hat{\theta}_t$, where $\hat{\theta}_t$ is the maximum likelihood (ML) estimate of θ_t at the observed outcome x. The best fitting specifications of H and K thus have equally good fit (in fact, equal maximum likelihoods). Let .01 be the mesh of the grid. Then for the average likelihood of K to be as high as that of H requires that, for each setting of the remaining parameters $\theta_1, ..., \theta_{t-1}$, the likelihood[6] $p_x(\theta_1, ..., \theta_{t-1}, \theta_t)$ summed over the 101 grid values of θ_t will be at least 101 times the ML at θ_t, or in symbols:

$$\sum_{\theta_t} p_x(\theta_1, ..., \theta_t) \geqslant 101 p_x(\theta_1, ..., \hat{\theta}_t).$$

I.e., it requires that the likelihood function be a constant. And that can only happen (for an informative experiment) if the sample size is zero. The argument does not depend on the choice of mesh, and so by letting the mesh tend to zero, we see that the integrals must stand in the same relation as their approximating sums. Therefore, the exact average likelihood of H will exceed that of K. (This proof, be it noted, is completely general; it is not restricted to multinomial models.)

We can say more. The likelihood function is asymptotically normal about the vector $(\hat{\theta}_1, ..., \hat{\theta}_t)$ of ML values, with covariance matrix inverse to Fisher's information matrix with typical entry $-nE(\partial^2 L/\partial\theta_i \partial\theta_j)$. For multinomial models,

$$(7) \qquad -nE(\partial^2 L/\partial\theta_i^2) = n \sum_{s=1}^{k} \frac{1}{p_s} (\partial p_s/\partial\theta_i)^2$$

$$(8) \qquad -nE(\partial^2 L/\partial\theta_i \partial\theta_j) = n \sum_{s=1}^{k} \frac{1}{p_s} (\partial p_s/\partial\theta_i)(\partial p_s/\partial\theta_j)$$

give, resp., the diagonal and off-diagonal entries of the information matrix, where $L = \ln p_x(\theta_1, \dots, \theta_t)$ is the log likelihood. The quadratic form Q that appears in the approximating normal density has as its coefficients the co-factors divided by the determinant of the covariance matrix (as in (2)). But these coefficients are just the elements of the inverse of the covariance matrix, and hence, of the information matrix. Therefore, the approximating normal density can be written:

(9) $\qquad (2\pi)^{-t/2} I^{1/2} e^{-Q/2}.$

where I is the determinant of the information matrix. As before, the quadratic form of the density has a chi square distribution, this time with t df, t being the number of parameters. Hence, $Q = \chi^2_{\alpha, t}$ is the equation of a $100(1 - \alpha)\%$ ellipsoidal confidence set (or *Bayesian confidence set*) for the parameters $\theta_1, \dots, \theta_t$. The volume of this set, as shown by (4), increases with the determinant of the covariance matrix, and so decreases with I, the determinant of the information matrix. I, in turn, increases with n (indeed, I is of the order of n^t). Hence, as the sample size n increases, the likelihood function becomes more and more concentrated in a small neighborhood of the ML vector (as well as more nearly normal). Given a grid of mash h, we can choose n so large that the 99.99% confidence set of the parameters contains only the ML vector (and the same will be true, *a fortiori*, of the 99.99% confidence set for the parameters of H). H will then have, at this mesh, an average likelihood roughly $1/h$ times that of K. But the mesh h is arbitrary, so we conclude that, as the sample size increases, the ratio of the average likelihood of H to that of its extension K will grow beyond all finite bounds. *In the limit, the simpler model becomes infinitely better supported.*

Before leaving this topic, some remarks on other approaches to the evaluation of models with many parameters are in order. The standard chi square test of fit does adjust for simplicity to the extent that one degree of freedom is lost for each parameter estimated from the data. But, as we will see, this test provides a rather crude adjustment for simplicity, and evaluations based on it often conflict sharply with those based on average likelihood.[7]

Another approach, yet cruder, is to compare models by their maximum likelihoods. After showing Felix Bernstein's data for the A-B-O blood groups (cf. Ex. 1) decisively supports the one factor model H_1, Edwards (1972), p. 42 writes: "As with H_2, two independent parameters have been evaluated in H_1, so that the hypothese are strictly comparable in 'simplicity'." The two models are *not* of equal simplicity[8], but the more interesting point I wish to raise here is this: What would Edwards have said if the two models differed in their numbers of free parameters? Edwards discusses the problem in Section 10.2 of his book and suggests (p. 200) as a rule-of-thumb that the addition of a parameter must increase the maximum of the log likelihood (which he calls 'support') by two units for the extended model to be accounted better supported. (The two-unit 'support' units are stressed because of their close correspondence, in the case of a normal data distribution, with the conventional 5% limits.) Now not only is Edwards' proposal *ad hoc* (as he recognizes himself), but we can see how to do better from a Bayesian point of view by employing average likelihood. (The approximation (10) below shows that the increase in maximum likelihood needed to increase average likelihood depends, in general, on the number of parameters and on the determinant of the information matrix.)

One might still contend that the maximum likelihood test is easier to work with than the more sensitive Bayesian analysis, especially when the likelihood function cannot be analytically integrated and it becomes necessary to carry out a tedious numerical integration. As matters turn out, though, the Bayesian method is not essentially harder to apply than the likelihood method, at any rate, for it proves possible to get a very satisfactory approximation to the average likelihood in terms of the maximum likelihood.[9]

As we remarked, the likelihood function is approximately normal in shape, but we have to apply a magnification A to the approximating normal density to make its maximum equal that of the likelihood function. (Equivalently, $1/A$ is the normalizing constant, multiplication by which transforms the likelihood into a true normal density integrating to one.) Consequently, we must solve $Ae^0(2\pi)^{-t/2}I^{1/2} = p_x(\hat{\theta}_1, ..., \hat{\theta}_t)$ for A to obtain an approximation to the likelihood. But the integral of the magnified normal density is just A times the integral of the unmagnified normal density, or A, so that the magnification factor itself should provide a good approximation to the area under the likelihood surface:

$$(10) \qquad A = (2\pi)^{t/2} I^{-1/2} p_x(\hat{\theta}_1, ..., \hat{\theta}_t) \doteq \int p_x(\theta_1, ..., \theta_t) \, d\theta_1 ... d\theta_t$$

writing $\doteq$ for approximate equality.

Example 3. Consider the quadrinomial with category probabilities

$$\tfrac{1}{2}\theta, \tfrac{1}{2}(1-\theta), \tfrac{1}{2}(1-\theta), \tfrac{1}{2}\theta$$

which the reader familiar with genetics will recognize as the linkage model for backcross data with the heterozygote in repulsion phase. (We allow θ to range from 0 to 1, instead of from 0 to $\tfrac{1}{2}$, as is customary, in order to simplify the integration.) The ML estimate is $\hat{\theta}=(x_1+x_4)/n$, the observed proportion of recombinants. Suppose that in a sample of $n=100$ we observed 20, 30, 30, 20 individuals in the four categories. Then $\hat{\theta}=0.4$, $p_x(\hat{\theta})=1.038\times 10^{-3}$, and the standard error of $\hat{\theta}$ is $(\hat{\theta}(1-\hat{\theta})/n)^{1/2}= =24^{1/2}\times 10^{-4}$ (using (7), and the normal approximation to the likelihood function). The approximation (10) gives $A=(2\pi)^{1/2}(24)^{1/2}(1.038)\times 10^{-3}\times \times 10^{-4}=127\times 10^{-6}$. The exact value of the average likelihood is $\binom{100}{20\ 30\ 30\ 20}2^{-100}\int_0^1 \theta^{40}(1-\theta)^{60}\,d\theta=128\times 10^{-6}$, using the tables of $\log N1$. Had we observed 16, 26, 34, 24, the ML estimate would be the same, the approximation would give $A=34.726\times 10^{-6}$ and the exact value would be $\binom{100}{16\ 26\ 34\ 24}2^{-100}\int_0^1 \theta^{40}(1-\theta)^{60}\,d\theta=34.467\times 10^{-6}$.

Example 4. Consider H_2, the two factor model of the A-B-O blood groups discussed in example 1, with category probabilities:

$$s(1-t), (1-s)\,t, \,st, \,(1-s)\,(1-t).$$

At a sample size $n=100$ we observe 64, 4, 16 and 16 in the categories A, B, AB and O. Then $\hat{s}=0.8$ and $\hat{t}=0.2$ (using $s=s(1-t)+st$ and $t=(1-s)\,t+st$ to obtain $\hat{s}=(x_1+x_3)/n$ and $\hat{t}=(x_2+x_3)/n$). The elements of the information matrix for this model, found from (7) and (8) above, are: $I_{ss}=n/t(1-t)$, $I_{tt}=n/s(1-s)$ and $I_{st}=0$. Hence $I=n^2/st(1-s)(1-t)$. Now $p_x(\hat{s},\hat{t})=2.406\times 10^{-3}$, whence our approximation to the average likelihood gives $A=2\pi\times(n^2/st(1-s)(1-t))^{-1/2}p_x(\hat{s},\hat{t})=2.41\times 10^{-5}$. The exact value of the average likelihood is $\binom{100}{64\ 4\ 16\ 16}\int_0^1 s^{80}(1-s)^{20}\,ds\times \times\int_0^1 t^{20}(1-t)^{80}\,dt=100!\,80!\,80!\,20!\,20!/64!\,4!\,16!\,16!\,101!\,101!= =2.39\times 10^{-5}$.

Let us next compare the average likelihood of H_2 to its special cases obtained by setting $t=0.2$ and then by setting both $t=0.2$ and $s=0.8$. For the former (one parameter) model the average likelihood is now $\binom{100}{64\ 4\ 16\ 16}0.8^{80}\,0.2^{20}\int_0^1 s^{80}(1-s)^{20}\,ds=2.40\times 10^{-4}$, while the average likelihood of the latter (no-parameter) model is $p_x(\hat{s},\hat{t})=2.41\times 10^{-3}$. At $n=100^2$, the average likelihoods at 6400, 400, 1600, 1600 of H_2 and its special cases can be calculated using Stirling's formula for the factorials, and we obtain 6.25×10^{-10} for H_2, 2.49×10^{-8} for the special case $t=0.2$, and 2.03×10^{-6} for the special case $t=0.2$ and $s=0.8$. The ratio of the average likelihood of the special case to that of its extension in each case is thus of the order of $n^{1/2}$. This example already exhibits the slow trend towards the infinitely greater support which a special case of a model enjoys at outcomes which fit it as well as its extension as the sample size becomes infinite.

Our approximation to the average likelihood can be usefully applied whenever a scientist contemplates complicating a model to improve its accuracy.

Example 3 (continued). Mendel treated the special case $\theta=0.5$ (independent assortment) of the linkage model H_θ. On this special case of H_θ, call it H, all expected category frequencies are equal. Is average likelihood increased for the data discussed earlier, viz. 20, 30, 30, 20, by introducing the linkage parameter θ? The likelihood for H is 3.64×10^{-4}. For average likelihood to increase, the maximum likelihood of the extended model H_θ (viz. the likelihood at $\theta=0.4$) must be roughly $(2\pi)^{-1/2} \times$ $\times (\hat{\theta}(1-\hat{\theta})/n)^{1/2} = 10^4/(48\pi)^{1/2}$ times the likelihood at $\theta=0.5$. In fact, $p_x(0.4) = 1.038 \times 10^{-3}$ which is not large enough (the average likelihood at H_θ being 1.28×10^{-4}). Even at a perfectly agreeing outcome, then, it requires a much larger sample size than 100 to make the average likelihood increase by adjoining the linkage parameter to H when $\hat{\theta}$ differs as little as 0.4 does from 0.5. By a chi square criterion, on the other hand, we would almost always be led to prefer a more accurate but more complicated model. For outcomes fitting the more complicated model exactly, this is obvious, since the chi squared deviation is then zero and the probability of exceeding zero deviation is one for any number of degrees of freedom. But even if we had observed 24, 28, 28, 20, an outcome which agrees numerically with the predictions of H_θ only very slightly better than it does with those of H, the chi squared deviation would have been 1.40 for H at 3 df and 0.36 for H_θ at 2 df. But at 2 df, $P(\chi^2 > .36) > 0.9$, while at 3 df, $P(\chi^2 > 1.40) < 0.7$, using the tables of chi square. The requirement that average likelihood increase is therefore far more stringent than the requirement that the probability of observing as large a chi squared deviation decrease.

Let us consider, on the other hand, some of the data that actually prompted introduction of the linkage parameter. Using *Drosophila Melanogaster*, T. H. Morgan backcrossed a wild-type heterozygote in repulsion phase to a double recessive having black body (bb) and vestigial wings (vv). The genotypic frequencies of the off-spring were:

TABLE II

	BbVv	bbVv	Bbvv	bbvv
Observed	338	1552	1315	294
Expected H:	874.75	874.75	874.75	874.75
Expected H_θ:	314.90	1434.60	1434.60	314.90

where we have shown the numbers expected on H and on H_θ (with the ML estimate $\theta = 0.18$). Clearly the numerical fit of H_θ is far superior to the numerical fit of H, its special case, but is the improved accuracy enough to compensate the loss of simplicity? We find the likelihood of H to be 3.14×10^{-346}. The standard error of $\hat{\theta}$ is about 0.0065, and so the maximum likelihood of H_θ must be $(2\pi)^{-1/2}(0.0065)^{-1}$, or roughly 61, times the likelihood of H for average likelihood to increase. In fact, the maximum likelihood of H_θ (viz. the likelihood at $\hat{\theta} = 0.18$) is 1.57×10^{-10}, which is about 5×10^{335} times as great as the likelihood at H. There is no question, then, that the introduction of the linkage parameter was fully justified by the lights of a Bayesian analysis. I am sure that similar results would be obtained for experiments that prompted various other complications of the Mendelian model, as, for example, those containing additional parameters for incomplete penetrance, differential viability, and so forth. It would be interesting to apply Bayesian analysis to cases of polygenic inheritance to determine the best supported hypothesis as to the number of genes actuating a polygenic effect like skin color or height in man.

We have just seen that the chi square test requires a less favorable trade-off between gain of accuracy per loss of simplicity than the Bayesian requirement that average likelihood increase (or, for that matter the very nearly equivalent requirement that OSC decrease). Further deficiencies of the chi square test include its insensitivity to the allowed ranges of the parameters and to constraints among the parameters (such as the

constraint $p + q \leqslant 1$ governing the parameters of the one factor model of
the A-B-O blood groups). Indeed, as we have remarked, a one-parameter
model can be simpler than a no-parameter model, and, in that event, the
chi square test, with a loss of one df for each estimated parameter, would
in effect penalize the simpler, rather than the more complicated model.
Even where two models have the same number of unconstrained parameters, it can happen that the fatter model has both a lower chi squared
deviation and a lower likelihood. E.g., let H be the trinomial with probabilities 0.4, 0.4, 0.2, while K is the trinomial with probabilities 0.1, 0.2, 0.7.
At $n = 50$ and observed counts 16, 10, 24, the chi squared deviations are
25.4 for H and 27.7 for K, while the log likelihoods are $0.6891 - 7$ for H
and $0.1025 - 6$ for K, so that the fatter model H, with the lower chi square,
also has lower likelihood.

6. THE SAMPLE COVERAGE RULE

In this section, we explore the validity of the sample coverage rule: that
OSC and (average) likelihood vary inversely. By the direct confidence
region of a model with many parameters we mean the union of the DCR's
associated with the special cases of the model. For multinomial models,
DCR's are ellipsoids, and the DCR of a model with many parameters can
equivalently be defined as the envelope of the associated family of ellipsoids. We will continue to approximate the number of sample points
inside a DCR by its measure. The approximation worsens with the simplicity of the model, but adjustments can be made. As to the measure of
the envelope itself, it can often be computed by multiplying the measure
of the hypersurface of predicted values by its average thickness (at least
where the surface does not curl back on itself). The average thickness is
the average distance from the center of an ellipsoid of the family to its
boundary in the direction of the normal to the tangent plane to the hypersurface of predicted values at the center of the ellipsoid.

Example 5. Let H be the one parameter trinomial with probabilities
$\frac{1}{2}\theta, \frac{1}{2}\theta, 1 - \theta$. At $n = 36$, the equation of the 95% direct confidence ellipse is:

$$(2 - \theta)(X - 18\theta)^2 + 2\theta(X - 18\theta)(Y - 18\theta) + (Y - 18\theta)^2 =$$
$$= 216\theta(1 - \theta)$$

after rounding off the upper 5% point of the chi square distribution with 2 df to 6. The hypersurface of predicted values is, in this case, the line of slope one joining the origin to the point (18, 18) of length $648^{1/2} = 25.45$. The normal to this line intersects the general ellipse in the point $(18\theta + \sqrt{54\theta}, 18\theta + \sqrt{54\theta})$, and the distance from this boundary point to the center of the ellipse is $(108\theta)^{1/2}$. The average length of this line is therefore $108^{1/2} \int_0^1 \theta^{1/2} \, d\theta = 6.93$, and hence the average thickness of the envelope is $2 \times 6.93 = 13.86$. When multiplied by 25.45, the length of the axis through the family of ellipses, we obtain 352.7 for the area of the envelope. The exact number of included sample points is 361. Since there are 703 sample points in all, the exact sample coverage is $\frac{361}{703} = 51.35\%$ and the approximation gives 50.21%.

Quite clearly the sample coverage rule holds when the simpler model is also better fitting. In the interesting case, a model H' is better fitting than a simpler model H'', so that $P'' < P'$ and $Q'' > Q'$, where the P's (the products of category probabilities, taken now at the best fitting specification of the model) measure simplicity (or, better, 'local simplicity') and the Q's measure the deviation between observed and predicted values (at the best fitting specification). Writing V', V'' for the measures of the associated DCR's, we seek the conditions under which $V' : V'' > 1$ and $P(x/H') : P(x/H'') > 1$, so that the fatter but better fitting model H' has both higher OSC and higher average likelihood.

Take first the case of no-parameter models. Using (2),

$$L = \ln \frac{P(x/H')}{P(x/H'')} > 0 \quad \text{iff} \quad \tfrac{1}{2}\ln \frac{P''}{P'} + \tfrac{1}{2}(Q'' - Q') > 0.$$

Since $\ln P''/P' < 0$, this requires $|\ln(P''/P')| < Q'' - Q'$. Likewise, $\ln(V' : V'') > 0$ iff $(k-1)|\ln(Q'/Q'')| < \ln P'/P'' = |\ln(P''/P')|$. Combining these results for the no-parameter case, a necessary and sufficient condition that H' have both higher OSC and higher likelihood (in violation of the sample coverage rule) is that the inequalities

$$(11) \qquad (k-1)\,|\ln Q'' - \ln Q'| < \left|\ln \frac{P''}{P'}\right| < Q'' - Q'$$

both hold.

Now (11) cannot hold when $P' = P''$ and the models are of equal simplicity. But where two- no-parameter models of the same experiment differ in simplicity, their predicted values will be widely separated, so that, at large sample sizes, (11) will hold only at outcomes which fit neither model well. This is also clear from the fact that when the middle term in (11) is large, $Q'' - Q'$ must be large and $\ln Q'' - \ln Q'$ must be small for (11) to hold. Given the concavity of $\ln x$, this means that both Q' and Q'' must be reasonably large, so that the outcome will be equivocal.

For a typical exception of this sort, take $k=2$, let the models be $H' : p' = 0.5$ and $H'' : p'' = 0.8$ and let $n = 10{,}000$. The standard deviations are $\sigma' = 50$ and $\sigma'' = 40$, and the expected values are $np' = 5000$ and $np = 8000$. Let $x = 6600$ be observed. This outcome is 32 σ' units from $np' = 5000$ and 35 σ'' units from $np'' = 8000$. Hence the fatter model H' has higher likelihood. But, in absolute units, x is closer to 8000 than to 5000, whence H' also has higher OSC.

When it is required that the outcome fit one of the models reasonably well, it is found that (11) can hold only for a small interval of sample sizes and for at most one or two outcomes at each sample size in the interval. In our binomial example, if x lies at the boundary of a 95% direct confidence interval of H', we have $|\ln(P''/P')| = \ln 25 - \ln 16 = 0.446$ and $Q' \doteq 2$. For (11) to hold requires that $Q'' \geqslant 2.446$ and $\ln Q'' \leqslant 0.446 + \ln 2 = 5.74$, or $2.45 \leqslant Q'' \leqslant 3.12$. Since $x = 0.5n + n^{1/2}$ is 2 σ units from $0.5n = np'$, we must have $0.8n - Q''(0.4)\, n^{1/2} = 0.5n + n^{1/2}$. At $Q'' = 2.45$, $n = 43.55$; at $Q'' = 3.12$, $n = 56.2$. Hence (11) holds only for n between 44 and 56. But it needn't hold even at these sample sizes, for the required x will usually be fractional, and the nearest integer values often fail to satisfy (11). E.g., when $n = 56$, $n^{1/2} = 7.4833$ and $x = 0.5n + n^{1/2} = 35.4833$. At $x = 35$, $Q' \times (\frac{1}{2} \times \frac{1}{2} \times 56)^{1/2} = 7$ implies $Q' = 1.87$, while $Q'' \times (0.16 \times 56)^{1/2} = np'' - 35 = 9.8$ implies $Q'' = 3.20 > 3.12$, so that (11) fails. Likewise, (11) fails at $x = 36$, as the reader can check. At $n = 56$, then, no outcome satisfies (11), and I think this situation more the rule than the exception.

Notice, finally, that due to the presence of the factor $k - 1$ in the left term, (11) becomes a more stringent condition (harder to satisfy) as the number of categories increases. We conclude that, with rare exceptions at small sample sizes, the sample coverage rule is violated only by equivocal outcomes in the no-parameter case.

The situation is more complex when we admit free parameters. Let H', H'' have t', t'' free parameters, resp., For general notation, write $P(x/H)$ for the average likelihood of H at x (stretching the usual notation somewhat) and $P_x(\hat{H})$ for the maximum likelihood of H at x. Using (10):

(*) $P(x/H) \doteq (2\pi)^{t/2} I^{-1/2} P_x(\hat{H})$

while from (2),

(**) $P_x(\hat{H}) \doteq (2\pi)^{-(k-1)/2} n^{-(k-1)/2} P^{-1/2} e^{-Q/2}$

where P is now the product of category probabilities at $\hat{H}$, the best fitting special case of H. Combining (*) and (**):

(12) $\ln \dfrac{P(x/H')}{P(x/H'')} \doteq \frac{1}{2}(t' - t'') \ln(2\pi) + \frac{1}{2} \ln \dfrac{I''P''}{I'P'} - \frac{1}{2}(Q' - Q'')$.

The determinant I of the information matrix is of the order of n^t, n the sample size, t the number of parameters. When $t' - t'' = s > 0$, $I''P''/I'P' \doteq \doteq n^{-s}$. The middle term on the right side of (12) is then numerically large relative to the first term, and so $Q'' - Q'$ must be large for (12) to be positive. I.e., when H' has more parameters than H'', it must fit the outcome much better in order to have higher average likelihood, the amount depending on the difference $t' - t''$. However, it may happen that the model with more parameters may be simpler (see below), and so this reasoning cannot be trusted too far.

Consider next the OSC's. In the general case, let H have $V = V_1$ at sample size n_1 and $Q = Q_1$. We wish to obtain an expression for the value of V (the measure of the DCR of H), at n_2 and Q_2. V is the product of the measure of the hypersurface of predicted values by its average thickness. (This is only an approximation if the hypersurface curls back on itself so that non-adjacent ellipsoids of the family overlap.) Now the measure of the hyper-surface at n_2 is just $n_2 : n_1$ times its measure at n_1. Its average thickness at n_2, Q_2 is $(Q_2 n_2 : Q_1 n_1)^{1/2}$ its values at Q_1, n_1. Hence:

(13) $V_2 = \dfrac{n_2}{n_1} \times \left(\dfrac{Q_2 n_2}{Q_1 n_1} \right)^{1/2} \times V_1$.

We are interested in the ratio $V' : V''$ for H', H''. Using (13), the sample sizes all cancel and we get:

$$V_2' : V_2'' = \left(\dfrac{Q_2' : Q_1'}{Q_2'' : Q_1''} \right)^{1/2} \times \dfrac{V_1'}{V_1''}.$$

This simplifies further if we set $Q_1' = Q_1''$, so that V_1' and V_1'' are the sample coverages of H' and H'' at the same confidence level and a given sample size n_1. Setting $R = V_1' : V_1''$ (when $Q_1' = Q_1''$) and writing V', V'' for the measures of the envelopes at current values of n, Q', Q'', we have:

$$V' : V'' = (Q' : Q'')^{1/2} \times R$$

whence

(14) $\ln \dfrac{V'}{V''} = \tfrac{1}{2} \ln \dfrac{Q'}{Q''} + \ln R.$

For H' to have both higher average likelihood and higher OSC, it is therefore necessary and sufficient that the right sides of (12) and (14) be positive.

This is not a very transparent condition. However, the problem simplifies considerably when $t' = t''$. For then the first term on the right side of (12) drops out. Moreover, I' and I'' will be of the same order. (Indeed, they are often identical; e.g., for all trinomial models whose category probabilities have the form $p\theta$, $(1-p)\theta$, $1-\theta$, $I = n/\bar{\theta}(1-\bar{\theta})$, and the result extends to higher numbers of categories.) Hence, the second term on the right side of (12) will simplify (at least approximately) to $\tfrac{1}{2}\ln(P''/P')$, and R, in (14) will likewise be close to P'/P''. The condition then reduces to (11). We conclude that the case of models with *the same number* of parameters – provided that both sets of parameters are unconstrained – is essentially no different from the no-parameter case covered by (11).

It is harder to see in detail what happens when $t' > t''$. We look at two examples that appear to provide an especially sensitive test of the sample coverage rule.

Example 6. The chi square statistic vanishes at perfectly agreeing outcomes so that, in effect, any model, however complicated, enjoys infinite support at perfectly agreeing outcomes by the lights of a chi square test. Our measure Q of deviation has this defect in lesser degree, for, at a perfectly agreeing outcome, Q will equal the number of predicted values that are possible sample points, as the parameters run through their range. At any rate, if we compare a model to a special case of itself at outcomes that fit the former perfectly, we should get a very sensitive test of the sample coverage rule.

Let H' be the trinomial with probabilities $\tfrac{1}{2}\theta$, $\tfrac{1}{2}\theta$, $1-\theta$, and let H'' be the special case $\theta = \tfrac{1}{2}$ of H' with probabilities $\tfrac{1}{4}$, $\tfrac{1}{4}$, $\tfrac{1}{2}$. Then, writing a, b for the numbers in the first two categories, $P(x/H'):P(x/H'')=2^n(a+b)! \times$ $\times (n-a-b)!/(n+1)!$, and, at $n=100$ and $a=b$, this ratio first exceeds 1 when $\hat{\theta}=0.38$ (being very close to 1 at $\hat{\theta}=0.40$). On the other hand, $V'=51$ at $a=b$, while $V''=55.5Q''$, which first exceeds 51 when $\hat{\theta}=0.44$. There is a slight discrepancy, then, between the requirements that average likelihood increase and that OSC decrease. However, if we allow even a minimal deviation, setting $a=\frac{1}{2}(a+b)+1$ and $b=\frac{1}{2}(a+b)-1$, the ratio of average likelihoods is unaffected, but the OSC of H' will be increased almost by a factor of five. In fact, both the ratio of average likelihood and the ratio of OSC's are now close to 1 when $\hat{\theta}=0.40$ (i.e., when $a=21$ and $b=19$), so that the discrepancy already disappears. Moreover, if we increase the sample size, the data will discriminate more sharply between H' and H'', so that the average likelihood of H' will come to exceed that of its special case H'' at a value of $\hat{\theta}$ closer to 0.50, while the OSC of H' will be larger (it is $\frac{1}{2}n+1$). Hence the discrepancy will likewise diminish. In fine, the requirement that OSC decrease is *practically* equivalent to the requirement that average likelihood increase. Both are sharply at odds with the chi square condition. At $n=100$ and minimally deviant outcomes $a=\frac{1}{2}(a+b)+1$ and $b=\frac{1}{2}(a+b)-1$, the chi square test favors H' as soon as $\hat{\theta}=0.46$ (viz., at $a=24$, $b=22$).

Example 7. The existence of cases in which the model with more parameters is simpler provides another sensitive check of the rule. In Example 2, the uniform trinomial with all probabilities equal to $\frac{1}{3}$ was found to have sample coverage 18% at a 95% confidence level ($Q=5.99\doteq6$) and $n=36$. At the same sample size and confidence level, the trinomial with probabilities $\frac{1}{36}\theta$, $\frac{35}{36}\theta$, $1-\theta$ is found to have sample coverage 13.79% (its DCR containing just 97 sample points.) Here, writing H' and H'' for the two models, $P(x/H'):P(x/H'')=36^{a+b}(n+1)!/3^{36}35^b(a+b)!(n-a-b)!$, and the common logarithm of this expression at $n=36$ is $30.8413-1.5441b$. From this, one easily finds that the ratio of average likelihoods first exceeds 1 at the point $a=5$, $b=19$, if we run through the points from $(0, 24)$ to $(12, 12)$ keeping $a+b=24$ (so that $\hat{\theta}=\frac{2}{3}$ throughout). I find that the OSC of H'' first exceeds that of H' at the same point, $(5, 19)$.

While much further investigation is needed, the evidence examined thus far does point to the conclusion that the sample coverage rule is rather satisfactory for practical purposes.

7. APPLICATIONS OF THE SAMPLE COVERAGE RULE

In this section some applications of the sample coverage rule are illustrated.

(a)   OSC's can be used to effect *global comparions*: viz., to compare the support of not necessarily exclusive hypotheses (which may even be drawn from disparate fields of science).

(b)   The *severity* of two tests of a theory can be compared by comparing the sample coverages of the theory for the two tests or experiments at a chosen confidence level.

(c)   Qualitative theories (with qualitative predictions) for which no data distribution can be obtained can nevertheless be tested if we can order sample points as agreeing more or less well with the theory in question.

(d)   The requirement that OSC decrease can be used to determine whether a theory's support is improved by adjoining a new parameter.

Fisher's example of the tea-tasting Lady illustrates both (b) and (c). His design, you recall, assigns each of the two treatments (milk-first and tea-first) at random to four of the eight cups. The probability that the Lady classifies a specified number of the eight cups correctly when she has some ability to discriminate the treatments cannot be calculated for this design (which is why Fisher chose it to illustrate significance tests). However, a larger number of correct classifications is surely more indicative of ability than a smaller number. Given this intuitive ordering of the sample points, the OSC of the outcome 'r of the 4 milk-first cups correctly classified is easily seen to be $\binom{4}{r}\binom{4}{4-r}/\binom{8}{4} + \binom{4}{r+1}\binom{4}{4-r-1}/\binom{8}{4} + \cdots + \binom{4}{4}\binom{4}{0}/\binom{8}{4}$. Fisher employed essentially the same analysis. He also observed that a more severe test of the Lady's ability is obtained by selecting the treatment to be assigned each cup at random. For $n=8$ cups there are now 256 possible assignments of the two treatments, and so the Lady's chance of correctly classifying all eight cups falls from $1/\binom{8}{4} = \frac{1}{70}$ to $\frac{16}{256}$. We briefly consider two more examples relevant to (b) and (c).

Example 8. In tables of empirical data, where the entries vary in length, such as one finds in an Atlas or Almanac, one might naively expect all nine digits $1, \ldots, 9$ to occur as first digit in a tabular entry with equal frequency. The physicist Frank Benford found, however, that the frequency P_k with which the digit k occurs as first digit is more accurately given by $P_k = \log(k+1) - \log k$, $k = 1, \ldots, 9$ (common logarithms). I looked up a table in my Atlas giving the areas of the world's largest islands. The observed and expected frequencies (based on Benford's law) follow:

TABLE III

	1	2	3	4	5	6	7	8	9
Obsvd	37	27	19	11	8	2	9	7	4
Expct	37.3	21.8	15.5	12.0	9.8	8.3	7.2	6.3	5.7

As you can see, the fit is excellent; a 25% confidence ellipsoid covers the observed values. The sample coverage at that confidence level is 0.0000014, or 1.4×10^{-6}. Also, agreement this good is virtually impossible if the nine digits occur with equal frequency as first digit. We are loath to believe that agreement this good – and this improbable – could be mere coincidence. R. S. Pinkham has shown that Benford's law is the only distribution invariant under change of the scale units in which the tabular entries are expressed. But even lacking such theoretical justification for the law, the evidence based on a single table of modest size is quite convincing.

Now suppose that Benford's law were weakened to require only that P_k be a decreasing function of k. If we rank the nine categories in order of highest occurring frequency as first digit for our table of areas, the expected ranks are $1, 2, 3, 4, 5, 6, 7, 8, 9$, and the observed ranks are $1, 2, 3, 4, 6, 9, 5, 7, 8$. We measure the deviation between observed and expected values for this qualitative hypothesis by the sum of the absolute deviations at the different categories, the sum in our example being 8. The number of sets of ranks for which the total deviation is $\leqslant 8$ is found (by counting) to be 238. Hence, the OSC is $238/9! = 6.56 \times 10^{-4}$ (as compared with an OSC of 1.4×10^{-6} for the quantitative law, $P_k = \log(k+1) - \log k$). There is, then, an appreciable loss of support in diluting a quantitative hypothesis to qualitative form, when the quantitative hypothesis fits the data reasonably well.

Example 9. Equating the severity of a test with its informativeness, we arrive at an alternative to various informational measures (e.g., the expected log likelihood ratio and its special case, Fisher's information, entropy reduction, and others) for comparing experiments. A standard example from genetics compares the single backcross $AaBb \times Aabb$ to the double backcross $AaBb \times aabb$ for the information about linkage which they yield. We assume here that genotypes within a phenotype are not distinguishable (at least not without further experimentation). We obtain multinomial models H' and H'' for the two experiments, each involving the linkage parameter (or 'recombination fraction') θ. Their category probabilities are listed in Table IV:

TABLE IV

	AB	Ab	aB	ab
H'	$\frac{1}{4}(2-\theta)$	$\frac{1}{4}(1+\theta)$	$\frac{1}{4}\theta$	$\frac{1}{4}(1-\theta)$
H''	$\frac{1}{2}(1-\theta)$	$\frac{1}{2}\theta$	$\frac{1}{2}\theta$	$\frac{1}{2}(1-\theta)$

where H' is the model for the single backcross. To ask which experiment is more informative for θ (on the average) is to ask which of the models H' or H'' is simpler. This is answered by comparing their respective products of category probabilities. It is found that $P'' = \theta^2(1-\theta)^2/16$ is smaller than $P' = (2-\theta)(1+\theta)\theta(1-\theta)/64$ at every value of θ in the open interval $(0, 1)$, while both are zero at the endpoints. Hence, H'' is *uniformly simpler* than H', and so the double backcross provides more information about θ (on the average) whatever the true value of θ (provided, we emphasize again, that genotypes within a phenotype are indistinguishable). For quantitative comparisons, we would compute the sample coverages of the two models at a given sample size and confidence level.

8. CONCLUDING REMARKS

Additional evidence for the intuitive acceptability of our explication comes from reflecting upon the simplicity of general laws. Consider first the problem posed by Hintikka (1967). Given a partition of the population into k possible kinds, we seek the number of non-empty kinds. (This is essentially a concept identification task.) Start with a law which designates w of the k kinds as non-empty. Any law which designates these w kinds and some additional kinds will be compatible with more possible samples of n items. For a law is compatible with a sample iff a (proper or improper) subset of its designated kinds appear in the sample. We are then led straightaway to conclude that the simplest law compatible with a sample designates all *and only* those kinds exemplified in the sample. And it is easy to show, as Hintikka does, that the fewer the number of kinds designated by a law, the lower its probability of holding in a finite disorderly universe (viz. one in which individuals are assorted among kinds at random). Finally, the simpler of two laws is found to have higher likelihood for a sample compatible with both. Indeed, if successively larger samples continue to exemplify but b of the k kinds, the posterior probability of the law designating those b kinds tends to one, while that of all other laws (including complications of the given law) tend to zero. This conclusion should follow for any reasonable assignment of prior probabilities to the competing laws or concepts, not just on Hintikka's assignment. The result is the counterpart of our earlier finding that special cases of a model come to enjoy infinitely greater support at data that fits both equally well as the sample size becomes infinite. One's intuition in the present context is *how improbable* that the remaining designated kinds should remain unexemplified in the successively larger samples when they are present in the population, the improbability mounting with the sample size and the number of additional unexemplified kinds.

We are now in a position to appreciate the sense in which the injunction to assume the simplest theory compatible with the data is a stricture against *arbitrary* extrapolations. Having observed a sample which exemplifies b of the k possible kinds, we may experience a temptation to posit additional kinds. But which kinds? No answer that is not completely arbitrary recommends itself (supposing nothing further is known). We would be right in thinking that a more complicated law stands a better *a priori* chance of proving compatible with larger samples, but without at all believing the more complicated law is more likely to be true of the population. E.g., in estimating the number of words (types, not tokens) in a children's book, higher estimates are more plausible than lower estimates up to a point, but beyond that ill-defined point become less probable. Hence we cannot conclude in general that simpler laws have lower prior probability than more complicated laws, as Popper has so often insisted.

Science is thought to rest on a principle of induction which, variously formulated, asserts the uniformity of nature, that the future will be like the past, that laws are made more probable by their positive cases, and so on. As I have argued elsewhere, there is an inherent vagueness in the notion of uniformity invoked here: any spatio-temporal slice of the world will exhibit no end of 'uniformities' or patterns, most of which there will be no reason to think characterize other slices.[10] Nevertheless, if we find that a theory appears adequately to represent a certain realm of phenomena, it is *simplest*, in general, to assume that it continues to hold outside the domain over which it has been thoroughly tested (supposing nothing to the contrary is known or inferable). Let me quote a very representative passage from a popular exposition of the steady-state theory by H. Bondi (in *Rival Theories of Cosmology*, Oxford, 1960, pp. 13–14):

Of course, it may be necessary to consider the very difficult problems of the variation of physics in a varying universe; but before we enter the enormous complication of the question, we first try to see whether our universe might not happen to be one that is the same everywhere and at all times when viewed on a sufficiently large scale. In examining this possibility, we by no means claim that this must be the case; but we do say that this is so straightforward a possibility that it should be disproved before we begin to consider more complicated situations.

This *principle of conservative extrapolation*, as we may call it, is closely related to E. T. Jaynes's *maximum entropy rule* for obtaining probability distributions from given constraints (e.g. the mean of the distribution): among all those distributions which satisfy the constraints, choose that which maximizes entropy or uncertainty. Otherwise put, the rule directs us to choose that distribution which goes *as little beyond what we know* as possible. Likewise, when complicating a theory to accommodate discrepant data, we strive to increase its sample coverage (over the range of its intended application) minimally. The principle of conservative extrapolation seems to me to capture the kernel of truth in traditional formulations of the 'principle of induction'. But there is no mystery at all about its justification. If the most conservative (and therefore simplest) extrapolation or theory turns out to fit additional experimental results, *those results will lend it higher support* than any rival theory.

Virginia Polytechnic Institute

NOTES

[1] First proposed explicitly, I believe, by Jeffreys and Wrinch (1921), though, as an intuitive yardstick, it must go back to the dawn of science.

[2] Many additional canons of inductive simplicity are captured by our explication, among them: (1) that it is simplest to assume that a law holds exactly (e.g., that energy is *exactly* conserved), (2) that a theory is simpler the more it determines, and the fewer the number (and kinds) of individuals it posits. But to explain the connection between these additional canons and sample coverage in depth will require another paper.

[3] Cf. Kemeny (1953) for this point. This interesting paper of Kemeny's shows that, by ordering hypotheses according to simplicity, and adopting the earliest hypothesis in the ordering that fits the data (in a direct confidence sense), then, assuming that one and only one of the considered hypotheses is true, the probability of choosing the correct hypothesis will tend to one with increasing sample size. However, it seems to me, many orderings will lead to this convergence to the truth, many of which may have a tenuous connection with any intuitive notion of simplicity. Our proposal imposes orderings for a variety of problems which are all based on a single principle (that of minimizing sample coverage). It would be interesting to investigate its convergence properties – whether, by choosing the simplest hypothesis in our ordering, one would always be led, on the average, most rapidly to the true hypothesis.

[4] This reasoning is considered in the following passage from Weyl (1949): "Since the concept of simplicity appears to be so inaccessible to objective formulation, it has been attempted to reduce it to that of probability.... If, for example, 20 corresponding pairs of values (x, y) of a functional connection $y = f(x)$, with the accuracy to be expected, lie on a straight line when plotted in a rectangular coordinate system, then a strict natural law will be surmised to the effect that y depends linearly on x. And this because of the *simplicity* of the straight line, or also because it would be so extremely *improbable* for the 20 pairs of observations to lie (nearly) on a straight line if the law in question were a different one" (pp. 155–6). But then Weyl goes on to object that the same claim of improbability could be made on behalf of any specific curve that fits. This is true of specific curves, like $3x^{1/2} + \ln x + 5$, but it is not true of curve families with adjustable parameters, like $ax^{1/2} + \ln x + b$. And it is curve families or theories rather than the special cases that result by fixing all the free parameters that are the proper objects of simplicity comparisons, just as argument *forms* are the proper objects of logical study.

[5] I owe this way of putting the point to a conversation with I. J. Good; cf. further his discussion of Bode's law in Good (1969).

[6] We use the notation '$p_x(\theta_1, ..., \theta_t)$' for the likelihood function, rather than '$P(x/\theta_1, ..., \theta_t)$', in order to stress that it is a function of the parameters indexed by the outcome x. Barry M. Loewer first suggested to me the possibility of analysing the affect of adding parameters to a model on its support in terms of average likelihood.

[7] For the relation between the average likelihood and the chi square criteria, cf. Smith (1969), Section 21.21. Smith provides a thorough discussion of tests for standard statistical problems (homogeneity, comparison of several means, etc.) using average likelihoods or 'Bayes factors'. Cf. also Jeffreys (1948) and Good (1950).

[8] Both models have the same number of free parameters, but while the parameters s, t of H_2 range independently through the interval from 0 to 1, those of H_1 are bound by the constraint $p + q \leqslant 1$. For a mesh of .01, the number of grid points (or predicted values) for H_2 is thus 101^2. But the number for H_1 is just the number of non-negative integer solutions of $p + q + r = 100$, or $\binom{100+3-1}{3-1} = 5151$, or roughly half as many. So we should expect H_2 to

have roughly twice the sample coverage of H_1. Actually, the gap is larger, for H_2 also has more symmetries, with the result that its predicted values are more spread out, thus minimizing opportunities for distinct specifications of it to fit the same sample point. At $n = 16$ and mesh $n^{-1/2} = 0.25$, I find the number of sample points with a sum-of-squared-deviations $\leqslant 2$ is 123 for H_1 and 217 for H_2, while the numbers giving a sum-of-squared deviations $\leqslant 4$ are 132 for H_1 and 276 for H_2, now more than twice as many.

[9] Cf. Smith (1969), Section 21.20.

[10] *Polya urn models* illustrate the point. Imagine an urn containing b black and r red balls. A ball is drawn at random and replaced, but in addition, c balls of the color drawn and d balls of the opposite color are placed in the urn. Now it should be clear that a given frequency of black balls on past drawings can connote either an increase, a decrease, or no change in the probability of drawing black on ensuing trials, depending on the values of the parameters c and d. But the traditional 'principle of induction' makes the inference depend only on the observed frequency, and not on what we take to be the kind of stochastic process under observation.

REFERENCES

Edwards, A. W. F., *Likelihood*, Cambridge 1972.

Fisher, R. A., *Design of Experiments*, ch. II, Edinburgh 1935.

Good, I. J., *Probability and the Weighing of Evidence*, London and New York 1950.

Good, I. J., 'A Subjective Analysis of Bode's Law and an 'Objective' Test for Approximate Numerical Rationality', *J. Amer. Statist. Assoc.* **64** (1969), 23–66 (with discussion).

Hintikka, K. J. J., 'Induction by Enumeration and Induction by Elimination', in *The Problem of Inductive Logic* (ed. by I. Lakatos), Amsterdam 1967.

Jaynes, E. T., 'Prior Probabilities', *IEEE Transactions on Systems Science and Cybernetics*, v. SSC-4, no. 3, Sept. 1968.

Jeffreys, H., *Theory of Probability*, Oxford, 1948 (3rd ed., Oxford, 1961).

Jeffreys, H. and D. Wrinch, 'On Certain Fundamental Principles of Scientific Inquiry', *Phil. Mag.* **42** (1921).

Kemeny, J., 'The Use of Simplicity in Induction', *Phil. Rev.* **62** (1953), 391–408.

Popper, K., *The Logic of Scientific Discovery*, London 1959.

Smith, C. A. B., *Biomathematics*, vol. 2, New York and London 1969.

Weyl, H., *Philosophy of Mathematics and Natural Science*, Princeton 1949.

DISCUSSION

Commentator: Bunge: You've defined the simplicity of a hypothesis in terms of the degree of its evidential support. Thus it is impossible for you to define a prior simplicity for hypotheses which have not yet been subjected to tests. Thus your notion of simplicity will not help the scientists to decide which among a set of hypotheses to subject to tests or to spend time making calculations with.

Rosenkrantz: I haven't *defined* simplicity in terms of support, but in terms of sample coverage (the proportion of possible outcomes of the relevant experiment in agreement with the hypothesis). But I have shown (in certain cases) that simpler hypotheses in equally good agreement with the data have higher likelihood or average likelihood and are thus better supported. Thus a scientist can say a priori that a simpler hypothesis will experience a larger increase in probability if it turns out to be in agreement with the experimental results. In short, simpler theories are more confirm*able*.

Bunge: How do you account for the history of science? It seems that, on the whole, as scientific knowledge advances, it gets more complex and at the same time better confirmed. For example, intuitively, I think we would all agree that quantum mechanics is far more complex than is classical mechanics and yet it is undoubtedly the case that quantum mechanics is far better supported than classical mechanics. Indeed, it is intuitively fairly clear (in the intuitive sense of the word 'complexity') that as our theories improve they get more complex but they also make more precise predictions and hence, when confirmed, become far more strongly confirmed than earlier theories.

Rosenkrantz: Theories which yield sharper predictions are *ipso facto* simpler in my sense, and so more confirmable. I take no issue, then, with your second point, nor, for that matter, with your first. Science *does* grow both more complex and better supported, but that is no difficulty for my view. We complicate a theory in order to improve its accuracy, and if the gain in accuracy is enough to off-set the loss of simplicity, as determined by a rise in average likelihood, then the complicated theory is of course better supported. There is another trap to be wary of here. An ever more complicated picture of the hereditary mechanism has emerged from the development of Mendelian theory, which now admits parameters for linkage, partial manifestation of a gene, and differential viability of recessives and dominants, and also countenances many-many relations between genes and their effects, complexly interacting systems of polygenes, multiple alleles, and so on. Yet we must be careful not to mistake complexity of the overall theory for complexity of the account it gives of particular experiments or the inheritance of particular traits in particular species. E.g., the tri-allelic model for A-B-O blood group data discussed in my paper is a rather simple model of those data. By contrast, Mendelian theory *qua particulate scheme* of inheritance is vacuous, or very nearly so, for it would be hard to imagine categorized data that could not be fitted by some particulate model or other. The term 'theory' does double duty here in a way which invites this confusion.

Good: I think it is somewhat ambiguous to say that the simplicity of a hypothesis H should be allowed for when considering the evidential *support* for that hypothesis. I think it is better to think of the simplicity of H as relevant to its prior probability and thus to separate simplicity from the evidence E based on observations of the world. My own views on support, or 'weight of evidence', for a hypothesis, provided by evidence,
· have been expressed in dozens of places and I shall not repeat them here except to mention that the weight of evidence does not depend on the prior probability of H. In fact any function of $P(E/H)$ and $P(E/\bar{H})$ that, together with the initial probability of H, determines its final probability, must be a monotonic function of weight of evidence. This was proved in my paper 'Corroboration, Explanation, Evolving [dynamic] Probability, Simplicity, and a Sharpened Razor', *Brit. J. Philos. Sc.* **19**(1968), 123–143. I should add that I have changed the definition of complexity given there, but the remainder of the paper remains unchanged: see my paper 'Explicativity, Corroboration, and the Relative Odds of Hypotheses' in the conference 'Methodologies: Bayesian and Popperian' (University of South Carolina, November 1973), in *Synthese* [**30**, 39]. Under some weaker desiderata I found all explicata of corroboration in *J. Royal Statist. Soc.* **22** (1960), 319–331; **30** (1968), 203; and these included both weight of evidence and the formula in Dr. Rosenkrantz's Note 2, $P(H/E) - P(H)$.

At one time I thought that the simplicity of a theory or hypothesis could be defined in terms of its prior probability, but my present view is that simplicity is merely *one* of the aspects of a theory that helps one to judge the prior probability. The complexity of a theory can be measured by the shortest description of that theory in a language, where the language is chosen to be efficient for communication in the broad field of interest. When judging whether to adopt one theory rather than another one, in the light of evidence, the prior probability (and hence the posterior probability) is more relevant than its simplicity. But even the posterior probabilities are not a sufficient guide for choosing between a pair of theories, because a theory can be more probable than another but can explain less. Thus a compromise is required between high probability (which depends on simplicity among other things) and the amount explained. A quantitative form of this compromise was proposed in my 1968 paper (the 'sharpened razor') and was developed further in the 'explicativity' paper. On another related topic, I wonder if Dr. Rosenkrantz has any comments on the notion of the complexity of assigned parameters? For example, if the inverse square law were replaced by an inverse xth power law, where $x = 2 + 7^{-77}$, I would say the theory had been made more complicated.

Rosenkrantz: I must concede the main thrust of Professor Good's last point: mathematical complications of a law seem intuitively to complicate it without necessarily complicating it in my sense. To vary his example, replacing the inverse square of Newton's law by a fractional or irrational exponent would certainly be held a complication of it. More generally, whole number relationships are conceded a higher degree of simplicity. It could be noted in my defense, however, that whole number relationships often owe their saliency to the discrete underlying model of the phenomenon which they suggest. One could cite in this connection the celebrated Mendelian ratios, the laws of definite and multiple proportions, and the Balmer formula, associated with the early development of Mendel's particulate scheme, Dalton's atomic scheme, and the older quantum mechanics, respectively. The point is that a 3:1 ratio of dominants to recessives in a hybrid cross suggests a simple discrete model in a way that, say, a 3.245:1.198 ratio does not. And discrete models, of course, impose strong constraints on multinomial data – they are simple in my sense. Moreover, we seldom complicate a law mathematically, in practice, without at the same time adding to its free parameters. Newton's law would be mathematically more complicated if the gravitational constant were demoted from that status and allowed to vary (as an ordinary parameter) from one application of the law to another. That would enormously increase the sample coverage of the gravitation law, as the measured or estimated value of G in one problem would no longer constrain the value of G in another (unrelated) problem. Mightn't it be the case that our feeling that mathematical complexity matters be due to its association with adjunction of new parameters?

Considering now Good's positive proposal, I am disinclined to follow him in viewing simplicity as a determinant of prior probability. The simplest hypothesis, as I am sure he would concede, is often quite absurd, or, at least, highly implausible. Having drawn an ace from a freshly opened pack of cards from a reliable firm, the simplest hypothesis, surely, is that all 52 cards in the deck are aces (indeed, aces of the suit drawn). But that hypothesis, under the envisaged circumstances, would have negligible prior probability. At the same time, I disagree just as emphatically with Popper's contention that probability and simplicity are inverse. In estimating the number of different words (i.e. word-*types*, not tokens) in a text, the prior probability increases with the size of the estimate up to an ill-defined point beyond which it declines. Given Hemingway's *A Farewell To Arms*, for example, I would assign higher probability to an estimate of 10,000 words than to one of 100,000, and more to an estimate of 100,000 than to one of 1,000,000. But, by the same token, 10,000 is a more plausible estimate than 100, and I wonder what Professor Good would say to that? For the hypothesis of 10,000 is more complicated than the hypothesis of 100 by his criterion (as well as by mine), if we take first-order logic as the relevant language. At the very least, the relation between prior probability and simplicity is not simple!

Settle: It seems odd that you discuss Popper, when it is clear that you wish to use the degree of simplicity to prefer hypotheses as *true*, whereas Popper wishes to use the same notion to prefer hypotheses as those to be *tested next*, not as being true. Thus there is a crucial difference in the reasons that people have for preferring a simple hypothesis.

Rosenkrantz: I couldn't agree more that Popper and I are up to very different things. He would use simplicity to gauge only the testworthiness of a theory, where I would use it to gauge the trustworthiness of a theory as well. Yet, there are resemblances equally worth emphasizing. If you take seriously an aside of his about relativising simplicity to a problem and construe his 'relatively atomic basic statements' as descriptions of simple experimental outcomes, our characterizations of simplicity come to look more and more alike. Moreover, just as simplicity is (for me) ingredient in support, so, too, for Popper, simplicity is ingredient in corroboration. For corroboration is the passing of severe tests, and, as Popper is well aware, the severity of a test has much to do with the simplicity of the hypothesis tested.

Lubkin: As a matter of fact, the quantum theory is a more coherent theory than classical mechanics (the number of states is smaller, or, the phase group is smaller) and in that sense it is actually a simpler theory to work with. Indeed, this is the motivation for the attempt to apply quantum logics, as opposed to classical logics, to situations outside atomic physics.

Robinson: I have been trying to think of examples of basic theories in the history of science which have been accepted or rejected on the grounds of their simplicity – and I cannot think of a single plausible example.

Rosenkrantz: If you mean theories are never accepted or rejected on grounds of simplicity alone – agreed! But insofar as simplicity is ingredient in support, it must play a critical role in evaluating a theory in the light of data. I would add it is only by appeal to simplicity that we can explain the otherwise mysterious fact that scientists sometimes regard an upstart theory as more promising than a demonstrably more accurate and entrenched rival. (The simplified heliocentric scheme of Copernicus – without epicycles – might be a case in point.) For if the upstart theory is radically simpler, then, even though less accurate, its average likelihood might be higher than that of its older rival. Copernicus himself made much of the heliocentric model's ability to determine the order of the inferior planets and the relative distances of all the planets from the sun, features which reflect the radical reduction of sample coverage in passing from an earth-centered to a sun-centered system. Kepler's laws effect an even greater reduction of sample coverage, especially as his third law relates the motions of different planets. The more a theory determines, the simpler it is, and the more it is confirmed by conforming data.

Van Fraassen: I want to try you on the tacking paradox. Suppose we compare two theories that are both solely about the structure of a given population. Then you wish to evaluate these two theories in part in terms of simplicity, which you relate to logical strength. It doesn't seem to me that you would simplify any such theory by conjoining it with the assertion, e.g., that all Australians are philosophical realists. As a matter of fact you could have a pair of theories, each with samples which they fit and others which they do not fit, and you could just conjoin them, so that the conjunction would fit just the samples in the intersection of the two sets of samples which each fitted separately, and yet, according to your account you would get a hypothesis which is simpler.

Rosenkrantz: The point is this: when you conjoin an hypothesis to a theory you increase its logical strength (reduce the number of possible worlds compatible with the theory), but you needn't thereby reduce the theory's sample coverage for a particular experiment. Thus, conjoining 'Australians are philosophical realists' to the tri-allelic model of the A-B-O blood types mentioned above presumably would not reduce the number of possible blood group frequencies fitted by that model, and, more generally, the conjoined but extraneous hypothesis in question would have nil affect on the relevant conditional outcome probabilities. (The latter feature is precisely what I mean by 'extraneous' in this context.) There is a difference, then, between Popper's formulation in logical terms, and my formulation in relativised and probabilistic terms. As to your schematic example, if the intersection is smaller, the conjunction of the hypotheses is indeed simpler, but I see nothing strange in that *per se.* Simpler theories result whenever disparate theories, like genetics and cytology, are conjoined or united. The union of genetics and cytology via the discovery of linkage added, for example, the powerful constraint that the number of linkage groups for a species must coincide with the (directly observable) number of chromosome pairs characteristic of that species. Of course, your point might be that there is more to union than mere conjunction, and I would tend to agree.

Note*: In the version of the paper delivered at the conference, three criteria of simplicity were discussed: (A) the paucity-of-parameters criterion, that the simplicity of a theory varies inversely with the number of its adjustable parameters, (B) a form of Occam's Razor asserting that the simplicity of an hypothesis about the number of kinds present in a population decreases with the posited number of kinds, and (C) the Popperian proposal that simplicity increases with logical strength.

Hooker: Let me attempt to extend van Fraassen's remarks. You have offered three criteria of simplicity and it does seem to me that these will often conflict and indeed criterion (C) is nearly worthless. Consider the following situation: you have two theories T_1, T_2, which have a subset of their parameters in common. Let us assume that both theories apply to overlapping ranges of subject matter. Each theory applied separately determines only interval ranges for the values of its parameters, but when conjoined, although the interval ranges for the parameters which are not in common between them remain unaltered, the interval ranges for the parameters in common are very likely restricted by their joint application. (Indeed, one could have imagined that the conjunction amounted to a complete theory of the intersection of the domains in question, and even on the union of the domains if the theories separately determine precise values for the parameters which were not in common between them.) Now there are more free parameters in the conjunction, so presumably, according to criterion (A), the conjunction is a more complex theory. But the conjunction is certainly a very much logically stronger theory, so according to criterion (C), it is actually simpler! But according to your criterion, the probability of some evidence E, given T_1 and T_2, might be higher or lower than E given T_1 alone, depending on what T_1 and T_2 say about the domain in question. Thus I see no connection between the three criteria whatsoever.

Rosenkrantz: I agree (A)–(C) are not equivalent, but they are surely connected. A model with parameters is logically the union of its special cases, and so logically weaker. This shows that (C) implies, not (A), but what I argued was the sound kernel of (A), viz. the weaker claim that you complicate a theory when you add a parameter to it. (The paper contains a counterexample to the stronger form of (A), namely, a one-parameter trinomial – whose one parameter is essentially a dead letter – which is simpler than the no parameter uniform trinomial.) Similarly, (B) is 'nearly' a special case of (A), if you think of the additional kinds as additional parameters. I did not set (A)–(C) up as criteria of adequacy; rather, my claim is that my explication of simplicity as sample coverage captures what I take to be the sound kernel of these three criteria, viz. that a theory is less simple than any of its special cases. The example Hooker sketched shows that a natural ceterus paribus clause is understood here: when you add a parameter to a theory *and do nothing else,* you complicate the theory. Obviously, if you add new parameters *and constrain old parameters,* a net simplification of theory may result.

PAUL TELLER

CONDITIONALIZATION, OBSERVATION,
AND CHANGE OF PREFERENCE*

INTRODUCTION

I take bayesianism to be the doctrine which maintains that (i) a set of
reasonable beliefs can be represented by a probability function defined
over sentences or propositions, and that (ii) reasonable changes of belief
can be represented by a process called conditionalization. Bayesians have
produced several ingenious arguments in support of (i); but the equally
important second condition they often seem to take completely for
granted. My main aim is to fill this gap in those bayesian positions which
characterize reasonable belief directly as a probability function. Thus,
what follows applies equally to the bayesian personalists' views which
characterize reasonable belief as having subjective sources and to views
such as that of Carnap which attempt to explicitly define a function which
characterizes the degree of belief it would be objectively reasonable for an
idealized rational agent to have in a given proposition in stated circumstances. Many frequentist views are also classifiable as bayesian, and I
will briefly discuss the justification of condition (ii) from the point of view
of a frequency interpretation of probability or reasonable degree of belief.
Two other topics will also come up along the way: the connection between change of belief and change of preference, and the connection between conditionalization and observation.

Throughout the discussion I will rely on several common bayesian
presuppositions. The object of study is a notion of belief, perhaps most
aptly described as degree of confidence, which can be ordered as to
strength and admits of quantatization. Such beliefs, or degrees of confidence, are assumed to reveal themselves in an agent's disposition to
make bets voluntarily or under duress. The agent whose beliefs are under
discussion is assumed to be an ideal logician, and the set of propositions
about which the agent has beliefs is assumed to be closed under all
logical operations. Also, the set of propositions for which the agent enter-
tains beliefs is assumed to be fixed. Quite clearly, when this assumption

*Harper and Hooker (eds.), Foundations of Probability Theory, Statistical Inference, and Statistical Theories of Science,
Vol. I, 205–259. All Rights Reserved. Copyright © 1976 by D. Reidel Publishing Company, Dordrecht-Holland.*

is violated, the bayesian model does not apply; and the most cogent arguments against conditionalization seem to turn on cases in which this assumption clearly fails (see, e.g., [10] *passim*). Finally, it is assumed that no such set of beliefs, taken together, are reasonable unless they can be described by a function satisfying the axioms of probability, where the numbers given by such a function can be taken to represent what the agent regards as fair betting quotients for those propositions on which it makes sense to bet. Clearly these assumptions are idealizations, and conclusions which depend on them can be applied to real cases only insofar as the idealizations are in relevant respects sufficiently close approximations to correct descriptions.

When an agent changes his beliefs his new beliefs may fail to be reasonable as a result of a number of different factors. For example, there may be nothing wrong in the way he changes his beliefs, but the old beliefs may have been unreasonable, and the new beliefs may inherit this irrationality. On the other hand, the old beliefs may have been perfectly reasonable, while the fault lies entirely with the pattern of change of belief. Preanalytically, there is a distinction to be made between the rationality of the method or pattern of change of belief, where the latter figures in but does not by itself determine the former. The relation is complex, as is illustrated by the example of an agent who starts with unreasonable beliefs and changes them by a method we would all applaud if his old beliefs had been reasonable. In such a case the new beliefs will still be correctly described as reasonable if the agent had no way of telling that his original beliefs were unreasonable, if he had every reasonable belief that, counter to fact, his original beliefs were perfectly reasonable. In view of this sort of complication, a precise account of the distinction between reasonable belief at a time and reasonable *change* of belief would require a precise way of treating an agent's second order beliefs about his beliefs and the second order rationality of the agent's belief about the rationality of his beliefs; and these tools are not yet available. However, the following rough and ready definition should make the distinction sufficiently clear for use in what is to follow:

0.1.1   A *change of belief is reasonable* if and only if
         (a) the new beliefs are reasonable,
         or

(b) both the new and the old beliefs are not reasonable,
but the new beliefs would have been reasonable
if (i) the old beliefs had been reasonable and (ii) both
before and after the change the agent has a high
reasonable degree of belief that his old degrees of
belief were reasonable.

A change from one set of beliefs to another will be said to be described by conditionalization on the proposition E if the old set of beliefs is described by the probability function P_0, and for every proposition A, in the domain of P_0, the new degree of belief in A is given by $P_n(A) = = P_0(AE)/P_0(E)$. Except in section 1.4, I will use $P(A/E)$ as a notational varient for $P(AE)/P(E)$; and throughout I will use 'is determined by', 'arises by', 'is given by' and 'takes place by' interchangeably with 'is described by', as stylistic varients in expressions such as 'change of belief is described by conditionalization'.

The problem of justifying conditionalization may now be loosely stated as the problem of supporting the claim that changes of belief described by conditionalization on some proposition which the agent has learned to be true (for example, by observation) are reasonable changes. More precisely, one might try to argue that all and only changes of belief by conditional are reasonal. Or one might try to show that all changes of belief by conditionalization are reasonable, while leaving it open whether or not other changes of belief might also be reasonable. Both of these options seem wholey unrealistic. Finally, one might argue that, under certain well specified conditions, only changes of belief by conditionalization are reasonable; that is that if any change of belief is reasonable, then such reasonable change of belief must be described by conditionalization on some proposition E. Most of the arguments which follow will be of this third form.

Finally it is to be noted that when change of belief is described by conditionalization on a proposition, E, the new degree of belief in E is 1, and it is often supposed that $P(E) = 1$ when the agent *knows* that E is true. I regard ascription of degree of belief of 1, and the associated concepts of certainty and knowledge, as at best idealizations; and fortunately there is a generalization of conditionalization which requires none of them. However, in Part I will assume that agents do gain knowledge which makes degree of belief of 1 reasonable, and that only knowledge makes degree of belief of 1 reasonable. I make these assumptions here because I wish to present the arguments in a form suitable for those who accept them and because the arguments of Part II, which do not require these assumptions, are easiest to present as straightforward generalizations of the arguments given in Part I.

PART I

1.1. *Introduction*

There are four well known types of arguments for the probability axioms: (1) Arguments from frequency definitions of probability; (2) 'Dutch Book' arguments which show that if an agent's belief function violates the probability axioms, his betting rates make him susceptible to forming collections of bets on which he will lose no matter what happens (e.g. [3], pp. 102–4); or for which no matter what happens he may lose but cannot gain (e.g. [9]); (3) Arguments from qualitative assumptions about belief and preference (e.g. [8], pp. 6–43); and (4) Arguments from assumptions about certain continuity properties of an agent's belief function (e.g. [2], [4]). In this part I will discuss arguments for conditionalization which are analogues of the first three types of arguments for conditionalization. (At the present time I have not investigated the possibility of analogues of the fourth kind of argument.) Sections 1.2 and 1.3 will very briefly examine the frequency and Dutch Book type of arguments. Sections 1.4 to 1.9 will develop in detail a pattern of argument which rests on a qualitative assumption about inductive reasoning. Sections 1.4 and 1.8 will also explore related questions concerning change of preference and questions concerning observation.

1.2. *Frequency Arguments*

If probability or reasonable degree of belief is defined in terms of relative frequency, the claim that new probabilities or reasonable degrees of belief are given by conditionalization on the old ones follows trivially from the definitions, though details differ with differing frequentist analyses. I will here give only a brief sketch of how such arguments proceed. Note that in this section capital letters are used as parameters for properties instead of, as elsewhere, propositions.

On a frequentist view the probability or reasonable degree of belief that a case has property A is defined as the frequency (observed or 'limiting') of cases that have property A among some non-empty reference class of cases which have property R. This will be written

$$P_R(A) =_{\text{def}} F(AR/R).$$

Suppose that the agent comes to know that (and no more than that) the case in question has property B as well as property R. Then, on any reasonable analysis of 'reference class' the reference class becomes the class of all cases which have both property R and property B. Thus the new probability or reasonable degree of belief is

(1.2.1) $P_{RB}(A) = F(ARB/RB).$

If the function F is being interpreted as a finite frequency we note that the reference class was assumed non-empty, and since the agent knows the case in question has property B, the class of cases which have both properties B and R, is non-empty. If F is interpreted as a limiting frequency we need to assume that $F(BR/R) \neq 0$. Using this assumption if needed it follows from (1.2.1) that

$$P_{RB}(A) = F(ARB/R)/F(BR/R)$$
$$= P_R(AB)/P_R(B).$$

Consequently, the new probability for the case having property A is given by conditionalization from the old, provided that the observed property B did not initially have probability zero, interpreted as a limiting relative frequency.

1.3. *The Dutch Book Argument*

In this section I report a version of the Dutch Book Argument devised by David Lewis.[1] Let P_0 and P_n be, respectively, the agent's old belief function at time 0 and his new belief function at time n. For this argument we must assume that the agent's belief function, P, represents his betting rates, so that, for any proposition A, in the domain of P, $P(A)$ is the price for which he would be indifferent between buying or selling the bet $\begin{bmatrix} 1 \text{ if } A \\ 0 \text{ otherwise} \end{bmatrix}$; and we must assume that the domain of P_0 and P_n includes a set $\{E_i\}_{i\in I}$ of mutually exclusive and jointly exhaustive propositions that specify, in full detail, all the alternative courses of experience the agent might undergo between time 0 and time n. Let $\{E_i\}_{i\in I'}$ be the subset of $\{E_i\}_{i\in I}$ such that $P_0(E_i)>0$ for $i\in I'$. A version of the Dutch Book Argument due to Abner Shimony [8] supports the claim that $I=I'$. If $I=I'$ is not assumed, the following argument supports change of belief by conditionalization for cases in which some E_i, $i\in I'$ occurs, while giving no information for cases in which some E_i, $i\notin I'$ occurs. The argument shows that if at time 0 the agent knows, for some $i\in I'$ what his new belief function, P_n will be if E_i should turn out to be true, and if for such an $i\in I'$ the function to be adopted, if E_i turns out to be true, is not $P_n(A)=$ $=P_0(A/E_i)=_{\text{def}}P_0(AE_i)/P_0(E_i)$, then a bookie who knows no more nor less than the agent can induce the agent to buy and sell bets on which he will have a net loss whatever happens. If one agrees that any plan for changing belief is unreasonable if it makes one vulnerable in this way to certain loss, the conclusion can be summarized thus: No explicitly formulated plan for changing beliefs in the face of new evidence is reasonable unless, for any $i\in I'$ for which the plan specifies the beliefs to be adopted should E_i occur, the plan calls for changing beliefs by conditionalization on E_i if E_i occurs.

The bookie's system for exploitation of the non-conditionalizing agent proceeds as follows: Suppose, for some $i\in I'$ and A, the agent plans to have new beliefs $P_n(A)<P_0(A/E_i)$ if E_i turns out to be true. Let $x=P_0(A/E_i)$ and $y=P_0(A/E_i)-P_n(A)$.
At time 0 the bookie sells the agent the bets

(a) $\qquad \begin{bmatrix} 1 \text{ if } AE_i \\ 0 \text{ otherwise} \end{bmatrix}$

(b) $\qquad \begin{bmatrix} x \text{ if } \bar{E}_i \\ 0 \text{ otherwise} \end{bmatrix}$

and

(c) $\qquad \begin{bmatrix} y \text{ if } E_i \\ 0 \text{ otherwise} \end{bmatrix}$

for the maximum price he will pay, namely $P_0(AE_i)+xP_0(\bar{E}_i)+yP_0(E_i)=$ $=P_0(A/E_i)+yP_0(E_i)$. If E_i turns out to be false the agent wins $P_0(A/E_i)$ on bet (b) and has a net loss of $yP_0(E_i)$. If E_i turns out to be true the bookie

buys back the bet $\begin{bmatrix} 1 \text{ if } A \\ 0 \text{ otherwise} \end{bmatrix}$ for the minimum price the agent will pay.

By hypothesis this is $P_n(A) = P_0(A/E_i) - y$. The agents wins y on bet (c). His total gain is $P_0(A/E_i) - y + y$, his total loss is $P_0(A/E_i) + yP_0(E_i)$, and he again has a net loss of $yP_0(E_i)$. Since, by assumption, $P_0(E_i) > 0$ and $y > 0$, the agent loses the positive amount $yP_0(E_i)$ whatever happens. If the agent plans new beliefs $P_n(A) > P_0(A/E_i)$ instead of $P_n(A) < P_0(A/E_i)$ the argument proceeds in exactly the same way except that the bookie buys (sells) where in the foregoing argument he sells (buys).

The argument can be rephrased by noting that bets (a) and (b), taken together can be viewed as a 'conditional bet' on which no one has net gain or loss if E_i turns out to be false and which becomes a bet on A if E_i turns out to be true. The bookie sells the agent this conditional bet and also (c), a bet at small stakes on E_i. If E_i turns out to be false the agent neither gains nor loses on the conditional bet, but loses on bet (c). If E_i turns out to be true, the bookie buys back the (now no longer conditional) bet on A at a reduced rate. The agent wins on bet (c), but the bookie has been careful to set the stakes small enough on (c) so that those winnings do not offset the agent's loss.

The bets (a) and (b) are just the ones used in the well known application of the Dutch Book argument which shows that to avoid vulnerability to certain loss one must use $P_0(AE_i)/P_0(E_i)$ as one's betting rate on the bet conditional on E_i, to be called off if E_i is false and to become a bet on A if E_i is true. Previous authors (e.g. Hacking [5], p. 315) concluded that this fact about conditional betting rates could not be applied in a Dutch Book type of argument to reach conclusions about change of belief. They were mistaken because they failed to consider application of the argument to a set of propositions meeting the special conditions specified above for $\{E_i\}_{i \in I'}$. Indeed it is easy to see that the above pattern of argument applies *only* to the propositions of $\{E_i\}_{i \in I'}$. Suppose a proposition F is incompatible with all the members of $\{E_i\}_{i \in I'}$. Then, by hypothesis $P_0(F) = 0$, and the fractions $P_0(AF)/P_0(F)$ needed for the argument don't exist. Suppose F implies E_i, for some $i \in I'$, but E_i does not imply F. If the bookie is to exploit the agent by making the relevant bets on F, he must be able to determine whether or not F is true. But since the E_i, $i \in I$ are assumed to describe, in full detail the various courses of experience which the agent might undergo between time 0 and time n, the agent learns only whether E_i is true while the bookie learns whether F is true. Exploitation by dint of such greater knowledge or keener powers of observation shows nothing derogatory about the agent's plan for change of belief. Suppose F is compatible with two or more members of $\{E_i\}_{i \in I'}$, but is not merely the disjunction of two or more members of $\{E_i\}_{i \in I'}$. Then there is at least one possible out come in which the bookie will not be able to determine whether F is true merely by knowing which member of $\{E_i\}_{i \in I}$ is true. Consequently the bookie cannot *whatever happens* take advantage of the agent without knowing more or having keener powers of observation. As the final possibility, F might be a disjunction, $F = \bigvee_{j \in J} E_j$, where $J \subset I'$. Then, either $P_0(A/E_j) = P_0(A/F)$ for all $j \in J$, in which case the argument applied to F yields no different results than when applied to members of $\{E_i\}_{i \in I'}$. Or, for some k, $l \in J$, $P_0(A/E_k) < P_0(A/F) < P_0(A/E_l)$. In this case, the bookie cannot inflict sure loss because he cannot tell for sure whether he should buy or sell the initial bets (a)–(c).

1.4. *Conditionalization and Change of Preference*

A prima facie different approach to the problem of justifying conditionalization is suggested by the personalist formulations which treat desirability and degree of belief simultaneously. I will outline the relevant considerations here in terms of Savage's presentation in *Foundation of Statistics* ([8], Chapeters 2 and 3 *passim*). Savage assumes that a decision problem in face of uncertainty can be described in terms of *possible states of the world, consequences,* and *acts.* A possible state of the world is specified by a conjunction of propositions which is not known to be false and which fully describes every condition which is believed possibly relevant to the problem situation. A consequence is, in Savage's words "anything that happens to a person;" consequences are, in some sense, "ultimate goods," and for our purposes they may be taken to be specified by conjucts, C, of propositions which are maximal in the following sense: for every contingent proposition, A, such that $C \& A$ is consistent with what the agent knows, the agent is indifferent to the prospect that C is true and the prospect that $C \& A$ is true. An act, in turn, is specified by a function, f, which maps every state of the world, s, into a consequence, $f(s)$, such that it is within the agent's power to make all propositions of the form 'If s, then $f(s)$' true. Savage supposes that the agent has preferences between these acts; if the agent prefers act g to act f, this is indicated by writing $f<g$. Savage presents six postulates for this preference ordering the relevant consequences of which are described below.

Most important for the present considerations is the definition of the relation g is *preferred to* f, *given* B, written $f<g$, *given* B. For any two acts, f, and g, and any proposition B, the postulates guarantee that there will also be two acts, f' and g', which have the following property: For every state of the world, s, which is in (i.e. implies) the proposition B, f' has the same consequences as f, and g' has the same consequences as g; but for all states of the world not in B, f' has the same consequences as g'. In short,

$$(1.4.1) \quad f'(s)=f(s) \quad \text{for} \quad s \text{ in } B$$
$$g'(s)=g(s) \quad \text{for} \quad s \text{ in } B$$
$$\text{and} \qquad f'(s)=g'(s) \quad \text{for} \quad s \text{ in } \bar{B}$$

The desired relation is then defined by

$$(1.4.2) \quad f<g \text{ given } B \quad \text{if and only if} \quad f'<g'.$$

The postulates also guarantee that definition (1.4.2) does not depend on which acts f' and g' satisfying (1.4.1) are used in definition (1.4.2), so that $f<g$ given B is always well defined.

The ordering $f<g$ can be used to define a qualitative ordering of beliefs. We say that *an act* f_A *constitutes a prize in case proposition A obtains*, if there are fixed consequences c and c', such that

$$f_A(s)=c \quad \text{for} \quad s \text{ in } A$$
$$f_A(s)=c' \quad \text{for} \quad s \text{ in } \bar{A}$$

and the agent prefers c to c'. Let f_B constitute the same prize in case of proposition B, i.e.

$$f_B(s)=c \quad \text{for} \quad s \text{ in } B$$
$$f_B(s)=c' \quad \text{for} \quad s \text{ in } \bar{B}$$

Then the relation $A<B$ (the agent has greater (qualitative) belief in B than in A) is defined by

$$A<B \quad \text{if and only if} \quad f_A<f_B.$$

The postulates guarantee that the relation $A < B$ is well defined for all propositions A and B, and that there is a unique quantitative ordering which corresponds to the qualitative ordering and which is a probability measure. That is, there is a unique function, P, from propositions to real numbers such that

> $A < B$ if and only if $P(A) < P(B)$, and P is a probability measure.

The postulates also guarantee that for any given proposition, C, such that $P(C) \neq 0$, there is a qualitative belief ordering among propositions, $A < B$, *given* C, which can be defined in terms of the preference relation among acts, $f < g$, given C, in exactly the same way that $A < B$ was defined in terms of $f < g$. Corresponding to the relation $A < B$, given C, there is a unique function, $P(A/C)$ from propositions A, to the real numbers such that

> $A < B$ given C if and only if $P(A/C) < P(B/C)$, and $P(A/C)$, as a function of A, is a probability measure.

Finally, it is a consequence of the postulates that $P(A/C) = P(AC)/P(C)$. It is important to bear in mind that in Savage's system this equation is not a definition: $P(A/C)$ is independently defined in terms of the qualitative conditional belief ordering, which is in turn defined in terms of the conditional preference ordering among acts.

Given any problem situation which is described in terms of Savage's framework, it is tempting to argue in the following way. Consider an agent who has a reasonable set of preferences and who is initially unsure of the truth of E. He then comes to know that E is true, as the direct result of an observation. And suppose that coming to have observational knowledge that E is true is the only epistemic change which takes place. Under these conditions what change is it reasonable for the agent to make in his preference ordering among acts? If before the observation the agent preferred g to f given E, it now seems reasonable for him to prefer g to f unconditionally. For $f < g$ given E meant that there were two acts, f' and g', such that g' was preferred to f'; such that for every state of the world in E, f, and f' had the same consequences and g and g' had the same consequence; and such that for every state of the world in $\bar{E}$, f' and g'

had the same consequences. But when the agent discovers that $\bar{E}$ is false, this should not affect his preference between f' and g', for these had the same consequences in $\bar{E}$. In other words, the result of the agent's coming to know E can be viewed simply as the agent's ruling out certain states of the world as possible, namely those states which are in $\bar{E}$; and ruling out these states should not affect his preference between f' and g' since these have indentical consequences on the states which are ruled out. Finally, it is to be observed that the new preference between f and g should be the same as the new preference between f' and g' since in all states which are now viewed as possible f and f' agree and g and g' agree. In sum, the new preference between f and g should be the same as the new preference between f' and g', and the new preference between f' and g' should be the same as the old preference between f' and g'; so the new preference between f and g should be the same as the old preference between f and g given E. And this holds for all f and g. However, the probability measure determined by the relation $f<g$ given E is the probability measure determined by the relation $f<g$, conditionalized on the proposition E. So if, as I have argued, it is reasonable for the agent's new preference ordering to be identical to the old preference ordering given E, then a reasonable set of new beliefs will be obtained by conditionalization on the former probability measure.[2]

However, there is a step in this argument which is fairly said to beg the the question at issue. It was claimed that if f' and g' agree on $\bar{E}$ (i.e. have the same consequences for the states in $\bar{E}$) then learning that $\bar{E}$ is false should not affect the agent's preference between f' and g', and the argument given for this claim was that learning E to be true is no more nor less than ruling out as possible the states in $\bar{E}$. But, the objection runs, it is possible to rule out the states of $\bar{E}$ in more than one way – they may be ruled out in such a way as to leave the relative degree of belief (as measured by ratios) in the states of E unchanged, and they may be ruled out in a way which shifts these relative degrees of belief. And it is easy to see that, on the assumption that the new degree of belief in E is 1, the new degrees of belief arise from the old by conditionalization on E if and only if the change of belief leaves unchanged all the ratios of degrees of belief of the states in E, or equivalently, all propositions which imply E. Consequently, the conclusion that the change of belief can be described by conditionalization on E follows immediately from the assumption of constancy of ratios of degrees of belief in the propositions which imply E, and making this assumption begs the question at issue.

It remains to complete these remarks by demonstrating the claimed equivalence of change of belief by conditionalization on E, and constancy of ratios of degrees of belief in propositions which imply E. More exactly let P_0 and P_n, respectively, be the functions representing the old and new degrees of belief, and let A, B, and E range over the propositions in the domain of the agent's belief function. We have to prove that

(1.4.3)   If $P_0(E)>0$ and $P_n(E)=1$, then
(a) For all A, $P_n(A)=P_0(A/E)$

if and only if

(b) For all A, B, each of which imply E,
if $P_0(A)=P_0(B)=0$, then
$P_n(A)=P_n(B)=0$; and if one, say
$P_0(B)\neq0$, then $P_0(A)/P_0(B)=P_n(A)/P_nB$.

Suppose that $P_0(E)>0$ and $P_n(E)=1$.
Suppose (1.4.3, b) that the ratios of degrees of belief in propositions which imply E remain constant. Then, for any A

$$P_n(A)=P_n(AE) \quad \text{(since } P_n(E)=1\text{)}$$
$$=P_n(AE)/P_n(E)$$
$$=P_0(AE)/P_0(E)$$
$$=P_0(A/E).$$

To prove the converse, let A and B be any two propositions in the range of P and P_n which imply E. Suppose (1.4.3, a) that P_n arises by conditionalization from P_0. If $P_0(A)=P_0(B)=0$, then

$$P_n(A)=P_0(A/E)=P_0(AE)/P_0(E)=0$$
and
$$P_n(B)=P_0(B/E)=P_0(BE)/P_0(E)=0.$$

Otherwise we may assume without loss of generality that $P(B)\neq0$. Since A and B each imply E, $P_0(AE)=P_0(A)$ and $P_0(BE)=P_0(B)$. So

$$P_0(A)/P_0(B)=P_0(AE)/P_0(BE)$$

$$= \frac{P_0(AE)}{P_0(E)} \bigg/ \frac{P_0(BE)}{P_0(E)}$$
$$= P_0(A/E)/P_0(B/E) =$$
$$= P_n(A)/P_n(B), \quad \text{by the assumption of } (1.4.3, \text{ a}).$$

Although this connection between conditionalization and ratios of degrees of belief is trivial, it deserves explicit mention because of the role it plays in undercutting the change of preference approach to the justification of conditionalization. And more importantly, as we shall see in the next sections, this trivial fact provides the pivot point for a new approach to the justification of conditionalization by appeal to qualitative assumptions on the nature of reasonable belief and change of belief.

1.5. *Inexact Sketch of an Argument*

Let us take the probablist's standard case par excellence, that of repeated tosses of a given coin. An individual event is the outcome H or T of a given toss. A possible state of the world (of this restricted problem situation) is a sequence of outcomes such as $HTTH...$[3]. These sequences may be of a fixed finite length n if only n tosses of the coin are planned, or they may be taken to be infinitely extendable. To raise the problem of conditionalization we suppose that a prior set of beliefs are given and that they are reasonable. What happens now when the coin is tossed for the first time and is observed to come up, say, heads? The agent learns by direct observation that certain sequences are excluded, namely all the initial sequences which begin with T as the outcome of the first toss. Furthermore, it is plausible to suppose that in such a case

> The only thing, or the only relevant thing the agent learns by direct observation is that H occurred on the first toss.

But if the agent learned that heads occurred on the first toss, his new degree of belief for heads on the first toss, to be reasonable, should be 1. And if this is the only thing, or the only relevant thing he learned then it seems that he acquires from his observation no reason to change his relative degree of belief in any of the initial sequences not ruled out by the observation, that is initial sequences beginning with H. However, if he acquires no reason to change any of these relative degrees of belief, then presumedly his new degrees of belief should, to be reasonable, exhibit the same relative degrees of belief. Since any proposition in the domain under consideration which implies 'H on the first toss' can be expressed as a disjunction of such initial sequences, the new relative degrees of belief among all propositions implying H should, to be reasonable, be the same as the old. But we have seen that, if the new degree of belief in H is 1, this condition is equivalent to the condition that the new degrees of belief be given by conditionalization of the old degrees of belief on the proposition that H occurs on the first toss.

I suspect that such considerations underly the felt obviousness of conditionalization. Though attractive, this argument sketch leaves much to be clarified and made explicit. In particular it makes assumptions about the nature of observation and qualitative principles of reasoning. It also assumes that relative degree of belief is to be measured by ratios, an assumption which turns out to be wholly eliminable. We will now make the argument both general and precise first by stating a qualitative condition on change of belief which, under certain background assumptions is equivalent to conditionalization; and then by describing the circumstances under which it is plausible to suppose that this condition holds.

1.6. *The Equivalence of Conditionalization and a Qualitative Condition on Change of Belief*

I turn now to characterizing conditionalization in terms of a qualitative equivalent. I will throughout suppose that a change of belief takes place, that P_0 describes the agent's beliefs before the change, and that P_n describes his beliefs after the change. I will also throughout suppose that propositions range over just those propositions which are in the agent's domain of beliefs. I will say that proposition E has condition C, or $C(E)$ for short, just in case the agent's belief in E changes from something greater than zero to unity, and furthermore, for any two propositions, A and B, each of which logically implies E, if A and B are believed equally before the change, then they are believed equally after the change. If we use '$\Rightarrow$' to mean 'logically implies', then $C(E)$ is defined precisely by

$$(1.6.1) \quad C(E) \equiv_{\text{def}} 0 < P_0(E) < 1 \ \& \ P_n(E) = 1 \ \& \ (\forall A)\,(\forall B)\,[(A \Rightarrow E \ \& \times \\ \times\, B \Rightarrow E \ \& \ P_0(A) = P_0(B)) \rightarrow P_n(A) = P_n(B)].$$

Note that although this condition has been stated in terms of the equality of two quantitative degrees of belief, no more is really required than that if certain of the agent's beliefs are qualitatively equal before the change, that is if neither is stronger than the other, then they are qualitatively equal at the conclusion of the change. This qualitative condition on change of belief, together with the assumption we are making throughout, that at a fixed time the agent's beliefs can be represented by a probability measure, will suffice for what follows.

I will use 'Cond(E)' to state that change of belief takes place by conditionalization on E; more exactly,

(1.6.2) $\text{Cond}(E) \equiv_{\text{def}} 0 < P_0(E) < 1 \,\&\, (\forall A)\,[P_n(A) = P_0(A/E)].$

I will prove that, in the presence of certain further assumptions, $C(E)$ is equivalent to Cond(E), for any E.

First I will show that

(1.6.3) For all E, if Cond(E), then $C(E)$

For an arbitrary E, assume Cond(E). By definition of 'Cond' $0 < P_0(E) < 1$ and $(\forall A)\,(P_n(A) = P_0(A/E))$. In particular, $P_n(E) = P_0(E/E) = 1$. And for any A, B, each of which implies E,

$$P_n(A) = P_0(A/E) =_{\text{def}} P_0(AE)/P_0(E) = P_0(A)/P_0(E).$$
$$P_n(B) = P_0(B/E) =_{\text{def}} P_0(BE)/P_0(E) = P_0(B)/P_0(E).$$

So if $P_0(A) = P_0(B)$, it follows that $P_n(A) = P_n(B)$.

To prove the converse, I first prove the lemma that for all E,

(1.6.4) If $0 < P_0(E) < 1$ and $P_n(E) = 1$ and
$$(\forall A)\,(P_0(A/E) = P_n(A/E)), \quad \text{then} \quad \text{Cond}(E).$$

Let A be an arbitrary proposition; and assume the antecedent of (1.6.4):

$$
\begin{aligned}
P_0(A/E) &= P_n(A/E) \\
&=_{\text{def}} P_n(AE)/P_n(E) \\
&= P_n(AE) \quad \text{(because } P_n(E) = 1 \text{ is assumed)} \\
&= P_n(A) - P(A\bar{E}) \\
&= P_n(A) \quad \text{(again because } P_n(E) = 1 \text{ is assumed).}
\end{aligned}
$$

To finish proving the equivalence of $C(E)$ and Cond(E) we need to prove that the antecedent of (1.6.4) follows from $C(E)$; and, as can be demonstrated by simple counterexamples, this can be done only in the presence of some further assumption. I will proceed by giving, in outline only, a very simple proof, which uses a strong further assumption. I will then provide a detailed proof which uses a weaker assumption. The first proof is mentioned because of its greater mathematical elegance and because it will enable those with background in mathematics to see very quickly what is 'really going on' in the second proof. The second proof is presented because it establishes the conclusion under an assumption weak enough to make the result useful to the bayesian characterization of treatment of change of belief.[4]

The first proof requires the definition

(1.6.5)   The agent's domain of beliefs will be said to be *full* if and only if for every number q and every proposition A in the domain such that $P_0(A) = r$ and $0 \leqslant q \leqslant r$, there is a proposition B, such that $B \Rightarrow A$ and $P_0(B) = q$.

Recall that the agent's domain of beliefs is assumed throughout to be closed under all logical operations and so forms a Boolian field. It is now easy to prove that for any E,

(1.6.6)   If the agent's domain of beliefs is full and if $C(E)$, then $(A) (P_0(A/E) = P_n(A/E))$.

Let E be given. Assume $C(E)$ and that the agent's domain of beliefs is full. In the presence of the fullness assumption the last conjunct of $C(E)$ is equivalent to the existence of a function, g, defined on $[0, P_0(E)]$ such that if $A \Rightarrow E$, then $g[p_0(A)] = P_n(A)$. Since P_0 and P_n are probability measures, g is a positive function and is additive, i.e., for positive x, y such that $0 < x + y < P_0(E)$, $g(x+y) = g(x) = g(y)$. It follows, though not completely trivially, that there is a constant, k, such that for all arguments, x, $g(x) = kx$ (the proof is a slight modification, devised by Arthur Fine, of the proof of Theorem 1, p. 34 in Aczél [1]). So for arbitrary A, and using the first two conjuncts of $C(E)$ which give $P_0(E) \neq 0$ and $P_n(E) \neq 0$, we have

$$
\begin{aligned}
P_n(A/E) &=_{\text{def}} P_n(AE)/P_n(E) \\
&= g[P_0(AE)]/g[P_0(E)], \quad \text{since} \quad AE, E \text{ each imply } E \\
&= kP_0(AE)/kP_0(E) \\
&= P_0(AE)/P_0(E) \\
&=_{\text{def}} P_0(A/E).
\end{aligned}
$$

The second proof will use the definition

(1.6.7) The agents domain of beliefs is *full enough at E* if and only if
    (i) If $B \Rightarrow E$ and $P_0(B) = (r/s) P_0(E)$, for integers r, s, then there is a sequence of propositions $\{X_i\}_{i=1}^s$ such that, for $1 < i < j < s$,
        (ia) $X_i X_j$ is logically false
        (ib) $\bigvee_{i=1}^s X_i$ is logically equivalent to E
        (ic) $P_0(X_i) = P_0(X_j)$.
    (ii) If $B \Rightarrow E$ and $P_0(B) = t P_0(E)$, t an irrational number, then there are four infinite sequences of propositions $\{X_i\}, \{X_i'\}, \{Y_i\}$ and $\{Y_i'\}$, each proposition of which implies E such that:
        (iia) $P_0(X_i) \to P_0(B)$ from below and $P_0(Y_i) \to P_0(B)$ from above, as $i \to \infty$
    and for all i,
        (iib) $P_0(X_i)$ and $P_0(Y_i)$ are rational multiples of $P_0(E)$
        (iic) $P_0(X_i \vee X_i') = P_0(B) = P_0(Y_i \bar{Y}_i')$.

I will now prove that, for all E.

(1.6.8) If the agent's domain of beliefs is full enough at E, and if $C(E)$, then $(A) (P_0(A/E)) = P_n(A/E))$.

Let E be given, suppose that $C(E)$, and that the agent's domain of beliefs is full enough at E. Since $C(E)$, $P_0(E) \neq 0$ and $P_n(E) \neq 0$. Finally, let A be given. The proof falls exhaustively into two cases:

Case 1: $P_0(AE) = (r/s) P_0(E)$, for integers r, s. Since the agent's domain of beliefs are assumed full enough at E, there is a sequence of propositions $\{X_i\}_{i=1}^s$ satisfying (1.6.7; ia–ic) with 'AE' substituted for 'B'. By the assumption of $C(E)$ and (1.6.7, ic) $P_n(X_i) = P_n(X_j)$ for $1 < i < j < s$. The desired result is now at hand because r of the s propositions of $\{X_i\}_{i=1}^s$ give the value for AE as r/s times the value for E both before and after the change of belief. More exactly, $\{X_i\}$ is a logically exclusive and exhaustive partition of E (1.6.7, ia, ib) and for each i, $1 < i < s$, $P_0(X_i) = (1/s) P_0(E)$ and $P_n(X_i) = (1/s) P_n(E)$. So

$$P_0\left(\bigvee_{i=1}^r X_i \right) = \frac{r}{s} P_0(E)$$

and

$$P_n\left(\bigvee_{i=1}^{r} X_i\right) = \frac{r}{s} P_0(E).$$

But by assumption of case 1, $P_0(AE) = (r/s) P_0(E)$. So

$$P_0(AE) = P_0\left(\bigvee_{i=1}^{r} X_i\right)$$

and by $C(E)$,

$$P_n(AE) = P_n\left(\bigvee_{i=1}^{r} X_i\right)$$

Putting the last four lines together we get $P_0(AE) = (r/s) P_0(E)$ and $P_n(AE) = (r/s) P_n(E)$. So

$$P_0(A/E) =_{\text{def}} P_0(AE)/P_0(E) = (r/s) = P_n(AE)/P_n(E)$$
$$=_{\text{def}} P_n(A/E)$$

which was to be shown.

Case 2: $P_0(AE) = t P_0(E)$, t an irrational number. Again, by the assumption that the agent's beliefs are full enough at E, there are four sequences of propositions, $\{X_i\}$, $\{X_i'\}$, $\{Y_i\}$ and $\{Y_i'\}$ satisfying the conditions of (1.6.7, iia–iic) with 'AE' substituted for 'B'. By the assumption of $C(E)$ and (1.6.7, iic) with 'AE' substituted for 'B' we have, for each i,

$$P_n(X_i \vee X_i') = P_n(AE) = P_n(Y_i \bar{Y}_i')$$

Using the inequalities $P_n(X_i) \leqslant P_n(X_i \vee X_i')$ and $P_n(Y_i \bar{Y}_i') \leqslant P_n(Y_i)$, the last line, and dividing through by $P_n(E)$, we get, for each i,

$$\frac{P_n(X_i)}{P_n(E)} \leqslant \frac{P_n(X_i \vee X_i')}{P_n(E)} = \frac{P_n(AE)}{P_n(E)} = \frac{P_n(Y_i Y_i')}{P_n(E)} \leqslant \frac{P_n(Y_i)}{P_n(E)}$$

By (1.6.7, iib), for each i $P_0(X_i)/P_0(E)$ and $P_0(Y_i)/P_0(E)$ are rational numbers; also X_i implies E, and Y_i implies E. So, by case 1, for all i,

$$\frac{P_0(X_i)}{P_0(E)} = \frac{P_n(X_i)}{P_n(E)} \quad \text{and} \quad \frac{P_0(Y_i)}{P_0(E)} = \frac{P_n(Y_i)}{P_n(E)}$$

By (1.6.7, iia), with 'AE' substituted for 'B' and the last line,

$$\frac{P_n(X_i)}{P_n(E)} \rightarrow \frac{P_0(AE)}{P_0(E)} \quad \text{from below as } i \rightarrow \infty \text{ and}$$

$$\frac{P_n Y_i}{P_n(E)} \rightarrow \frac{P_0(AE)}{P_0(E)} \quad \text{from above as } i \rightarrow \infty.$$

This together with the one but last line gives

$$P_0(A/E)/P_0(E) = P_n(AE)/P_n(E),$$

i.e. $P_0(A/E) = P_n(A/B)$, which was to shown.

Putting together (1.6.3), (1.6.4), and (1.6.8) it follows for every proposition, E, in the agent's domain of beliefs that, if the agent's domain of beliefs is full enough at E, then Cond(E) if an only if $C(E)$. Henceforth I will assume that the agent's domain of beliefs is full enough at every E in his domain of beliefs, and I will call this the limited fullness assumption. Under this assumption it follows from (1.6.3), (1.6.4) and (1.4.8) that

(1.6.9)   For every E in the agent's domain of beliefs Cond(E) if and only if $C(E)$.

The limited fullness assumption seems to me to be a less severe assumption than it will at first appear to the reader. Let E be given and consider any proposition B which implies E and such that $P_0(B) = = (r/s) P_0(E)$ for integers r and s. Suppose that there is some randomizing device such as a coin or a die which has outcomes $\{U_i\}_{i=1}^s$ which the agent regards as equiprobable and independent of the truth of B before and after he comes to know that E is true. Then the sequence $\{X_i\}_{i=1}^s$ which must exist for the agent's beliefs to be full enough at E according to definition (1.6.7) are just the propositions $\{BU_i\}_{i=1}^s$. As for sequences of propositions as described in (1.6.7, ii), the limited fullness assumption requires them only in case there is a proposition, B, in the agent's domain of beliefs such that $P_0(B)/P_0(E)$ is irrational; and for a rational agent this will be an exceptional situation. It seems plausible to suppose that this will happen only when there is some operation, such as taking a logarithm or square root or calculating the area of a circle, which is entering into the considerations of the agent. But if some such specific operation is in question, it again seems plausible to suppose that it can be employed to set up certain randomizing devices with outcomes described by the sort of propositions required for limited fullness. It is hard to see how this contention could be supported in general, but I will illustrate it in terms of a specific example. Suppose that there are propositions B and E in the agent's domain of beliefs such that B implies E and $P_0(B)/P_0(E) = 1/\sqrt{2}$. Let k_i be a sequence of rational numbers that converge to $1/\sqrt{2}$ from below; for example, k_i could be taken to be the ith decimal expansion of $1/\sqrt{2}$ rounded down. For a given i, mark off a segment of length $k_i\sqrt{2}$. This can be done, for example, by constructing a square with side of length k_i and taking the square's diagonal as the segment. Then take a roulette wheel with unit circumference and fashioned with a pointer, which the agent regards as balanced and the outcomes of which the agent regards as independent of the truth of B, both before and after E is found to be true. Color the constructed line segment green and lay it on the circumference of the roulette wheel, and color the remaining portion of the circumference red. Spin the roulette wheel. Then X_i is the proposition that B-and-the-pointer-comes-to-rest-on-the-green-portion-of-the-circumference. X_i' is the proposition that B-and-the-pointer-comes-to-rest-on-the-red-portion-of-the-circumference. To obtain the propositions Y_i and Y_j', let j_i be a sequence that approaches $1/\sqrt{2}$ from above, where j_1 is sufficiently close to $1/\sqrt{2}$. For a given j_i, use geometrical methods to construct line segements of length $(j_i - 1/\sqrt{2})/(1 - 1/\sqrt{2})$. Color this line segment green and lay it along the circumference of a roulette wheel as in the previous case. Spin the wheel. Then Y_i is the proposition that B-or-($\bar{B}$-and-E-and-the-pointer-comes-to-rest-on-the-green-portion-of-the-circumference). Y_i' is the proposition that $\bar{B}$-and-E-and-the-pointer-comes-to-rest-on-the-green-portion-of-the-circumference.

Although these existence assumptions are weak, they still involve a considerable degree of idealization. Most of the required randomizing devices do not actually exist, and certainly not infinitely many of them. So real agents do not have most of the propositions describing the outcomes produced by such devices in their actual belief structures. However, the strength of the assumptions can be drastically reduced in the following way. All we really need to assume is the truth of certain counterfactual conditionals. For every B and E such that B implies E and $P_0(B)/P_0(E)$ is irrational, and for every integer n, we need to assume that

> if there were n randomizing devices of the kind described, and $\{X_i\}_{i=1}^n$ were in the agent's belief structure, the agent's relative degrees of belief $P_0(B)/P_0(E)$ and $P_n(B)/P_n(E)$ would

be unaffected either by the existence of the devices or by
learning the truth values of the propositions in $\{X_i\}_{i=1}^n$; and
$C(E)$ would hold in the expanded belief structure if it held in
the original belief structure.

and similarly for the propositions in $\{X_i'\}$, $\{Y_i\}$, and $\{Y_i'\}$. Since it is easy
to imagine randomizing devices which, if they were to exist, would have
the required properties of independence from the agent's beliefs, all such
counterfactuals seem quite clearly to be true.

Finally, it is to be remarked that the existence assumptions for the case
in which $P_0(B)/P_0(E)$ is rational can be similarly exchanged for the
assumption of the truth of corresponding counter-factuals.

1.7. *A Qualitative Principle of Inductive Reasoning and the Justification of
Conditionalization.*

How might the equivalence of condition C and conditionalization be of
help in justifying conditionalization? In the first place the equivalence
should help us see to what we are committed when we embrace conditionalization as describing reasonable change of belief, for condition C is,
psychologically speaking, a much simpler condition than conditionalization. I say C is simpler because (1) it is a qualitative rather than a
quantitative condition on change of belief, (2) it specifies how new beliefs
are related to old only in the highly restricted case in which, initially, two
beliefs are of equal strength, and (3) this specification is itself highly
simple, stating simply that if equal before the change, two beliefs are
equal after the change.

In setting out to use the equivalence to justify conditionalization, we
should first note that conditionalization on a given proposition E surely
does not always describe reasonable change of belief even when the agent
comes to know that E is true. Clear cut exceptions are to be found among
cases in which E does not cover everything relevant that the agent comes
to know in the process of changing beliefs and among cases in which a
degree of belief equal to 1 is changed. On the other hand, I doubt that
anyone wants to deny that conditionalization describes reasonable
change of belief in any circumstances. Surely, in at least some of the highly
regimented situations studied by statisticians, conditionalization gives a
correct description. So our problem is clearly one of stating the circum-

stances under which it seems that reasonable change of belief is described by conditionalization. And since condition C is psychologically simpler than conditionalization, it should be easier to single out such circumstances in terms of condition C. Condition C could be used for this purpose in a great variety of ways. What follows is merely one, I think quite conservative such attempt.

I will proceed by first singling out circumstances in which new beliefs are not reasonable unless condition C holds for a given proposition E. By definition, $C(E)$ holds only if the agent moves from a state of doubt about the truth of E to a state of certainty that E is true. Since such certainty seems reasonable only if the agent comes to know E is true, we will look only at circumstances in which the agent comes to have such knowledge. Since it is unclear what new beliefs are reasonable when the agent's initial beliefs are unreasonable, we will restrict the circumstances under consideration to those in which the agent's initial beliefs are reasonable. When we turn to consider reasonable *change* of belief, this restriction will in large measure drop out. Finally, reasonable new beliefs often do not seem to arise by conditionalization on E when E does not cover all of the agent's relevant new knowledge; so we should also restrict the circumstances to those in which E satisfies some sort of a total evidence requirement.

These suggestions are captured by the following qualitative principle of inductive reasoning:

P: Let E be any proposition such that

     (a) The agent's initial degrees of belief are reasonable.

     (b) Initially the agent is unsure of the truth of E.

     (c) The agent comes to know that E is true.

     (d) After coming to know that E is true, any reasons the agent might have which in fact make reasonable or justify changes in other beliefs are either directly given by or included in his new knowledge that E is true; or such reasons indirectly rest on his new knowledge that E is true.

Then for any two propositions A and B, such that

     (e) A and B each logically imply E

     (f) The agent's initial degree of belief in A and in B are the same it is also the case that

> (g) The agent's new degrees of belief in A and in B are
> reasonable only if after coming to know that E is true
> they continue to be the same.

Several of the terms used in clause (d) need to be explained. I take a reason
(of the sort a person may be said to *have*) to be a belief. I will say that a
belief whose object is proposition X is directly given by a belief (or, in
particular, knowledge) whose object is proposition Y, just in case $X = Y$ or
$X = \bar{Y}$. Thus, for example, the agent's knowledge that $\bar{X}$ is false is directly
given by his knowledge that X is true. Next, I will say that a belief whose
object is proposition X is included in the belief (or, in particular, knowledge) whose object is proposition Y if X is a conjunct of Y or X is the
negation of a conjunct of Y. Finally, I will say that a belief whose object is
proposition X indirectly rests on a belief (or again, knowledge) whose
object is proposition Y just in case the agent has arrived at his belief that
X is true by a chain of reasoning whose initial premises are all directly
given by or included in his belief (or knowledge) that Y is true.

Note that principle P does not assume that degrees of belief can be
quantized, it assumes only that beliefs can be qualitatively ordered as to
strength, and for any two beliefs, one will be found to be stronger than the
other or they will be found to be the same in strength.

It is easy to check that, if principle E is true, and if conditions (a)–(d)
hold true of a proposition E, then the new beliefs are all reasonable only
if $C(E)$. Since $C(E)$ if and only if change of belief is described by conditionalization on E, and since by definition (0.1.1), if the initial beliefs are
reasonable, *change* of belief is reasonable only if the new beliefs are
reasonable, it follows that if principle P is true, and if conditions (a)–(d)
hold true of a proposition E, then change of belief is reasonable only if it is
described by conditionalization on E.

Principle P also applies, as follows, to certain cases in which (a) fails.
Assume

(1.7.1) P is true, (b)–(d) hold true of E, and (a) is false.

Then the foregoing argument sustains the following counter-factual
conditional:

(1.7.2)   If the initial beliefs (the ones the agent in fact held) had been reasonable, the new beliefs would have been reasonable only if they arose from the old beliefs by conditionalization on E.

Assume further that

(1.7.3)   Had the initial beliefs (the ones the agent in fact held) been reasonable, then *some* new overall set of beliefs would have been reasonable.

Then it follows, from (1.7.2) and (1.7.3) that

(1.7.4)   If the initial beliefs (the ones the agent in fact held) had been reasonable, then new beliefs arising from the old by conditionalization on E, and only these, would have been reasonable.

Finally, consider the assumption that

(1.7.5)   Both before and after the change of belief, the agent has a high, reasonable degree of belief that his initial beliefs were reasonable.

It follows from (1.7.4), (1.7.5) and the definition of reasonable *change* of belief (0.1.1) that a *change* of belief arising from the old beliefs by conditionalization on E is a reasonable change.

In summary, if principle P is true, and if conditions (b)–(d) hold true of a proposition E, then

(i)   if (a) holds, change of belief by conditionalization on E is a reasonable change (and the only reasonable change), if any change is reasonable.

and

(ii)   if (a) fails, but assumptions (1.7.3) and (1.7.5) hold, then by definition (0.1.1), change of belief by conditionalization on E is reasonable.

Thus barring situations in which no change is or would be reasonable, and situations in which the agent fails to have considerable reasonable confidence in the rationality of his original beliefs, if principle P is true, knowledge that (b)–(d) hold true of a proposition E provides substantial justification for change of belief by conditionalization on E.

But is principle P true? I think the principle is intuitively plausible, it is interesting in its own right, and it merits both critical scrutiny and efforts to derive it from still more obvious principles of non-deductive reasoning. But for the moment I can defend it only by offering some examples, which the reader is invited to multiply, and by examining some general arguments which some will think to advance against it.

Suppose two men are going to race and the agent has equal strength of belief in the propositions A, that the first man wins, and B, that the second man wins. The agent is unsure of the truth of E, the proposition that one of the men wins, because he recognizes that the race might be called off or might result in a tie. The agent now learns (and learns no more than) that E is true: the race was successfully completed and did not result in a tie. Under these conditions it would be absurd for him now to shift his beliefs so that he is rather more confident in A than in B or B than A.

To present another example, suppose that five tosses of a given coin are planned and that the agent has equal strength of belief for two outcomes, both beginning with H, say the outcomes $HTTHT$ and $HHTTH$. Suppose the first toss is made, and results in a head. If all that the agent learns is that a head occurred on the first toss it seems unreasonable for him to move to a greater confidence in the occurrence of one sequence rather than another. The only thing he has found out is something which is logically implied by both propositions, and hence, it seems plausible to say, fails to differentiate between them.

This second example might be challenged along the following lines: The case might be one in which initially the agent is moderately confident that the coin is either biased toward heads or toward tails. But he has as strong a belief that the bias is the one way as the other. So initially he has the same degree of confidence that H will occur as that T will occur on any given toss, and so, by symmetry considerations, an equal degree of confidence in $HTTHT$ and $HHTTH$. Now if H is observed on the first toss it is reasonable for the agent to have slightly more confidence that the coin is biased toward heads than toward tails. And if so it might seem he now should have more confidence that the sequence should conclude with the results $HTTH$ than $TTHT$ because the first of these sequence has more heads in it than tails. However, there is more to be said. Consider that nothing has happened to cast doubt on the agent's (assumed reasonable) original opinion that the coin is biased one way or the other and that $HTTH$ is a relatively unlikely occurrence under either the hypothesis of bias towards heads or the hypothesis of bias towards tails. It is true that, after the observation, $TTHT$ seems less likely than, say, $HHTH$; but the evidence will not have fully convinced the agent that bias towards tails is to be rulled out. Indeed, in this example, after the observation the agent will have only slightly more confidence that the bias is toward heads rather than tails. Insofar as the agent continues to believe that there is a bias one way or the other, both $TTHT$ and $HHTH$ should seem more likely than a sequence like $HTTH$. Whether or not these qualitative considerations exactly balance out cannot be decided without, at least implicitly, opting for some quantitative principle for change of belief. And since an argument for or against such a principle is precisely what is at issue in examining conditionalization, we must, on pain of begging the question, leave these competing qualitative considerations to compete inconclusively.

I turn now from specific examples to a general argument which might be put forward against principle P. On the hypothetico-deductive account of confirmation, a theory T is said to be confirmed by the observations it entails. But it is now widely agreed that of two theories, T_1 and T_2, both of which imply all the observations which have in fact been made, one may be better confirmed than the other. One might be tempted to conclude that a counter example to our principle can be found in some conjunction O, describing all performed observations entailed by T_1 and T_2 where O confirms one of the theories more than the other. But to be a counter example such a case would have to be one in which both T_1 and T_2 commanded equal reasonable confidence before the observations and in which the final difference in confirmation cannot be accounted for by factors besides the entailed observations, such as such as simplicity or explanatory power. Careful examination of specific examples suggests that, when the observational evidence is the same, differences in final confirmation can always be attributed to differences in initial confirmation or other considerations extraneous to the observations. These claims are born out particularly clearly in the simple case of Goodman's green hypothesis, that all emeralds are green, and grue hypothesis, that all emeralds are grue. Both hypotheses entail the observational instantial evidence of heretofore observed green emeralds. Yet the green hypothesis is to be counted as the better confirmed of the two. But for this case to provide a counter example to principle P one would have to establish the claims that if no emeralds had been observed the two hypotheses would command equal degrees of reasonable belief and that the observations confirmed the green hypothesis but not the grue hypothesis or that the observations confirmed the first more than the second. The first claim is patently false. As for the second, Goodman and others often say that only the green hypothesis is confirmed by emeralds observed to be green, but it has never been argued that the difference in final confirmation of the two hypotheses is to be accounted for by the differential bearing of the observations themselves rather than other considerations. Indeed, when viewed as part of the overall problems of confirmation it becomes at least as plausible to say that the difference in the final status of the hypotheses is wholly attributable to the difference in the hypotheses before observations are taken into account. (This is argued in [11], pp. 234–7 and *passim*.)

1.8. *On Observation*

If principle P is accepted, we have an interesting specification of a wide range of circumstances in which change of belief by conditionalization can be justified. However, this specification is really valuable only if we can ascertain when the conditions (b)–(d) of principle P are met; and this might be found to be difficult. Often, when change of belief by conditionalization seems appropriate, the change appears to originate in the agent's making a perceptual observation. Thus it seems plausible to suppose that some of the conditions (b)–(d) of principle P might be illuminatingly tied to an analysis of observational knowledge. In this section I propose a partial analysis of observation which will allow us to connect observation with conditions (c) and (d) of principle P. The following section will give the details of the connection. The material presented here will focus on the connection between observation and conditionalization, and will not deal with independently existing problems in the analysis of observational knowledge.

I will use the term 'strict observation' to refer to any event satisfying the following conditions:

(1.8.1)

(i)   There is a non-empty finite set of propositions $\{A_i\}_{i \in I}$ such that at the time of or during the occurrence of the event the agent's degree of belief in each of these propositions changes.

(ii)   For each $i \in I$, the agent's change of belief in A_i and in $\bar{A}_i$ is caused by the environment's effects on the agent's sense organs.

(iii)   For each $i \in I$, the agent's change of belief in A_i and in $\bar{A}_i$ takes place without conscious inference or reasoning of any kind.

(iv)   For any proposition B which is not in the set $\{A_i\}_{i \in I}$ or in the set $\{\bar{A}_i\}_{i \in I}$ and for which the agent's degree of belief changes, the change of belief in B is not both caused by the environment's effects on the agent's sense organs and also not the result of conscious reasoning of any kind (i.e., the conditions of (ii) and (iii) do not both hold).

(v)   Conditions are such that for each A_i, $i \in I$, the agent's new belief in A_i is reasonable, and in particular counts as knowledge that A_i is true.

For any strict observation the proposition $A = \bigwedge_{i \in I} A_i$ will be called the observations observed proposition.

The importance of this definition lies in the fact that many, perhaps most, of the events we commonly refer to as observations seem to satisfy the conditions for being a strict observation, with one qualification, to be discussed below. When I draw the blinds in the morning and, blinking in the sunlight, observe that the sun is shining, the sunlight striking my eyes under those conditions causes me to believe that the sun is shining. Though, conceivably, I might later revise my opinion, I have no choice when I first look; upon looking I am caused, willi nillie, to believe that the sun is shining. Nor, in the usual case, does any form of conscious inference accompany this change of state of belief. I look and I am caused to believe that the sun is shining and that, from the point of view of my conscious rational processes, is all there is to it. Similarly, in ordinary circumstances, when I observe a flipped coin to come up heads, the visual pattern which is presented to my eyes causes me to believe, without conscious inference and without choice, that the coin has come up heads. Many more recondite cases are also to be described this way: the practiced archeologist looking at a chipped piece of stone, may be caused by the visual pattern to come to believe that the stone is an artifact. Ordinarily he will do so without conscious inference, even though the unpracticed archeology student, in the same circumstances, may come to the same belief only deliberately and as a result of considerable conscious reasoning. What a man is able to observe strictly will often depend on skills acquired by practice and training.

The reservation in the claim that many cases of ordinary observation constitute strict observations lies in the condition that the new beliefs in the propositions, $A_i, i \in I$ constitute knowledge. I have been assuming throughout Part I, that knowledge involves certainty and that certainty is analyzable within a bayesian framework only as degree of belief equal to 1. But it seems unrealistic to suppose that people often or even ever have degree of belief of 1 in a proposition, or that any such belief is reasonable. Furthermore, even if some observations result in knowledge, clearly many do not. That most ordinary observations constitute strict observations as defined here seems clearly to be an idealization which is both complementary to and on the same sort of footing as the idealization that reasonable change of belief often takes place by conditionalization on some proposition, E, which the agent has come to know to be true. We will continue in both these idealizations here and dispose of them in Part II.

1.9. *Strict Observation and the Conditions for Conditionalization*

We have seen in Section 1.7 that conditionalization on a proposition E can be justified by appeal to principle P when it is known that conditions (b)–(d) of principle P are satisfied. And in Section 1.8 we raised the question of how it can be known that these conditions hold. Condition (b) is not particularly problematic. And we can now make use of our definition of strict observation in specifying circumstances in which (c) and (d) hold.

Before examining conditions (c) and (d), we need several preliminary definitions and remarks. Let us say that an agent's beliefs are *stable* if none of his beliefs constitute reasons which would make reasonable or justify changes in the degree to which he believes other propositions. Clearly (d) holds only if the agent's beliefs are initially stable. Moreover, reasonable change of belief will not in general be given by conditionalization when the initial beliefs are not stable. Consequently we need to require stability of initial beliefs for condition (d) to hold. Let us call a degree of belief, $P_n(R)$, after the agent comes to know that E is true a *new (degree of) belief* if $P_n(R) \neq P_0(R)$, where $P_0(R)$ is the degree of belief the agent had before coming to know E is true. If $P_n(R)$ also constitutes a reason the agent has, which in fact makes reasonable or justifies his changing other degrees of belief, we will also call $P_n(R)$ a *new reason*. Next, we shall say that

> Proposition X is a *particle* of proposition Y if and only if $X = Y$ or $X = \bar{Y}$ or X is a conjunct of Y or X is the negation of a conjunct of Y.

I said in Section 1.7 that if $X = Y$ or if $X = \bar{Y}$, the agent's belief in X was directly given by his belief in Y; and if X is a conjunct of Y, or the negation of a conjunct of Y, the agent's belief in X was included in his belief in Y. Consequently, for any particle X, of Y, $P_n(X)$ is included in or given by $P_n(Y)$. In particular for any particle, X, of Y, the agent's knowledge that X is true (or false) is directly given by or included in his knowledge that Y is true. Finally, I shall assume, which I hope is obvious, that an uncaused, unreasoned new belief (if such a thing exists) is not reasonable. Furthermore an unreasonable new belief cannot in fact make reasonable or justify changes in other beliefs. So an unreasonable new belief cannot be a new reason.

I will now argue that, if the agent's initial beliefs are stable and if he makes a strict observation with observed proposition E, conditions (c) and (d) of principle P hold for E. Assume initial stability and the occurrence of a strict observation with observed proposition $E = \bigwedge_{i \in I} A_i$. Assuming that if an agent (assumed throughout to be an ideal logician) knows the conjuncts of a proposition, he knows the proposition, it follows from condition (v) of the definition of strict observation (1.8.1) that condition (c) holds true of E. We next show that condition (d) holds for E. Since the initial beliefs are assumed to be stable, any reason the agent might have which in facts makes reasonable or justifies changes in other beliefs must be a new reason. Let $P_n(R)$ be any such new reason. If R is a particle of E, $P_n(R)$ is directly given by or included in the agent's knowledge of E. So we have only to consider R's which are not particles of E. Assume R is not a particle of E. $P_n(R)$ is either consciously reasoned or not consciously reasoned. Let us consider unreasoned $P_n(R)$ first. If $P_n(R)$ is also uncaused, it is not reasonable, and so, as argued above it is not, after all, a new reason. If $P_n(R)$ is caused as well as unreasoned, it is, by the definition of strict observation, after all, a particle of E. So we have left to consider only R's such that $P_n(R)$ is reasoned. Since the agent's beliefs are assumed to be initially stable $P_n(R)$ must be reasoned on the basis of new reasons $\{P_n(R_j)\}_{j\in J}$. If a $P_n(R_j)$ is unreasoned, as before it must be caused and so a particle of E. If an $P_n(R_j)$ is reasoned the argument applies again as it did in the case of the reasoned $P_n(R)$. Assuming that chains of reasoning are finite, such reapplications of the argument must come to an end in a case in which the basis of reasoning contains only particles of E. Hence *all* the original premises which form the basis of reasoning for $P_n(R)$, are particles of E, used at one or another stage of the reasoning in support of $P_n(R)$; and $P_n(R)$ indirectly rests on the agent's knowledge that E is true, as 'indirectly rests on' was explained in Section 1.7.

1.10. *Further Remarks on Change of Preference and Change of Belief*

We saw in Section 1.4 that, within the framework of Savages system, preference orderings uniquely determine degrees of belief and conditional preference orderings uniquely determine degrees of conditional belief. This connection did not itself provide a non-question begging argument for conditionalization; but the connection is worth remarking because it has as an immediate consequence

(1.10.1)   Assume $P_0(E) \neq 0$: then,
            If (i) The agent changes his preference ordering among
                 acts so that the new preference ordering is identical
                 with the old preference ordering given E
         then (ii) The agent changes his beliefs by conditionalization
                 on E.

(1.10.1) is of interest because, whenever change of the preference described in (1.10.1) (i) is independently known to be reasonable (1.10.1) applies to justify change of belief by conditionalization on E. Furthermore, it may be possible to find some independent general conditions under which change of preference as described in (1.10.1) (i) can be justified. If so, (1.10.1) would apply to such conditions to yield a justification for change of belief by conditionalization.

It is easy to show that if the agent's preference among consequences are constant (an assumption on which Savages wholé system heavily depends), the agent's belief function uniquely determines his preference ordering among acts and his conditional belief function, conditional on E uniquely determines his preference ordering among acts, conditional on E. This has as an immediate consequence the converse of (1.10.1):

(1.10.2)   If (i) The àgent changes his beliefs by conditionalization on E.

            then (ii) The agent changes his preference ordering among acts so that the new preference ordering is identical with the old preference ordering given E.

(1.10.2) makes it possible to bring all the information about change of belief of Sections 1.6–1.9 to bear on the heretofore untouched problem of change of preference. Whenever it is possible to justify change of belief by conditionalization the argument will apply together with (1.10.2) to provide, within Savages framework of assumptions, a justification for the corresponding change of preference.[5]

1.11. *Summary and Concluding Remarks to Part I*

In this part I have briefly remarked on frequentist and Dutch Book arguments in support of change of belief by conditionalization. I have shown that change of belief by conditionalization is equivalent to a psychologically simpler qualitative condition on change of belief. I have used this condition in developing, P, a qualitative principle of inductive reasoning, which can be used to justify change of belief by conditionalization when the conditions of application of the principle are met. And finally, I have presented a partial analysis of observation and used this analysis in further specifying the conditions of application of principle P. Residual problems include that of finding other conditions, if there are any, under which the conditions of application of principle P apply to a proposition E, and the problem of further clarifying the conditions under which an event constitutes a strict observation. Finally, we still have the problem of generalizing our results to enable us to describe change of belief which does not require the agent to have either degree of belief of 1 in any proposition or the sort of knowledge and certainty which seem to be describable only in terms of such absolute degrees of belief. This last problem will be the subject of Part II.

2.1. *Generalized Conditionalization*

To present the present topic, it will help to have at hand several examples of observations which are not strict and changes of belief which cannot be described using the methods of Part I.

Case 1: The agent has a piece of cloth which he knows is either brown (B) or green (G), and is either dyed with a natural (N) or a synthetic (S) dye. Initially the agent's degree of belief are

$$P_0(B)=\tfrac{1}{3}, \qquad P_0(S/B)=\tfrac{1}{3}, \qquad P_0(SB)=\tfrac{1}{9}$$
$$P_0(G)=\tfrac{2}{3}, \qquad P_0(N/B)=\tfrac{2}{3}, \qquad P_0(NB)=\tfrac{2}{9}$$
$$P_0(S)=\tfrac{5}{9}, \qquad P_0(S/G)=\tfrac{2}{3}, \qquad P_0(SG)=\tfrac{4}{9}$$
$$P_0(N)=\tfrac{4}{9}, \qquad P_0(N/G)=\tfrac{1}{3}, \qquad P_0(NG)=\tfrac{2}{9}$$

The venn diagram is

	B	G
S	$P_0(SB)=\tfrac{1}{9}$	$P_0(SG)=\tfrac{4}{9}$
N	$P_0(NB)=\tfrac{2}{9}$	$P_0(NG)=\tfrac{2}{9}$

The agent observes the cloth by candle light, so that he cannot see the color very clearly. As a result, he comes to have new degrees of belief about the color:

$$P_n(B)=\tfrac{2}{3}, \qquad P_n(G)=\tfrac{1}{3},$$

but the conditional degrees of belief remain unaltered:

$$P_0(S/B)=P_n(S/B), \quad P_0(N/B)=P_n(N/B), \quad P_0(S/G)=P_n(S/G),$$

and

$$P_0(N/G)=P_n(N/G).$$

We assume that the change of belief is reasonable.

Case 2: The agent had the same initial degrees of belief as in case 1, but this time he observes the cloth in broad daylight and sees quite clearly that it is brown. However, the circumstances are such that his observation causes him to shift his belief that the dye is synthetic to $\frac{2}{3}$. His new degrees of belief are

$$P_n(B)=1 \qquad P_n(S)=\tfrac{2}{3}$$
$$P_n(G)=0 \qquad P_n(N)=\tfrac{1}{3}$$

We assume that neither we nor the agent can describe any special quality of the brown hue which causes this shift, so that there is no further proposition available in terms of which the problem could be redescribed allowing the change of belief about the dye to be characterized as arising by conditionalization. We assume the agent's change of belief to be rational, as might be the case if he has wide ranging but untutored experience in judging dyes.

These cases exemplify observation which are not strict and changes of belief which cannot be described by conditionalization. But the changes of belief can be characterized by a generalized form of conditionalization suggested by Richard Jeffrey ([5] pp. 157–63). Let capital letters range over propositions in the domain of the agents belief function, and suppose that the agent changes his belief from the original function, P_0, to the new function P_n. Suppose $\{E_i\}_{i \in I}$ is a set of propositions such that

(2.1.1) (a) $\bigvee\limits_{i \in I} E_i$ is logically true and

$E_i E_j$ is logically false, $i \neq j$; $i, j \in I$
(b) If $P_0(E_i)=0$, then $P_n(E_i)=0$, $i \in I$
(c) If $P_0(E_i) \neq 0$ and $P_n(E_i) \neq 0$, then for all A,
$P_0(A/E_i)=P_n(A/E_i)$, $i \in I$.

(Henceforth the index variable i will always be assumed to be in index set, I; and explicit reference to I will be omitted when no confusion will result.) If these conditions are met it follows immediately that, for all A,

(2.1.2) $P_n(A)= \sum\limits_{\substack{i \\ P_0(E_i) \neq 0}} P_0(A/E_i)\, P_n(E_i)$.

If a set of propositions $\{E_i\}$ meets conditions (2.1.1) for a change of belief from P_0 to P_n we will say that the change *originates* in the set of propositions $\{E_i\}$, and we will refer to any change of belief which can be described by the corresponding (2.1.2) as a change of belief described by, determined by, or arising by *generalized conditionalization*. To illustrate, in case 1, the conditions are satisfied for saying that the change of belief originates in the propositions B and G. The new degree of belief in S is

$$P_n(S) = P_0(S/B) \, P_n(B) + P_0(S/G) \, P_n(G)$$
$$= \left(\tfrac{1}{3}\right) \left(\tfrac{2}{3}\right) + \left(\tfrac{2}{3}\right) \left(\tfrac{1}{3}\right)$$
$$= \tfrac{4}{9}.$$

Henceforth the form of conditionalization discussed in Part I and defined by the conditon $\mathrm{Cond}(E)$ will be called *strict conditionalization*. Strict conditionalization is obviously a special case of generalized conditionalization.

We have strong reasons for seeking to describe reasonable change of belief by generalized rather than strict conditionalization. Strict conditionalization requires description of the agent as coming to have a perfect degree of belief in a proposition, a degree of belief which cannot be altered by conditionalization within the bayesian frame work, and which commits one to betting on the proposition at arbitrarily risky odds. Rarely, if ever, is it reasonable to have such a perfect degree of belief. To be sure, reasonable degrees of belief are often close to one, in which case strict conditionalization can be used as an idealization or useful approximate description. But in such a situation generalized conditionalization offers a more accurate and less idealized description of what change of belief is reasonable. Moreover, in situations like case 1, strict conditionalization does not offer even an approximate description of the change of belief. This is so in such cases unless some intervening proposition, perhaps describing something like 'sense data' can be found. In case 1, for example, it might be suggested that after all, the agent comes to have a perfect or near perfect degree of belief in a proposition describing the way the data appeared to him in the dim light, and his degree of belief in propositions B, G, S, and N can be described by conditionalization on this proposition. However, such attempts to make strict conditionalization applicable to this kind of situation seems to me entirely gratuitous. In most such cases our language and that of the agent does not include the sentences with which the required propositions can be specified. And there does not seem to be any other, perhaps indirect way of singling out the required propositions other than question begging descriptions of the sort, 'The proposition in terms of which the change of belief can be described by strict conditionalization'. Furthermore, in view of the strength of present arguments against a sense data account of human perception, there is no reason to expect that future advances in our understanding of perceptual processes will present us with well confirmed theories according to which agents do after all entertain beliefs in propositions of the sort required to make strict conditionalization applicable to cases like case 1.

While the foregoing remarks show generalized conditionalization to be an attractive refinement of strict conditionalization, it remains to be shown how change of belief by generalized conditionalization can be justified. And there might seem to be a problem with appeal to generalized conditionalization as a means of describing reasonable change of belief, for almost *any* change of belief can be characterized as one which takes place by generalized conditionalization orginating in some set of propositions. In case 2 above, the change can be characterized as one originating in the set of propositions *SB*, *NB*, *SG*, and *NG*, as can any other change from this original set of beliefs. With one class of exceptions, if the propositions of a problem situation include a finite basic set, that is a set of logically exhaustive, mutually exclusive propositions in terms of which all the other propositions can be expressed as truth functional combinations, then any change of belief can be characterized by generalized conditionalization originating in the basic set. The exceptions are cases in which an initial belief of degree zero or one changes. If the propositions of a problem situation cannot be characterized in terms of a basic set, because the problem situation includes an infinite number of logically independent propositions, the propositions of the problem situation can in all interesting cases still be characterized to an arbitrarily set degree of accuracy by a basic set; and with the exceptions mentioned above, any change of belief can be characterized to within an arbitrarily set degree of accuracy by generalized conditionalization originating in such a basic set.

Since virtually all changes of belief can be characterized, either exactly or to within an arbitrarily set degree of accuracy, as changes by generalized conditionalization, it might seem impossible to use generalized conditionalization to distinguish reasonable from unreasonable change of belief. This conclusion is unwarranted if situations of the following sort may arise. Suppose that a change of belief from P_0 to P_n originates in the set $\{E_i\}$ nontrivially, in the sense that at least one proposition A, in the domain of the agent's belief function cannot be expressed as a truth functional combination of the members of $\{E_i\}$ nor in any relevant sense can A be approximated by such a truth function. Suppose that it can be independently established that the agent's change of belief for the members of $\{E_i\}$ is reasonable. For example, this might be done by appeal to facts about an observation. Finally, suppose it can be argued that, under the circumstances, if change of belief in the members of $\{E_i\}$ is reasonable, no overall change of belief is reasonable except the one described by generalized conditionalization originating in the set $\{E_i\}$. Under such conditions one may conclude that only this change is reasonable.

The following sections will show how change of belief by generalized conditionalization may be justified by developing an argument following the pattern of the kind sketched above. This is accomplished by a straightforward generalization of the contents of Sections 1.4–1.7. I will omit arguments which are largely repetitious of ones given in Part I.

2.2. Equivalence of Generalized Conditionalization and a Qualitative Condition On Change of Belief

Suppose again that a change in belief takes place, that P_0 describes the agent's beliefs before the change, and that P_n describes his beliefs after the change. Let '$\Rightarrow$' mean 'logically implies'. A set $\{E_i\}$ will always be understood to be a set of at least two propositions satisfying condition (2.1.1, a), that is, a set of logically exhaustive and mutually exclusive propositions. Such a set will henceforth be called a *partition*. We now define condition GC applying to sets $\{E_i\}$, by the following generalization of condition C of Section 1.4:

$$(2.2.1) \quad GC(\{E_i\}) \equiv_{\text{def}} (\forall i) \left[(P_0(E_i) = 0 \to P_n(E_i) = 0) \, \& \right.$$
$$(\forall A)(\forall B)((A \Rightarrow E_i \, \& \, B \Rightarrow E_i \, \& \, P_0(A) = P_0(B)$$
$$\left. \to P_n(A) = P_n(B)) \right].$$

As in the case of condition C, condition GC has been stated in terms of quantitative degrees of belief for convenience of the following proofs. But the conditions is really of a qualitative nature, since each clause is stated only in terms of equality of beliefs or the condition of a belief having an extremum value.

Next we define G-Cond $(\{E_i\})$ as the appropriate generalization of

Cond (E) of Section 1.4:

(2.2.2) G-Cond$(\{E_i\}) \equiv_{\text{def}} (\forall i)\,(P_0(E_i)=0 \to P_n(E_i)=0)$ &

$$(\forall A)\,\Big(P_n(A)= \sum_{\substack{i \\ P_0(E_i)\neq 0}} P_0(A/E_i)\,P_n(E_i)\Big)$$

I will now show that, under the limited fullness assumption described in Section 1.6,

(2.2.3)   For any partition, $\{E_i\}$, $GC(\{E_i\})$ if and only if G-Cond$(\{E_i\})$.

Let a partition, $\{E_i\}$, be given. First I show that if G-Cond$(\{E_i\})$, then $GC(\{E_i\})$. Suppose G-Cond$(\{E_i\})$. Let A and B be any two propositions and E_j a member of $\{E_i\}$ such that $A \Rightarrow E_j$ and $B \Rightarrow E_j$, and $P_0(A)=P_0(B)$. By the assumption of G-Cond$(\{E_i\})$,

(2.2.4) $$P_n(A)= \sum_{\substack{i \\ P_0(E_i)\neq 0}} P_0(A/E_i)\,P_n(E_i)$$

$$P_n(B)= \sum_{\substack{i \\ P_0(E_i)\neq 0}} P_0(B/E_i)\,P_n(E_i)$$

Since, by assumption $A \Rightarrow E_j$ and $B \Rightarrow E_j$, and since the members of $\{E_i\}$ are logically incompatible, $P_0(A/E_i)=0$ if it exists and $i \neq j$ and $P_0(B/E_i)=0$ if it exists and $i \neq j$. Consequently, if $P_0(E_j)=0$, (2.2.4) gives $P_n(A)=P_n(B)=0$. If $P_0(E_j)\neq 0$, then (2.2.4) gives

$$P_n(A)=P_0(A/E_j)\,P_n(E_j)$$
$$\qquad = P_0(A)\,P_n(E_j)/P_0(E_j), \quad \text{since} \quad A \Rightarrow E_j$$

and

$$P_n(B)=P_0(B/E_j)\,P_n(E_j)$$
$$\qquad = P_0(B)\,P_n(E_j)/P_0(E_j), \quad \text{since} \quad B \Rightarrow E_j.$$

But, by assumption $P_0(A)=P_0(B)$. So $P_n(A)=P_n(B)$. This completes the proof that if G-Cond$(\{E_i\})$, then $CG(\{E_i\})$.

Next we turn to the proof of the converse. Assume $GC(\{E_i\})$. It follows that

(2.2.5) $(\forall i)\,[P_0(E_i)=0 \to P_n(E_i)=0].$

First we need to show, on the assumption of (2.2.5) that if

(2.2.6) $(\forall i) \left[(P_0(E_i) \neq 0 \ \& \ P_n(E_i) \neq 0) \rightarrow (A) \left(P_0(A/E_i) = P_n(A/E_i) \right) \right]$

then

(2.2.7) $(\forall A) \left(P_n(A) = \sum\limits_{\substack{i \\ P_0(E_i) \neq 0}} P_0(A/E_i) \, P_n(E_i) \right.$

Assume (2.2.6). Let A be an arbitrary proposition.

(2.2.8) $P_n(A) = \sum\limits_{i} P_n(AE_i)$

$= \sum\limits_{\substack{i \\ P_n(E_i) \neq 0}} P_n(AE_i)$

$= \sum\limits_{\substack{i \\ P_n(E_i) \neq 0}} P_n(AE_i) \, P_n(E_i)/P_n(E_i)$

$= \sum\limits_{\substack{i \\ P_n(E_i) \neq 0}} P_n(A/E_i) \, P_n(E_i)$

By (2.2.5) if $P_n(E_i) \neq 0$, then $P_0(E_i) \neq 0$. And by the assumption of (2.2.6), if $P_n(E_i) \neq 0$ and $P_0(E_i) \neq 0$, then $P_0(A/E_i) = P_n(A/E_i)$. Consequently (2.2.8) gives

(2.2.9) $P_n(A) = \sum\limits_{\substack{i \\ P_n(E_i) \neq 0}} P_0(A/E_i) \, P_n(E_i)$

The constraint $P_n(E_i) \neq 0$ has no effect, except, together with (2.2.5) to guarantee that the terms of the sum exist. So (2.2.9) gives

$P_n(A) = \sum\limits_{\substack{i \\ P_0(E_i) \neq 0}} P_0(A/E_i) \, P_n(E_i)$

This concludes the proof of (2.2.7) from (2.2.6) and (2.2.5).

Since (2.2.7) follows from (2.2.6) and (2.2.5), and (2.2.5) follows from $GC\{E_i\}$, we need only prove (2.2.6) from $GC\{E_i\}$ to complete the proof of G-Cond$(\{E_i\})$ from $CG(\{E_i\})$. This final step requires the assumption that, for each i, the agent's domain of beliefs is full enough at E_i (definition 1.6.7); and as in Part I, I will assume that the agent's domain of beliefs is full enough at each proposition in the domain. Under this assumption, (2.2.6) is seen to follow from $GC(\{E_i\})$ as an immediate consequence of (1.6.8). It need only be remarked that although $C(E)$, which appears in the antecedent of (1.6.8), incluces the conjuncts $0 < P_0(E) < 1$ and $P_n(E) = 1$, only the weaker antecedents $P_0(E) \neq 0$ and $P_n(E) \neq 0$, corresponding to the antecedents of (2.2.6), were used in the proof of (1.6.8).

2.3. *Generalization of Principle P, and the Justification or Generalized Conditionalization*

The equivalence of generalized conditionalization and condition GC should help us to understand what we are committed to when we accept generalized conditionalization, originating in a given set of propositions, as describing reasonable change of belief. The reasons for this claim are quite the same as the reasons for the parallel claim in the case of strict conditionalization and condition C. Again, the generalized equivalence might be exploited in many ways. I present here a straightforward generalization of principle P and the use to which the new principle may be put. Corresponding to the case of strict conditionalization, generalized conditionalization on $\{E_i\}$ clearly does not describe a reasonable change of belief when $\{E_i\}$ does not cover everything relevant the agent comes to know in changing beliefs or when the agent changes a belief of degree 0 or 1. Hence, requirements of total evidence and of preservation of extremum degrees of belief in the propositions of $\{E_i\}$ are needed in the conditions of application of the principle.

The important step in generalizing principle P is to replace the characterization of the agent as coming to have new knowledge of a proposition E with the characterization of the agent as coming to have new reasonable degrees of belief in the propositions of a set $\{E_i\}$ of logically exhaustive and mutually exclusive propositions. Incorperating this change results in principle PG:

PG: Consider any partition (a set of at least two propositions satisfying (2.1.1, a)) $\{E_i\}$ and any change of belief such that

(a)   The agent's initial degrees of belief are all reasonable.

(b)   For any E_i, if the agent is certain that E_i is false before the change, then he is certain that E_i is false after the change.

(c)   The agent's strength of belief in at least one of the propositions E_i changes, and after the change the agent's belief in E_i is reasonable for each i.

and (d)   After the change of belief, any reasons the agent might have which in fact make reasonable or justify changes in belief in any proposition $A \notin \{E_i\}$ are beliefs whose objects are propositions in $\{E_i\}$ or disjunctions of these propositions; or else such reasons indirectly rest on his beliefs in the E_i, or their disjunctions.

Then, for any two propositions A, and B such that

(e)   There is an i such that A and B each logically imply E_i.

and (f)   The agent's initial strength of belief in A and in B are the same.

It is also the case that

(g)   The agent's new beliefs in A and in B are reasonable only if after the change of belief his strength of belief in A and in B continue to be the same.

As in the case of principle P, I take a reason (of the sort a person may be said to *have*) to be a belief; and I say that a belief whose object is proposition X indirectly rests on beliefs whose objects are the propositions in $\{Y_j\}_{j \in J}$ just in case the agent has arrived at his belief in X by a chain of reasoning whose initial premises are among his beliefs in the propositions in $\{Y_j\}_{j \in J}$. In examining condition (d), note that our general assumption that the agent's beliefs are, at any given time, described by a probability function implies that the agent's degrees of belief in the propositions in $\{E_i\}$ fully determine the value of his degrees of belief in the disjunctions of these propositions. As in the case of principle P, principle PG is a purely qualitative principle of inductive reasoning.

I believe that the acceptability of principle PG is much on the same footing as that of principle P; considerations advanced for or against P carry over to the case of PG. I will illustrate briefly by generalizing on the first example given in support of principle P. The agent is a spectator at a race between two men. Initially he has equal strength of belief in the proposition A, that the first man wins and B, that the second man wins. E is the proposition that one of the men wins, and $\{E_i\}$ is the set $\{E, \bar{E}\}$. The agent initially has a much stronger belief in E than in $\bar{E}$. The sprinters have been neck and neck from the start, and as they burst through the tape the hot dog vender passes through the agent's line of vision so he cannot see the finish. The announcer informs the spectators that he could not see who, if either, finished first and so he is waiting for the judge's verdict. Under these circumstances the agent's strength of belief in $\bar{E}$ increases; but it would be absurd for him to shift his beliefs so that he is now more confident in A than in B or B than in A.

Continuing the parallel to the development of Section 1.6, it is easy to check that, if principle PG is true, and if conditions (a)–(d) hold true for a partition, and if certainty in the falsehood of a proposition A is interpreted by setting $P(A)=0$, then the new beliefs are all reasonable only if $GC(\{E_i\})$. This fact, together with the equivalence of conditions GC and G-Cond can be applied in an exact repetition of the argument given in Section 1.6 which links these facts to the definition of reasonable *change* of belief (0.1.1). Since this argument changes in none of its details it need not be repeated here. The conclusion is as follows:

> If principle PG is true and if conditions (b)–(d) of principle PG hold true for a change of belief and a partition $\{E_i\}$, then
>
> (i) if (a) of principle PG holds, change of belief by generalized conditionalization originating in the set $\{E_i\}$ is a reasonable change, and the only reasonable change, if any change is reasonable.
>
> and
>
> (ii) if (a) fails, but assumptions (1.7.3) and (1.7.5) hold, then change of belief by generalized conditionalization originating in $\{E_i\}$ is reasonable.

Thus, again in parallel to the case of strict conditionalization, barring situations in which no change is or would be reasonable, and situations in which the agent fails to have considerable reasonable confidence in the rationality of his original beliefs, if principle PG is true, reasonable belief that (b)–(d) hold true for a partition $\{E_i\}$ provides substantial justification for change of belief by generalized conditionalization originating in $\{E_i\}$.

1.4. *Generalized Observation*

As in the case of principle P, principle PG provides an interesting specification of a wide range of circumstances in which change of belief by generalized conditionalization can be justified. But the specification is really useful only if we can ascertain when the conditions (b)–(d) of principle *PG* are met. And again, to aid in this task, I propose here a generalization of the analysis of observation presented in Section 1.8 which avoids commitment to observational knowledge. As before, the analysis is partial and leaves aside independently existing problems in the analysis of observation.

I will use the term 'G-observation' to refer to any event satisfying the following conditions:

(i) There is a partition $\{E_i\}$ (a set of at least two propositions which satisfy condition (2.1.1, a) i.e., they are logically exhaustive and mutually exclusive), such that at the time of or during the course of the event the agent's strength of belief in at least one of the E_i changes.

(ii) For each i, if the agent's degree of belief in E_i changes, the agent's new degree of belief in E_i is caused by the environment's effects on the agent's sense organs.

(iii) For each i, if the agent's degree of belief in E_i changes, the agent's change of belief in E_i takes place without conscious reasoning of any kind.

(iv) For any proposition A which is not in $\{E_i\}$ or a disjunction of member of $\{E_i\}$, if the agent's degree of belief in A changes, the change in belief in A is not both caused by the environment's effects on the agent's sense organs and also not the result of conscious reasoning of any kind (i.e., the condition of (ii) and (iii) do not both hold for A).

and

(v) Conditions are such that at the end of the event the agent's degrees of belief in the members of $\{E_i\}$ are all reasonable.

If a partition $\{E_i\}$ satisfies the conditions (i)–(v), $\{E_i\}$ will be called the (generalized) observation's *observation set*.

The definition of G-observation does away with strict observation's requirement of observational knowledge, so that we may now say without reservation that many, perhaps most of the events we commonly refer to as observations seem to constitute G-observations. All strict observations with non-conjunctive observed propositions are strict observations. If a strict observation's observed proposition is the conjunctive proposition $A = \bigwedge_{i \in I} A_i$ it will either constitute a g-observation taking the observation set to be $\{A, \bar{A}\}$; or else the strict observation can be redescribed as a sequence of G-observations, with observation sets $\{A_i, \bar{A}_i\}$ occurring in sequence or simultaneously. If, as I have suggested, what are commonly taken to be strict observations are merely approximations to strict observations because the agent achieves strong reasonable belief but not the certainly required of knowledge, such observations are nonetheless correctly described as G-observations or sequences of G-observations. Finally, the kind of clear cut departures from strict observations described in Section 2.1 constitute G-observations. This kind of case may occur more frequently than we commonly suppose. Often we look or listen, our sense organs are effected by objects of perception, and we are caused to shift our beliefs without conscious inference and without arriving at new beliefs which approach anything like certainty. Such new beliefs are often reasonable because the agent's perceptual capacities, both innate and learned, are reliable.

At the same time, it must be born in mind that not all events which satisfy conditions (i)–(iv) in the definition of G-observation will qualify as G-observations. A crack on the head might put a man's mind in a deranged state in which the visual pattern presented to him when he looks at an ordinary tree will cause him to believe, without inference, that money grows on trees. This new belief will not qualify as reasonable, and the event will not qualify either as a strict observation that money grows on trees or as a generalized observation giving a reasonable and high degree of belief that money grows on trees.

The reader should be dissatisfied with my definitions of observation because they appeal to conditions for which no analysis is at hand. Not just any causally necessary condition nor just any part of a casually sufficient condition for a change of belief will count as the, or a, cause of a given change of belief. More obscure yet are the conditions under which a new belief caused by the environment's effects on the agent's sense organs will count as a reasonable belief. But the need to put such matters straight is an independently existing problem in the analysis of observation which need not detain us here. I take it to be a simple fact that many cases of observation satisfy the conditions of G-observations. If this is correct, we may appeal to our characterization of observation in explaining how information obtained by observation is to be brought to bear on our overall set of belief. And as long as we do not rely on any unstated assumptions about the conditions used in the definition, we may do so however these conditions are to be further analyzed.

2.5. *G-observation and the Conditions for Generalized Conditionalization*

We turn now to applying the definition of G-observation in the task of determining when conditions (c) and (d) of principle *PG* hold for a partition $\{E_i\}$. 'Stability', 'new belief' and 'new reason' are defined as in Section 1.9. We shall say that a proposition X *stems from* the set $\{Y_j\}_{j \in J}$ just in case it is a member of $\{Y_j\}_{j \in J}$ or it is a disjunction of members of $\{Y_j\}_{j \in J}$. The argument proceeds much as it did in Section 1.9, with the relation of stemming from here playing the same role formerly played by the relation of being a particle of.

Suppose that a G-observation takes place with observation set $\{E_i\}$. We want to argue that conditions (c) and (d) of principle *PG* hold for $\{E_i\}$. Conditions (i) and (v) in the definition of G-observations explicitly say that (c) holds for $\{E_i\}$. Turning to (d), clearly (d) holds only if the agent's beliefs are initially stable. So I will now argue that if a G-observation has occurred and the agent's beliefs are initially stable, condition (d) holds for $\{E_i\}$, the observation set of the G-observation.

Given the definition of 'stems from', above, the definition of 'indirectly rests on' given in 2.3, and the assumption of Section 2.3 that the reasons we are concerned with may be taken to be beliefs, condition (d) can be restated as:

(2.5.1)   After the change of belief any reasons the agent might have which in fact make reasonable or justify changes in belief in any proposition $A \notin \{E_i\}$ are either
   (i) beliefs whose objects are propositions which stem from $\{E_i\}$

or   (ii) beliefs at which the agent has arrived by a chain of reasoning whose original premises are beliefs whose objects are propositions which stem from $\{E_i\}$.

Since the agent's initial beliefs are assumed to be stable, any reason the agent might have which in fact makes reasonable or justifies changes in other beliefs must be a new reason. Let $P_n(R)$ be any such new reason. To prove (2.5.1) we have to prove that

(2.5.2)   Either (i) R stems from $\{E_i\}$
          or (ii) The agent has arrived at the belief $P_n(R)$ by a chain of reasoning whose original premises are beliefs whose objects are propositions which stem from $\{E_i\}$.

I will prove (2.5.2) by proving

(2.5.3)   If not (2.5.2, i), then (2.5.2, ii).

Assume that not (2.5.2, i), i.e. that R does not stem from $\{E_i\}$. $P_n(R)$ is either consciously reasoned or not consciously reasoned. Let us consider unreasoned $P_n(R)$ first. If $P_n(R)$ is also uncaused it is not reasonable, and so, as noted in Section 1.9, not after all, a new reason. If $P_n(R)$ is caused as well as unreasoned, it follows from (iv) in the definition of G-observation that it stems from $\{E_i\}$ after all. Since we are supposing that R does not stem from $\{E_i\}$, $P_n(R)$ must be reasoned. Since the agent's beliefs are assumed to be initially stable, $P_n(R)$ must be reasoned on the basis of new reasons $\{P_n(R_j)\}_{j \in J}$. If a $P_n(R_j)$ is unreasoned, as before it must be caused and so it must stem from $\{E_i\}$. If a $P_n(R_j)$ is reasoned the argument re-applies as it did in the case of the reasoned $P_n(R)$. Assuming that chains of reasoning are finite, such reapplications of the argument must come to an end in a case in which the basis of reasoning contains only beliefs whose objects are propositions which stem from $\{E_i\}$. Hence *all* the original premises which form the basis of reasoning for $P_n(R)$ are beliefs whose objects are propositions which stem from $\{E_i\}$, used at one or another stage of reasoning in support of $P_n(R)$. This completes the proof of (2.5.3), and so of (2.5.2), which was to be shown.

2.6. *Further Remarks on Change of Belief and Change of Preference*

To complete our generalizations of the results of Part I, we note the following generalization of (1.10.1) and (1.10.2), taken together. Within the framework of Savage's system the following holds:

(2.6.1)   Suppose that the agents beliefs change from P_0 to P_n and that $\{E_i\}$ is a partition. Suppose also, that if $P_0(E_i)=0$ then $P_n(E_i)=0$, $i \in I$. Then
          (i) The agent's conditional preference orderings, $f<g$ given

E_i, remain unchanged for all i such that $P_0(E_i) \neq 0$ and $P_n(E_i) \neq 0$

if and only if

(ii) The agent's overall change of belief is given by generalized conditionalization originating in $\{E_i\}$.

Corresponding to the remarks of Section 1.10, whenever the change of preference described in (2.6.1, i) is independently known to be reasonable, and whenever $P_0(E_i) = 0$, $P_n(E_i) = 0$ also, then (2.6.1) applies to justify change of belief by generalized conditionalization originating in $\{E_i\}$. If some general, independent, conditions could be found under which change of belief as described in (2.6.1, i) can be justified, (2.6.1) would apply to such conditions to yield a justification of change of belief by generalized conditionalization. Finally, within Savage's framework of assumptions, the results of Sections 2.2–2.5 apply, together with (2.6.1), to provide justifications of change of preference described by (2.6.1, i).[6]

2.7. *Final Remarks*

I have shown in this paper that change of belief by conditionalization, both strict and generalized, can be justified in terms of qualitative principles of inductive reasoning, and that the difficult conditions of application of these principles can be ascertained by application of a corresponding analysis of observation. The conditions of application of the principles might also be taken to suggest, at least provisionally, limits on the correct application of conditionalization in describing reasonable change of belief. I have also shown that change of belief by conditionalization can be justified when certain changes of preference are independently known to be reasonable. But the critic may charge that the arguments presented here accomplish no more than shifting the problem of justifying conditionalization to the problems of justifying principles P and PG, of further clarifying the analysis of strict and G-observation, and of giving independent specification of the conditions under which change of preference as described by (1.10.1, i) and (2.6.1, i) are reasonable. One may agree with this criticism, and still maintain that we have made progress toward a complete account of reasonable change of belief. If not universally correct, principles P and PG must be agreed to hold in many circumstances. It seems clear that many events constitute G-observations;

and, if we ever really come to have observational *knowledge*, many events constitute strict observations. Changes of preference as described by (2.6.1, i) and (if we ever attain the relevant knowledge) as described by (1.10.1, i) are often reasonable. And insofar as we have succeeded in reducing the problem of justifying conditionalization to problems about the analysis of observation, the justification of principles of change of preference, and the justification of plausible qualitative principles of inductive reasoning, we have moved some way toward solving outstanding difficulties in epistemology.

Univ. of Illinois, Chicago

NOTES

* Portions of this work were carried out with the support of a National Science Foundation post-doctoral fellowship and a University of Illinois summer faculty fellowship. Many thanks to David Lewis for supplying the argument of Section 1.3. The argument is entirely his, though the presentation and possible defects in detail of formulation are mine. Arthur Fine is responsible for much clarification of the proof in Section 1.6. I am also indebted throughout to Richard Jeffery for many of the ideas developed here.
[1] Lewis reports that the argument was suggested to him by remark's of Hillary Putnam (p. 113 in [7], reprinted from a Voice of America Forum Lecture). Others have examined the possibility of carrying out essentially the same idea but wrongly concluded (e.g., Hacking [5] p. 315; also this author) that it could not be done.
[2] This argument has been noticed by Richard Jeffrey. In *Logic of Decision* [6] Jeffrey takes propositions, rather than acts, to be the objects of desire. In an excercise (p.80) he asks the reader to show that if desirabilities shift, so that, for a fixed proposition B, Prob $(B) \neq 0$, and arbitary proposition X, the new desirability for X is equal to the old desirability for BX, then the new beliefs are given from the old by conditionalization on B. (The suggested proof does not work for propositions, X, toward which the agent is indifferent in certain ways; but this difficulty is easily resolved by taking some proposition, C, toward which the agent is not indifferent, and calculating the new degrees of belief in XC and $X\bar{C}$.) Jeffrey claims that "when you learn that a proposition B is true, where prob $(B) \neq 0$" your desirabilities change as described above; and that thus "your new probability assignment to A [for an arbitrary A] after learning that B is true ought to be... prob AB/prob B."

The connections between change of preference or desirabilities and change of belief is a good bit easier to present in Jeffrey's system than in Savage's; and Jeffrey's axioms avoid many unrealistic features of Savages's assumptions. However, these very advantages make the intuitive ideas underlying the conclusions presented here harder to grasps. Hence the choice of Savage's system as the framework for this presentation. The difficulties discussed below apply equally to the argument whether presented in terms of Savage's or Jeffrey's system.
[3] I shall use symbols such as '$HTTH...$' both as names of sequences of outcomes and as names of propositions which are used to assert that such a sequence occurs.
[4] My original proof was a slightly weaker form of the second one presented here. Arthur

Fine discovered that the mathematically illuminating way to regard the proof was as indicated in the outline of the first proof. His version of the proof led to strengthening and shortening of my original formulation, resulting in the second proof below.

[5] The appropriate analogue of (1.10.1) holds in Jeffrey's system as remarked in note 2, so that the comments made about (1.10.1) are as applicable from the point of view of Jeffrey's system as from Savage's. However, the appropriate analogue of (1.10.2) does *not* hold in Jeffrey's system. This is because Jeffrey does not assume, as Savage does, that there are 'ultimate goods' which are of fixed desirability to the agent.

[6] As in the case described in (1.10), in the framework of Jeffrey's system (2.6.1, ii) follows from (2.6.1, i), but not conversely.

BIBLIOGRAPHY

[1] Aczél, J., *Lectures on Functional Equations and Their Applications*, Academic Press, New York, 1966.

[2] Cox, Richard, *The Algebra of Probable Inference*, John Hopkins Press, Baltimore, 1961.

[3] de Finetti, Bruno, 'Foresight: Its Logical Laws, Its Subjective Sources', in *Studies in Subjective Probability* (ed. by H. Kyburg and H. Smokler), John Wiley and Sons, New York, 1964.

[4] Good, I. J., *Probability and the Weighing of Evidence*, G. Griffin, London, 1950.

[5] Hacking, Ian, 'Slightly More Realistic Personal Probability', *Philosophy of Science* **34** (1967), 311–25.

[6] Jeffrey, Richard C., *The Logic of Decision*, McGraw-Hill, New York, 1965.

[7] Putnam, Hilary, 'Probability and Confirmation', in *Philosophy of Science Today* (ed. by S. Morgenbesser), Basic Books, New York, 1967.

[8] Savage, Leonard, J., *Foundations of Statistics*, Wiley, New York, 1954.

[9] Shimony, Abner, 'Scientific Inference', in *The Nature and Function of Scientific Theories* (ed. by R. Colodny), University of Pittsburgh Press, Pittsburgh, 1970.

[10] Suppes, Patrick, 'Concept Formation and Bayesian Decisions', in *Aspects of Inductive Logic* (ed. by J. Hintikka and P. Suppes), North-Holland, Amsterdam, 1966.

[11] Teller, Paul, 'Goodman's Theory of Projection', *British Journal for the Philosophy of Science* **20** (1969), 219–38.

DISCUSSION

Rosenkrantz: I don't think the principle of total evidence to which you appeal is strong enough, for the following reasons: we assign our probabilities on the basis of some kind of background knowledge, now the conditioning event might be implied by both of two initially equally probable hypotheses but it might bear on the background knowledge, it might in fact significantly alter that background knowledge; in this case hypotheses which began equally probable are not going to finish up equally probable after all. This, I believe, is exactly what happens in your coin test....

Teller: The question we have to ask ourselves is this: Suppose that, *were* we to find your conditioning event to have occurred, we *would* give the two hypotheses unequal new probabilities. Would we then really, on reflection, rest content with the suggestion that the hypotheses start off with equal prior probabilities? You say, "Sometimes yes." Examination of cases, such as the coin case of my section 1.7, incline me to think that the answer is always "no". But really, I am in a kind of a circle here because, against the background assumptions of my paper, Bayes rule stands or falls with the answer to this question. This equivalence, however, is important; for by appeal to it we can transfer our disagreements about Bayes rule to disagreements about the equivalent, but psychologically much more transparent, qualitative condition. This should be a useful tool both for those who seek to justify Bayes rule and those who seek to challenge it.

Lindley: I challenge you to give me a single instance where Bayes theorem has been challenged.

Teller: Challenging Bayes' theorem and challenging Bayes' rule are two entirely different things. Of course Bayes' theorem is a valid theorem, it follows from the probability axiom as an elementary exercise. But Bayes' rule is a rule for changing one's belief in the light of accumulating evidence. Bayes' theorem says nothing whatever about this problem in itself. It merely expresses a relation among certain probabilities. That we decide to use this relation in order to give a rule for the rational change of belief under accumulating evidence is a further, substantial claim which is controversial and needs to be argued. This is what I am about here.

Harper: Perhaps you would relate what you have said to the justification for probability axioms based on a characterization of betting functions.

Teller: De Finetti showed that if you are placing traditional bets, conditionalized on E (i.e. bets on certain hypotheses, for example, which are on if the statement E turns out to be true and are called off if E turns out to be false), then the ratio of the bets on the hypotheses had to be calculated according to conditionalization. De Finetti evidently thought that this rule alone was enough to determine how your beliefs should change, that once you had discovered that E was true your new beliefs ought to have been shifted to the value $P(H/E)$. Lots of people seem to have thought that this 'Dutch Book' argument does justify this rule for change of beliefs. The trouble is that nowhere in the argument is there a principle of total evidence, nowhere is it assured that all the relevant evidence has been taken into account. It is easy to cook up cases (e.g. Ian Hacking, 'Slightly More Realistic Personal Probability', *Philosophy of Science* **34** (1967), 311–25 where a statement E is admitted to be true but is obviously not the total evidence. In this case you obviously should not change your beliefs by conditionalization on that statement alone. This discovery led Hacking to believe that you could not apply the Dutch Book argument at all to change of belief. David Lewis, however, has come up with a nice formulation of the Dutch Book argument, which essentially imports an additional condition that E be the total evidence. (Lewis's argument is the one I presented in Section 1.3 of my paper.)

Kyburg: Comments on Paul Teller's Paper

Suppose that H and H' are two laws. If Teller's Qualitative Principle of Inductive Reasoning holds, if they logically imply E, and they are assigned the same degree of belief, then if the agent comes to know E, his degree of belief in them will continue to be the same. Thus evidence implied by each of two hypotheses cannot discriminate between them inductively. We already know that it cannot discriminate between them deductively (since both H and H' imply E). But this is counterintuitive, for clearly the evidence, though implied by both might be the outcome of a far more crucial test of the one hypothesis than the other. To be sure, one would prefer to test a consequence of H whose denial is a consequence of H'; but we may not be in such a fortunate situation. An historian of science would surely be able to come up with examples where the content of some bit of evidence, implied by each of two hypotheses, was clearly more relevant to the truth of one than to the truth of the other.

For a made-up example, consider the theory T and the theory T': Theory T has E as a straightforward consequence – we might even take T to be E itself and tack on some irrelevant stuff to arrive at the right probability. T', on the other hand, is very tightly organized, and almost everything in the whole theory is called upon in deriving the consequence E. Now surely we can find or construct a T as described whose probability overall is the same as that of T', prior to the observation of $E : P_0(T) = P_0(T')$. Surely, after coming to know E, the degree of belief that it is appropriate to have in T' is greater than that appropriate to have in T: E in some sense represents, by depending on almost everything in T', a representative sample of T', while E is merely an isolated fragment of T.

As a trivial example, consider the theory $H : \bigwedge_x (Fx \rightarrow Gx)$. Let a_0 be 'the first F to be examined', a, be 'the second F to be examined,'... A_{1000} be 'the $1000^{\text{th}} F$ to be examined', and let K be the conjunction $Ga_0 \wedge Ga_1$, $\wedge ... \wedge Ga_{1000}$. Clearly the prior probability of K is less than that of H; so we can find a formula S such that the probability of $K \wedge S$ is the same as that of H. Let E be the conjunction K. There are surely circumstances under which E (i.e., K) would be strong evidence for H, but not for $K \wedge S$ – despite the fact that the conditional probabilities are the same.

Professor Teller's treatment of the grue and green emeralds in the written version of his paper simply doesn't hold water. He says that 'both hypotheses entail the observational instantial evidence, of heretofore observed green emeralds.' [ms. p. 36] (Of course, this is false; what it entails is that of every emerald $Ex \rightarrow Gx$ is true – and also that that is true of every raven. In a detailed analysis, the difference might turn out to be important. But no matter here.) He agrees that the green hypothesis is confirmed better than the grue hypotheis. But he says it is 'patently false' that the two hypotheses would command equal degrees of belief prior to the observation (he says, 'of emeralds' but he must surely mean) of anything. It may well be false, but it is surely not *patently* false, and it will be false at all only relative to some specific technique for assigning a priori probabilities to statements.

This counterintuitive consequence of the qualitative principle of inductive reasoning follows from the principle of conditionality in even stronger form: If both H and H' imply E, the ratio of the prior probability of H to the prior probability of H' will be equal to the ratio of the posterior probability of H to the posterior probability of H': E cannot discriminate between H and H'. Furthermore, this is true on Jeffrey's Principle of Conditionalization, where E does not become known, but merely changes its probability.

The importance of justifying the principle of conditionality that Professor Teller is concerned about is clear already in F. P. Ramsey, although it seems to have been taken largely for granted by modern Bayesians. When Ramsey defines "degree of belief in p given q", he points out, "This is not the same as the degree to which he would believe p, if he believed q for certain; for knowledge of q might for psychological reasons profoundly alter his whole system of beliefs." Harper and May are approaching exactly this problem, but from the logical point of view: If the agent's degree of belief in q is changed from one value to another, how *ought* this alter his whole system of beliefs? By conditionalization? Or in some other way? Ramsey seems to have seen that the question could be regarded as open.

Teller: Response to Comments of Professor Kyburg

On a trivial point Professor Kyburg and I are in complete agreement: My qualitative principle of inductive reasoning is an immediate consequence of the rule of conditionalization. If the qualitative principle leads to counterintuitive consequences, then conditionalization is in some kind of serious trouble. But putting things this way misleads us as to what I believe is most usefully taken to be the 'problem of conditionalization'. I already agree that there are situations in which conditionalization does not correctly apply – for example, when the underlying domain of propositions changes, or when the conditionalizing proposition does not cover the total evidence. On the other hand, surely there are some situations in which conditionalization is the right thing to do, for example in certain highly structured situations encountered by practicing statistions. Consequently, the interesting problem is, as I have said in my paper, to discover the conditions under which conditionalization is appropriate and to state these conditions so clearly that we can give a clear, even trivial argument to the conclusion that *under these conditions* only change of belief by conditionalization is reasonable. I hope that my paper is a useful contribution to this program, but I should be very foolish to think that I have carried it out to its conclusion. One, and I would guess the most profitable way to develop the program further would be to find counterexamples to my principle; and, by examination of these counterexamples, formulate further conditions which have to be met for conditionalization to apply. To this end examples of the kind Kyburg indicates might turn out to be very useful.

In keeping with the spirit of the program I have described, I also looked keenly for counterexamples to my qualitative condition. In so doing I found intuitions of my own to be similar to some of those of Professor Kyburg. But to make progress with the program it is not enough to become convinced that there are counterexamples of such and such a general form, convinced that "surely there are circumstances under which" (as Kyburg says of his schematic example) a case of such and such a general form provides a counterexample. We must formulate sufficiently concrete counterexamples so as to reveal wherein conditionalization breaks down, so that further restrictive conditions will become apparent. And in trying to make counterexamples sufficiently concrete to serve this purpose, I found that they always slipped through my fingers. My coin example was the closest I came. I can in a schematic way, and in reference to Kyburg's sketch of an example, indicate the kind of difficulty which I always encountered. Let H be Kyburg's "coherent" hypothesis where H implies K; and let $H' = K \wedge S$, where S is some formula picked so that H' has the same prior probability as H. Kyburg says, "There are surely circumstances in which $[K]$ would be strong evidence for H but not for $K \wedge S$ [i.e., for H']." However, H can also be viewed as a conjunction, $K \wedge S^*$, where S^* is, for example $(H \wedge \sim K)$ (concrete cases usually suggest more natural choices for S^*, as in the example below); and for K to provide strong evidence for H but not for H', K must be more positively relevant to the truth of S^* than to the truth of S. (In simple examples, K is just irrelevant to S but positively relevant to S^*.) Whenever I formulated concrete pairs of the form $K \wedge S^*$ and $K \wedge S$, where K is more positively relevant to the truth of S^* than to the truth of S, it seemed clear that

$K \wedge S^*$ commanded a higher prior degree of belief than did $K \wedge S$, which rendered the status of the example as a counterexample to my principle at least unclear. Goodman buffs will easily find a concrete instance of these general remarks in Kyburg's example of the generalization $\bigwedge_x(Fx \to Gx)$, with S^* being a formula which says that all Fs to be examined from the 1001st on are Gs, whatever the first 1000 Fs were. Of course, as Kyburg remarks in response to my treatment of Goodman's case, there are 'techniques' for assigning prior probabilities which will give $K \wedge S^*$ and $K \wedge S$ equal priors. But unless such a 'technique' assigns the two hypotheses in question priors which most of us will agree are intuitively plausible (or at least not intuitively wildly implausible) we still do not have the sort of clear-cut counterexample to my qualitative principle which will enable us to correct it. This fact also underlies my confidence that my rejection of Goodman's example as a counterexample to my principle holds more water than Kyburg allows. Of course there are 'techniques' for assigning priors which give the green and grue hypotheses equal priors. But these 'techniques' are, intuitively, absurd. So Goodman's case provides no clear-cut counterexample to my principle.

Secretely, I feel sure that my principle is wrong, in the sense that it will turn out to need considerable, even drastic correction. But in public I continue to maintain that my principle is correct and that the vague, general characterizations of kinds of possible counterexamples which I have been offered in no way underline it. In so doing I hope to madden someone into constructing a counterexample so clear-cut that even a very obstinate man will have to concede. When we have such an example we will also be able to see how to correct the principle.

BAS C. VAN FRAASSEN

PROBABILITIES OF CONDITIONALS

> 'He says he'll pay me every pfennig if he gets
> this job as barman at the Lady Windermere
> ... if, if...' Frl. Schroeder sniffs with intense
> scorn: 'I dare say! If my grandmother had
> wheels, she'd be an omnibus!'
>
> Christopher Isherwood, *Goodbye to Berlin*

Both conditionals and probabilities have been the subject of lively philosophical debate. Lately their interaction has been in the limelight,
through the disputed thesis that $P(A \rightarrow B) = P(B/A)$, *the probability of the
conditional is the conditional probability* (of consequent on antecedent).
This thesis is tenable for the Stalnaker conditional if nesting of arrows
is not allowed; for nested arrows I have weaker results. For ease of
reading, I have limited the body of this paper to an exposition of the
philosophical disputes, while the technical results are collected in a many-
sectioned appendix.[1]

Both for probabilities and conditionals, I shall make a distinction
between interpretations and paradigms. An interpretation, in this context, is a full-fledged account of the subject. A paradigm is something that
guides our attempts to arrive at an interpretation: it is an idea that
explains 'clear' cases, breaks down immediately for more complex cases,
but is returned to for inspiration whenever more formal attempts at
interpretation run into their own difficulties.

1. CONDITIONALS

The first paradigm that guided the explication of conditionals was the
idea of this situation: a person asserts $A \rightarrow B$ (read 'if A then B') to signify
that the argument with premise A and conclusion B, is valid. The first
obvious extrapolation of this idea is the assertion that it is the exact and
sole function of a conditional to express the statement that a certain

*Harper and Hooker (eds.), Foundations of Probability Theory, Statistical Inference, and Statistical Theories of Science,
Vol. I, 261–308. All Rights Reserved. Copyright © 1976 by D. Reidel Publishing Company, Dordrecht-Holland.*

corresponding argument is valid. And this extrapolation then yields the following principles for reasoning with conditionals.

(1)   If A and $A{\rightarrow}B$ are true, then B is true
       (Symbolically: $A, A{\rightarrow}B \Vdash B$) *(Modus Ponens)*
(2) $A{\rightarrow}B \Vdash (A$ & $C){\rightarrow}B$ (Weakening)
(3) $A{\rightarrow}B, B{\rightarrow}C \Vdash A{\rightarrow}C$ (Transitivity).

There are many logics of conditionals that incorporate these principles, for the above paradigm led to a number of different (detailed) interpretations of the conditional.

However, these principles can be held only at the cost of ignoring a large class of cases of conditional assertion, which apparently do not fit the paradigm. These cases were discussed in the forties by Goodman and Chisholm, under the heading of *contrary-to-fact* or *counterfactual conditionals*. A typical example is the statement 'If this match were struck (now), it would light', said with reference to a match held up for inspection. Agreement that this statement is true does not allow the inference that if this match were a burnt match, and struck now, it would light. Similarly, it seems that I could truly say of a drinking glass: 'Were this glass dropped now, it would break' but not 'Were this glass dropped, and were the floor covered with foam, the glass would break'. These examples contravene principle 2 above (Weakening), and once you see the trick, you can also provide examples that contravene Transitivity *(though not Modus Ponens)*.

The trick, as everyone saw at once, is no trick at all: many conditional statements in English carry a tacit *ceteris paribus* clause on their antecedent. There are two paradigms for this case: what I shall call the Ramsey paradigm and the Sellars paradigm. The Ramsey paradigm centres on a person with body of information K. This person asserts $A{\rightarrow}B$ in each of two cases: A is compatible with K, and the argument from K and A to B is valid; A is not compatible with K, but minimal changes in K yield an alternative $K(A)$ compatible with A, such that the argument from $K(A)$ and A to B is valid. Goodman's original arguments seem to establish pretty well that we cannot give anything like a general recipe for finding the correct body $K(A)$, or even for saying what changes in K are more minimal than others. But the logician adhering to the Ramsey paradigm is not worried by this, since he seeks generality: he asks what principles of reasoning remain correct whatever the recipe be.

The second paradigm is due to Wilfrid Sellars, who considers the relevant class of conditionals $A \rightarrow B$ to have a restricted syntactic form: A is an 'input' statement and B an 'output' statement. The person asserting the conditional has a general background theory with principles of form

(4) ϕ-ing a thing of kind K in conditions C makes it ψ

The conditional (4) has no *ceteris paribus* clause; it fits the original 'valid argument' paradigm. The tacit *ceteris paribus* clause in conditionals that violate principles of Weakening and Transitivity is a tacit specification of kind K and/or circumstances C for the antecedent of, say,

(5)   If X be ϕ-ed, it will (would) ψ.

To a logician this paradigm may seem very limited, because of the restricted syntactic form. But it is very likely that many problems are artificially created in logic through syntactic generality, going beyond the original context of philosophical problems.

There are two main theories of conditionals in which the interpretation is sufficiently detailed to allow a complete characterization of the logic of reasoning with conditionals. These are due to Robert Stalnaker and David Lewis; they are cast in the terminology of the current semantic analysis of modal logic.

Simplifying a bit, Stalnaker's theory is this: every statement is true or false in each possible world (or possible situation, or set-up, if you like); there is for each world α a *nearness ordering* of worlds such that α is nearest α, and if there are any worlds in which A is true, there is a nearest world to α among those in which A is true ('the nearest A-world to α'); $A \rightarrow B$ is true in α if and only if: B is true at the nearest A-world to α, or there are no A-worlds.

Lewis accepts the basic approach but denies that there must be a unique nearest A-world to α. Hence he says: $A \rightarrow B$ is true in α if and only if B is true at all the nearest A-worlds to α. Besides their more basic agreements, we find therefore that both Stalnaker and Lewis reject Weakening and Transitivity, and accept *Modus Ponens* and a kind of weakened transitivity:

(6) $A \rightarrow B, B \rightarrow A, A \rightarrow C \vdash B \rightarrow C$

But Stalnaker alone, and not Lewis, accepts

(7) $(A \rightarrow B) \vee (A \rightarrow \neg B)$ must always be true
        (Symbolically: $\Vdash (A \rightarrow B) \vee (A \rightarrow \neg B)$)

where $\vee$ is the sign of inclusive disjunction ('and/or') and $\neg$ the sign of negation ('not').

To the philosophical status of the 'possible world' talk and its intelligibility I shall return below. But first I must discuss a technical question: the question of nestings of arrows, of conditionals whose antecedents or consequents are themselves conditionals.

2. NESTED CONDITIONALS

The preceding section does not presuppose that nesting of arrows, as in $(A - B) \rightarrow C$ or $A \rightarrow (B \rightarrow C)$, yields syntactically well-formed sentences, or that similar constructions in English are meaningful before rephrasing them. One might hold, for example, that $A \rightarrow (B \rightarrow C)$ is to be understood as equivalent to $(A \& B) \rightarrow C$, and that when it sounds as if some-one is saying $(A \rightarrow B) \rightarrow C$ he really intends the metalinguistic $(A \rightarrow B) \Vdash C$. More abstruse constructions might be considered totally useless or unintelligible.

Much to my regret, the facts of discourse do not seem to allow of such simplifications. I was convinced of this by Richmond Thomason, with such examples as

(8)   If the glass would break if thrown against the wall, then it
        would break if dropped on the floor.

This cannot be construed as of form

(8a) $(A \rightarrow B) \Vdash (C \rightarrow D)$

which asserts a relation between two statements, but must be accepted as having the form

(8b) $(A \rightarrow B) \rightarrow (C \rightarrow D)$

because it suffers from typical conditional trouble, in that (8) does not imply

(9)   If this glass would break if thrown against the wall, and the floor were covered with foam rubber, then it would break if dropped on the floor.

Of course, it is still possible to react that one is not interested in the facts of discourse *per se*, and that such assertions as (8) and (9) are of no interest for more substantive reasons either. Much of what follows in this paper can be read while ignoring nested arrows, or abhorring them; only the last result I shall discuss benefits from restricting our attention to statements less complex than (8).

In any case, both Stalnaker and Lewis consider it a virtue of their account that the truth-conditions for statements involving nested arrows are provided automatically. For example, let β be the nearest A-world to α, and γ the nearest B-world to β; then clearly, on their account, $A \rightarrow (B \rightarrow C)$ is true at α exactly if C is true at γ.

To give our discussion somewhat more precision, I shall now state, in simplified form, the semantic account (essentially) due to Stalnaker.[2] A *model structure* is a couple $M = \langle K, s \rangle$ with K a non-empty set (the *possible worlds*), and s a map such that for each member α of K, and each subset X of K, $s_\alpha(X)$ is a subset of K also; subject to the conditions

(10a) $s_\alpha(X)$ is included in X

(10b) $s_\alpha(X)$ contains at most one member

(10c) $s_\alpha(X) = \{\alpha\}$ if α is in X

(10d)   If $s_\alpha(X) \subseteq Y$ and $s_\alpha(Y) \subseteq X$ then $s_\alpha(X) = s_\alpha(Y)$.

The set $s_\alpha(X)$ is the set of nearest worlds in X to α. Lewis denies (10b), but adds other clauses.

An *interpretation* (this technical usage of the term is to be distinguished from my earlier use of it) is a function I which assigns to each sentence A a subset of K (intended meaning: $I(A)$ is the set of worlds in which A is true), subject to the clauses

(11a) $I(\neg A) = K - I(A)$

(11b) $I(A \,\&\, B) = I(A) \cap I(B)$

(11c) $I(A \vee B) = I(A) \cup I(B)$

(11d) $I(A \rightarrow B) = \{\alpha : s_\alpha(I(A)) \subseteq I(B)\}$

We say that A is *valid* ($\Vdash A$) exactly if $I(A)=K$ for each interpretation I on each model structure; and that $A_1, ..., A_n$ *semantically imply* $B(A_1, ..., A_n \Vdash B)$ exactly if $I(A_1) \cap ... \cap I(A_n) \subseteq I(B)$ for each interpretation I on each model structure.

A set of possible worlds is called a *proposition*; to say that sentence A 'expresses' the proposition X means that X is the set of worlds in which A is true. The family $\{I(A): A$ is a sentence$\}$ is called an algebra of propositions with Boolean set operations, plus binary operator

$$X \to Y = \{\alpha : s_\alpha(X) \subseteq Y\}.$$

All this is general for model structures of any conditional logic; in the present case these algebras of propositions may suitably be called *Stalnaker algebras*. The question: what is the logic, i.e. what principles govern valid inferences for this language? is clearly answered if and only if we can give an exact account of the class of Stalnaker algebras. This was done axiomatically by Stalnaker and Thomason; see Appendix Section 1 for a simplified account.

3. A PHILOSOPHICAL DIGRESSION

To the question what principles govern deductive reasoning involving conditionals, Stalnaker nad Lewis give exact replies. But the validity of an argument does not depend on whether its premises are true; and indeed, Stalnaker and Lewis have not notably increased our ability to decide whether particular conditionals are true or false.

I opened this paper with a quote from Isherwood; his Frl. Schroeder seems to challenge the whole world to refute even the wildest consistent counterfactual conditional. Possible world discourse may seem to meet the challenge in principle: even if we cannot tell whether the dear lady's grandmother would have been an omnibus if she had been endowed with wheels, there is, as a matter of objective fact, a set of nearest worlds to the actual one in which the grandmother is so endowed. And the question whether in those worlds, she is an omnibus, has an objective answer.

However, one needs to swallow a great deal of metaphysics to take this seriously. Let me say at once that I do not; there are no possible worlds except the actual one, in the literal sense of 'there are'. I see the possible world machinery just as Duhem saw the rope-and-pulley models of the

English physicists: such fictions are useful when giving an account of the surface phenomena – and there is, in reality, nothing below the surface. In our case the phenomena are the inferential relations among statements, attested in the inferential behaviour of those engaged in such discourse. Within the model structure, what is meant to mirror these phenomena is the algebra of propositions. We introduce possible worlds, and relations on them, because that yields an intuitively simple, but formally indirect, way of defining that proposition algebra.

Let me return to the truth-values of conditionals. If there are facts only about this world, and no counterfacts or facts about other possible worlds, sentences involving conditionals cannot be evaluated by asking whether they correspond to the facts. For they are about what is not a fact. So except for ones that cannot be true – like $[A \& (A \to \neg A)]$, which just has to be false; and other limiting cases – the truth-value of such sentences seems to be indeterminate. Stalnaker and also Thomason have indicated how this may be accepted without imperiling the logic of inference involving conditionals; and I have elsewhere shown how this yields a perspicuous way to relate Lewis' and Stalnaker's theories to each other.[3]

However, this is a bit of an uncomfortable position, for one would like to say that all sorts of contingent conditionals are true. If this butter were heated to 150 °C, it would melt. Well, these may be the cases in which the context makes very clear the exact content of that tacit *ceteris paribus* clause. But the semantic analysis is in terms of possible worlds, and an alternative semantics based on the ceteris paribus idea yields a logic that is a *proper* part of both Lewis' and Stalnaker's logics, and indeed also of any other logic which we shall encounter in this paper.[4] So we feel the inclination to say that even if the truth-values of some conditionals are indeterminate, there are yet many non-trivial ones that we are entirely prepared to assert.

I must now state my own position: counterfactual conditionals do not have the function of stating facts, and in a strict sense, none deserve to be called true or false. Their function is different, just like the function of sentences in the interrogative or imperative mood is different.

Before I elaborate this position, I may as well mention at once that David Lewis considers such positions (in his discussion of Adams) and rejects them. Lewis writes:[5]

I cannot think of any conclusive objection to the hypothesis that indicative conditionals are non-truth-valued sentences, governed by a special rule of assertability.... I have an inconclusive objection, however: the hypothesis requires too much of a fresh start. It burdens us with too much work still to be done, and wastes too much that has been done already. So far we have nothing but a rule of assertability for conditionals with truth-valued antecedents and consequents. But what about compound sentences that have such conditionals as subsentences?

I assume that Lewis would voice these same admirably conservative sentiments in response to similar hypotheses about any conditional or modal connective. Yet he probably does not mean that it is not a virtue of a philosophical position, if it spurs its adherents on to new ventures, such as the development of a pragmatics-oriented alternative to truth-value semantics. What Lewis intends, presumably, is a challenge: do not claim any advantages for your hypothesis until you have substantiated its feasibility in detail. And Lewis assumes that, since the hypothesis must also eventually yield a characterization of the logical laws of inference, substantiation will require a full-fledged alternative to the usual semantics, in which concepts like *assertability* replace concepts such as truth.

This assumption accompanying the challenge I deny. Clearly, if anyone claims that conditionals, and sentences with conditionals as parts, have a different function from statements of facts, then he must give an account of that function. But the account I shall now give implies that, to ferret out the logic of discourse involving conditionals, it is appropriate exactly to engage in the usual (possible worlds & truth-values) semantics.

Let us return for a moment to the Sellars paradigm.[6] A person asserts that this pat of butter would/will melt if heated. He knows very well that his experience of heat melting butter in the past does not warrant his assertion; he has no such simple faith in straight rule induction as Russell's chicken. But in asserting the conditional he signifies his allegiance to a certain background theory, in which relevant principles of form 4, about butter and heating, hold. He may be more or less vague on the theory and the theory in question will be more or less sophisticated; it might surprise you to find what vague and unsophisticated theories some of your nearest and dearest have about butter. But his allegiance is strong, and he will reiterate the conditional with rising volume and timbre if pressed. For when a man enters a commitment, whether to a scientific theory or an ideology, he assumes *ipso facto* the office of explainer. He undertakes to answer questions *ex cathedra, qua* adherent of the theory.

And his use of conditionals has the function of signalling his commitment.

It would be nice to have a formal model for this situation, so that we could calculate what a man's *ex cathedra* answers would be, given his theoretical commitments, plus observations. I admit this, and that I have no such model to give. But I add that I can nevertheless describe informally how such a situation develops, and that in a way which shows the correct approach to the central question: what is the logic of inferences involving conditionals?

Your run of the mill conditional-user has no exact account of how theories plus facts warrant the assertion of conditionals or complex sentences involving them. So how does he keep straight how he is to behave in argument, carried on in such discourse? The situation is similar to that of a physicist unacquainted with the axiomatic basis of his discipline. He is willingly bewitched by a heuristic picture, and this picture guides his inferences. It is also similar to the pictorial way in which we give set-theoretic or algebraic proofs without reference to any axiomatic basis. (In a more serious vein, Milton chose to use the language of an earlier cosmography than that of his own day, when writing *Paradise Lost*: and today too, a scientifically educated person can give expression to his religious commitments through texts or hymns talking of the seven days of creation and four corners of the earth.) If there is an axiomatic basis in existence, the heuristic pictures or pictorial talk has no *theoretical* interest any more. But when there is no axiomatic basis and we try to produce one, we proceed by carefully examining the pictures. Recall for example Helmholtz's pictorial 'axiom of free mobility' and Lie's formalization in terms of geometric transformations.

By what picture, then, is the conditional-user willingly bewitched, when he uses such discourse? If pressed, he will talk about imagining alternative possibilities, and such; he may even give such elegant arguments as Lewis'

I believe that there are possible worlds other than the one we happen to inhabit. If an argument is wanted, it is this. It is uncontroversially true that things might be otherwise than they are. I believe, and so do you, that things could have been different in countless ways. But what does this mean? Ordinary language permits the paraphrase: there are many ways things could have been besides the way they actually are. On the face of it, this sentence is an existential quantification. It says that there exist many entities of a certain description,

to wit 'ways things could have been'. I believe that things could have been different in countless ways; I believe permissible paraphrases of what I believe; taking the paraphrase at face
value, I therefore believe in the existence of entities that might be called 'ways things could
have been'. I prefer to call them 'possible worlds'.[7]

The point to which I agree is that logician's possible World machinery is a
passable explication of the picture that bewitches users of modal and
conditional discourse. Since that picture is what guides inference, the
logical catalogue of valid patterns of inferences can only be searched out
by exploring this picture, or the logician's explication thereof. Hence the
success of the usual form of semantics.

4. PROBABILITY

As every knows, there are five Schools or interpretations of probability:
the logical, frequency, subjective, propensity, and Kyburg. I do not wish
to discuss these interpretations, nor align myself with one, but to discuss
paradigms. There seem to me to be two paradigms guiding the interpretation of probability, and their use cuts across the five Schools (though
at least among philosophers if not among statisticians, some Schools
show a distinct preference for one or other). These paradigms I shall call
the *epistemic ticker tape* and the *finite state machine*. Each of these is a
conceived situation, or type of situation, in which some probability talk
at least appears eminently intelligible.

A philospher adhering to the first paradigm imagines a subject who has
both knowledge and degrees of belief. His knowledge comes to him on a
ticker tape bearing simple sentences, and his total knowledge at time t is
the content of the tape at t. The tape delivers one message per unit time,
and each time it does, the subject's knowledge is increased, and he revises
his degrees of belief. And the subject asserts 'Prob$(A) = r$' if and only if his
degree of belief that A, equals r.

A philosopher adhering to the second paradigm imagines a subject (or
nature) feeding inputs into a black box with finitely many states. Call the
possible inputs $I_1, ..., I_m$ and the possible states $B_1, ..., B_n$. The box
comes with an instruction sheet giving for each state B_i a matrix $[p(i, j, k)]$
which purports to mean that if the machine is in state B_i and input I_j is
applied, then the machine transits to state B_k with probability $p(i, j, k)$ –
the *transition probability*. The philosopher likes to refer to this situation as

a *chance set-up*, and sees it as his task to explain what the instruction sheet really means.

Conditional probabilities have a place in both paradigms, but the guidance which the paradigms give for defining conditional probability is distressingly less than complete. Consider first the ticker tape. Bayesians established early on that, if to be rational is to be such that no one can make book against you, then the degrees of belief of a rational subject at a given time t must be, mathematically speaking, an assignment of probabilities. Professor Teller's paper in this volume proceeds in adherence to the epistemic ticker tape paradigm, and asks whether the conditional probability $P(A/B)$ of A given B must be defined by

$$(12) \qquad P(A/B) = P(A \& B)/P(B)$$

which is the usual formula.

The basic constraints on $P(A/B)$, namely that $P(B) = 1$, and that the function $P(A/B)$ is itself a probability assignment, are nowhere near enough to deduce (12). Mathematically at least, there are many alternatives to (12); and David Lewis has defined one for Stalnaker models. However, Teller's paper gives a proof, also due to Lewis, that the Bayesian requirements of rationality also require (12). That is, if the ticker tape subject revised his degrees of belief other than by conditionalizing to his increasing knowledge by formula (12), it would be possible to make book against him. This is a very nice result. Its philosophical significance, however, is somewhat marred, to my mind, by the lack of realism in the rational ticker tape subject as model for the scientific inquirer. The latter will frame theories, suggested but not entailed by his evidence, commit himself to them, and conditionalize his degrees of belief accordingly. Although these commitments are open to revision in the light of new evidence, he will nevertheless make his bets in accordance with these theoretical commitments in the meanwhile. So a real scientific pilgrim's progress violates the rationality conditions on which Lewis' proof is predicated.[8]

It might seem that the finite state machine paradigm is in good shape to guide the definition of conditional probability. For after all, is the transition probability not a conditional probability

$$(13) \qquad p(i, j, k) = P(B_k \text{ next} \mid B_i \text{ now \& } I_j \text{ now})$$

of the next state given present state and input?

Indeed; and it is possible to reconstruct the $p(i, j, k)$ from other probabilities so that (13) follows from (12) (see Appendix, Section 2). But the moment we start varying the conditions, we run into trouble. How are we to interpret

(14) $P(B_k \text{ next} \mid \text{one of } B_k, ..., B_m \text{ next})$

for example? Well, the transition matrix describes tendencies of the machine due to its physical make-up; hence the condition in (14) is to be conceived of as a condition imposed on the physical make-up of the machine. We put a 'damper' on the machine, 'closing off' possible states $B_1, ..., B_{k-1}$. Thus the machine has an offspring, the 'dampened' machine, with new transition probability matrices $[p^*(i, j, k)]$. What are these new matrices to be like? Well, that will depend on exactly what the damper does; the only real constraint is that after the damper is put on, P (one of $B_k, ..., B_m$ next)$= 1$. But there must be lots of dampers having that effect, and only one kind of damper will allow us to calculate $p^*(i, j, k)$ using formula (12). This kind of damper can be described (see Teller's proof that (12) must be true if, roughly speaking, $P(A/B)$ is to be functionally determined by $P(A \& B)$ alone, and not by other factors). But in the present context, this is still only one case, and it is hard to see why it should have such a preferential status.

In both paradigms, there are simple, clear cases, in which formula (12) appears the correct one. If I am a relatively dull subject, not given to much theorizing, like Conrad's Winnie Verloc of the conviction that things don't bear looking into much, then I shall have reason only to revise my degrees of belief by conditionalizing on new knowledge via formula (12). If I consider a finite state machine, and look only at those conditional probabilities listed in the transition probability matrices, again I shall see formula (12) satisfied. Outside these simple realms, all is grey. However, having noted that (12) cannot be considered sacrosanct, I shall now accept the usual extrapolation, and henceforth consider conditional probability to be defined by formula (12) everywhere.

5. The Stalnaker Thesis

The English statement of a conditional probability sounds exactly like that of the probability of a conditional. What is the probability that I throw a six if I throw an even number, if not the probability that: if I throw an even number, it will be a six? And if we do not allow nesting of conditionals, nor conjoining conditionals with other sentences, '$P(B \to A)$' is surely no more than a harmlessly rewritten '$P(A/B)$'. But the rewriting may not be so harmless if we regard $B \to A$ as a full-fledged sentence and make logical claims about the arrow – even if we do not allow nesting of arrows. But even if a thesis is not harmless or trivial, it may be true; and Robert Stalnaker advanced

(15)   Stalnaker's Thesis.
         $P(A/B) = P(B \to A)$ whenever $P(B)$ is positive

in conjunction with the usual (formula 12) definition of $P(A/B)$. I shall refer to this briefly as the Thesis. It must be distinguished from Stalnaker's secondary claim that the Thesis holds with $\to$ being the Stalnaker conditional.[9]

This secondary claim goes well beyond the Thesis, which supposes only that we are speaking of some logically respectable conditional. In the Appendix (Section 2) I shall show that chance set-ups may be regarded as Stalnaker models; but there is then in general no *obvious* way of getting the Thesis to hold, for the probabilities $p(i, j, k)$ of the transition probability matrix. Later on (Section 6) it will appear that there is an unobvious way to do that, involving the fictional introduction of infinitely many further possible states not observationally distinguishable from the ones originally countenanced. In any case, the matter is not obvious even for simple probability statements.

David Lewis has an argument to show that instead, the Thesis is untenable, on pain of triviality. He deduces, starting from the Thesis, that no probability assignment can have more than four distinct values. This demonstration he gave at the Canadian Philosophical Association in June 1972, and it was a veritable bombshell; for months afterward everyone believed that the Thesis was defunct.

But Lewis' demonstration has subtle auxiliary assumptions that go into his formulation of the problem. He introduces his formulation with the following persuasive commentary:[10]

What needs explaining is the fact that for all speakers at all times... the assertability of indicative conditionals goes by conditional subjective probability. If we hope to explain this general fact by the hypothesis that the probabilities of conditionals always equal the corresponding conditional probabilities, and if we assume that → means the same for speakers with different beliefs, then the fixed interpretation of → must be such as to guarantee that those equalities hold ho matter what the speaker's subjective probability function may be.

It may not be immediately obvious how this commentary affects the formal reasoning. The key phrase is 'the fixed interpretation of →'. What Lewis is actually demanding is that the model structures be such that any probability assignment (of a large enough class) can be superimposed without violating the Thesis, and without corresponding changes elsewhere in the model structure.

If persuasive English lacks some perspicuity, so does the formalistic jargon in which I have just restated it. Let us imagine the following situation. A certain person, who has a certain amount of information, and a commitment to certain theories, represents the world to himself by means of a model structure with probability measure P on its set of possible worlds. In this way he properly allows for his partial ignorance: there are many ways things might be compatible with his information and theories. His structure also has a nearness relation among worlds, so that he can tell in which worlds $A \rightarrow B$ is true. But he does not know which world is the actual one; however, if X is a set of these possible worlds in the domain of P, then $P(X)$ is the probability, according to him, that the actual world is in X.

Now a revision occurs; this person's information and/or theoretical commitment changes in a certain way. His ideas about what the world is like change; and also the degrees of belief he attaches to the sentences in his language. So he revises his model structure cum probability measure.

And here, Lewis introduces the requirement that it should be possible to make this revision by changing the probability measure alone – and not the constitution of the possible worlds or the nearness relation on them. What inspires this requirement, which is crucial to Lewis' reductio? Would it not seem rather, that our probabilities are inextricably involved in the way we represent the possibilities, and nearness relations among them, to ourselves? In the finite state machine paradigm, the transition probability matrix presumably reflects the machine's physical structure; if our ideas about the one change, will we not revise our modelling of the other?

The inspiration for the requirement must doubtlessly be Lewis' metaphysics, according to which one should always be able to say: let the possible worlds in my model structure be those which there actually are, let the nearness relation on them be the one reflecting their actual and objective similarities. In this scheme, the probability measure is nothing but a device to picture our ignorance. Hence it has nothing to do with the internal constitution of the model structure, which is reality itself. For this reality of possible worlds exists independent of the mind, its evolution flows on in its own even tenor;

The Moving Finger writes; and, having writ,
Moves on: nor all thy Piety nor Wit
Shall lure it back to cancel half a Line,
Nor all thy Tears wash out a Word of it.

How very different it looks to those of us who locate all of reality in the actual world and the representing subject, seeing nothing but manipulable fictions in the possible world menagerie!

So the logical disaster was precipitated not by Stalnaker's Thesis, but by the Thesis coupled with Lewis' metaphysical realism.

Lewis also has another demonstration about the Thesis, in which it is considered independently of formula (12), the usual definition of conditional probability. Assume the Thesis, and assume that $P(-/B)$ is a revision of P such that $P(B/B) = 1$, and the revision is 'in some sense minimal'. Then Lewis derives the conclusion that $\rightarrow$ must be the Stalnaker conditional. The same results would follow *a fortiori* with (12) assumed; and this was also proved by William Harper.[11] But in this demonstration too a crucial role is played by the assumption that the probability measure on the model structure may be revised, without violating the Thesis, and without changing any other aspect of the structure. (The specific revisions used are to the zero-one measures P_α giving probability 1 to the set $\{\alpha\}$, where α is any one of the possible worlds.) But this is exactly the assumption found in the preceding demonstration, justified only by metaphysics, and not acceptable to me.

In conclusion then, I see the state of the issue as follows: the Thesis coupled with the usual definition of conditional probability, and the assumption that $\rightarrow$ is a logically respectable conditional, does not reduce to absurdity, and also does not imply that $\rightarrow$ is the Stalnaker conditional. There are two questions I want now to explore: *first*, what is the minimal logic of conditionals such that the Thesis holds non-trivially, and *second*, is it tenable to assert the Thesis for the Stalnaker conditional?

6. A MINIMAL LOGIC

Let us now assume the Thesis, and the usual definition of conditional probability, and the respectability of → as a conditional connective. What logical principles must hold then? I shall need to introduce some Auxiliary Assumptions which are meant as explicating the respectability of the arrow. Henceforth capital letters denote propositions, i.e. sets of possible worlds; & is intersection, → a certain binary operation, ∨ is union and $\neg A$ or $\bar{A}$ denotes $K - A$ where K is the set of all possible worlds.

(16)   *Assumption.* If two propositions must always have the same probability, then they are identical.

This is not logically precise: I mean that if $P(\phi(X_1, ..., X_n)) = P(\psi(X_1, ..., X_n))$ for every model structure with probability measure, and all propositions $X_1, ..., X_n$ in that structure, then $\phi(X_1, ..., X_n) = \psi(X_1, ..., X_n)$ for every model structure etc.

(17)   *Theorem.* $A \to A = K$, the set of all possible worlds.

For $P(A \to A) = P(A/A) = 1 = P(K)$, using the Thesis and Assumption (16).

(18)   *Assumption.* If A is not the empty proposition then $A \to B$ and $A \to C$ are disjoint if B and C are disjoint.

(19)   *Theorem.* $A \to (B \vee C) = (A \to B . \vee . A \to C)$.

The proof, for the case $P(A) \neq 0$, is as follows.

$$P(A \to . B \vee C) = P(B \vee C/A) =$$
$$= P(B \& C . \vee . B \& \bar{C} . \vee . \bar{B} \& C/A) =$$
$$= P(B \& C/A) + P(B \& \bar{C}/A) + P(\bar{B} \& C/A) =$$
$$= P(A \to . B \& C) + P(A \to . B \& \bar{C}) +$$
$$+ P(A \to . \bar{B} \& C).$$

By Assumption (18), the propositions involved are disjoint, so

$$= P[(A \to . B \& C) \vee (A \to . B \& \bar{C}) \vee (A \to \bar{B} \& C)]$$

Now by (16), we conclude

(20) $[A \rightarrow B \vee C] = [(A \rightarrow B \ \& \ C) \vee (A \rightarrow B \ \& \ \bar{C}) \vee (A \rightarrow \bar{B} \ \& \ C)]$

However, by exactly similar reasoning we get, because $B = (B \ \& \ C) \vee \vee (B \ \& \ \bar{C})$, the following

(21) $A \rightarrow B = (A \rightarrow B \ \& \ C) \vee (A \rightarrow B \ \& \ \bar{C})$
(22) $A \rightarrow C = (A \rightarrow B \ \& \ C) \vee (A \rightarrow \bar{B} \ \& \ C)$

and noting that the first disjunct on the right in (20) can be repeated, we derive Theorem (19) by substitution of equals for equals in (20). Again, the case of $P(A) = 0$, if allowed to be defined at all, is obvious.

The last theorem has a corollary that $(A \rightarrow B. \vee .A \rightarrow \bar{B}) = K$, which is the peculiar Stalnaker principle first denied by Lewis. To this extent, at least, Stalnaker's Thesis is intrinsically connected with Stalnaker's theory of conditionals. A second corollary is that, if $P(A) \neq 0$, then, because $A \rightarrow B. \vee .A \rightarrow \bar{B} = K$, $\overline{A \rightarrow B} = A \rightarrow \bar{B}$. And with this second corollary in hand, we derive that $P(A \rightarrow B \ \& \ C) = P(A \rightarrow B. \ \& . \ A \rightarrow C)$ in all cases, because of De Morgan's law and Theorem (19). Hence by (16) again,

(23)   *Theorem.* $A \rightarrow (B \ \& \ C) = A \rightarrow B. \ \& . \ A \rightarrow C$.

We now have three theorems, but do not know yet that modus ponens holds – as it must, or the name 'conditional' is inappropriate. Indeed, in both Stalnaker and Lewis's systems, one finds the stronger law

(24)   *Assumption.* $(A \rightarrow B) \ \& \ A = (A \ \& \ B)$

to which I hereby agree.

This is where I stop; let me call a *proposition algebra* (*tout court*, as opposed to 'algebra of propositions in model structure *M*') any field of sets with unit K and binary operation $\rightarrow$ (possibly a partial operation, e.g. not defined for $\Lambda \rightarrow X$), such that

(I) $(A \rightarrow B) \ \& \ (A \rightarrow C) = (A \rightarrow .B \ \& \ C)$
(II) $(A \rightarrow B) \vee (A \rightarrow C) = (A \rightarrow .B \vee C)$
(III) $A \ \& (A \rightarrow B) = (A \ \& \ B)$
(IV) $A \rightarrow A = K$

hold where defined. The minimal logic of conditionals suitable for probabilification I shall call *CE*; it is adequately described by saying that in any acceptable model structure with probability, the algebra of propositions must be a proposition algebra in the above sense.

So the Thesis requires logic *CE*; but does the Thesis allow non-trivial probability measures for the model structures countenanced by *CE*? Indeed, as I shall prove in the Appendix, Section 3,

(25)   *Theorem.* Any antecedently given probability measure on a countable field of sets can be extended into a model structure with probability, in which Stalnaker's Thesis holds, while the field of sets is extended into a proposition algebra.

which theorem is as essential as the soundness and completeness theorems for *CE*, with respect to the criteria of adequacy for our present problem area.

Principle (III) (Assumption (24)) was introduced somewhat cavalierly perhaps, as being previously agreed to by all contestants. Even if it is only reasonable to approach the problem with a strong an agreement with Stalnaker and Lewis as seems possible, it may be well to look, momentarily, at the status of the Thesis purely *vis-à-vis* this area of agreement.

First, assume the Thesis, and also the nesting-reducing principle

(26) $A \rightarrow (A \rightarrow B) = (A \rightarrow B)$

also common to all contestants. Then we derive

(27) $P(A \rightarrow B) = P(A \rightarrow (A \rightarrow B)) = P(A \rightarrow B/A)$

that is, that the conditional is stochastically independent of its antecedent. Conversely, assume principle (III), plus (27), and derive

(28) $P(A \rightarrow B) = P(A \rightarrow B/A) =$
$$= P(A \rightarrow B. \ \& \ A)/P(A) =$$
$$= P(A \ \& \ B)/P(A) =$$
$$= P(B/A)$$

which is the Thesis. So the agreement, by all concerned, about the Thesis is that it is equivalent to the stochastic independence between the conditional and its antecedent. This might provide a new fulcrum for the application of a philosophical critique or defence of the Thesis.

However that may be, in Theorems (17)–(25) we have a complete solution to the problem of the minimal logic CE of conditionals required by the Thesis. Also, since nesting of arrows plays, at most, an inessential role in all the demonstrations involved, these results remain if such nesting is disallowed. I turn now to the Stalnaker logic of conditionals.

6. STALNAKER BERNOULLI MODELS

I may as well say at once that I cannot prove an analogue to Theorem (25) for Stalnaker algebras. (See note (12)). From the Appendix, Section 1, it appears that a Stalnaker algebra is a proposition algebra satisfying two further principles:

(V) $(A \rightarrow B) \& (B \rightarrow A) \& (A \rightarrow C). \supset .(B \rightarrow C)$

(VI)   if A is included in B, then $B \rightarrow \Lambda$ is included in $A \rightarrow \Lambda$.

The second of these is about conditionals with impossible antecedents, which is like probabilities, conditionalized on a condition with zero probability, an I shall ignore this (as a mainly technical matter). And (V), I consider debatable, and not established by philosophical argument; but I shall not debate it here.

In this section, I mean to show that one can adhere to the use of Stalnaker models and maintain the Thesis for all conditionals of a few simple kinds. (If no nesting of arrows is allowed, this solves the problem entirely; but I cannot so restrict myself with good conscience.) The kinds of conditionals I shall handle have the forms:

$$A \rightarrow B$$
$$A \rightarrow (B \rightarrow C)$$
$$(A \rightarrow B) \rightarrow C$$

in which A, B, and C themselves contain no arrows.

Imagine the possible worlds are balls in an urn, and someone selects a ball, replaces it, selects again, and so forth. We say he is carrying out a series of Bernoulli trials. Now it is possible, is it not, that the selects first α, then the nearest world to α, then the nearest to that, and so on. In that case, $A \rightarrow B$ is true at α exactly if the first A-ball he selects is a B-ball.

This suggests a way of constructing Stalnaker models, in which probabilities enter very straightforwardly. Indeed, the proofs I shall then give about the Thesis in such models, proceed by quite simple probability calculations. (They were not simple to me until Ian Hacking provided me with a crucial lemma; but they seem simple now.)

I must interject here that, from the point of view of Stalnaker's semantics, these models are of a very special variety. (Hence, although I do not know that the Thesis fails in them for more complicated conditionals, I expect it does.) To explain this limitation, it is easiest to refer to Lewis' reformulation of Stalnaker's semantics. In Lewis' version, a Stalnaker model is a couple $M = \langle K, \leqslant \rangle$ in which, for each α in K, $\leqslant_\alpha$ is a total pre-ordering of K. (For all present purposes, $\leqslant_\alpha$ may be taken to be a discrete linear ordering.) Then the nearest A-world to α is the minimal element by the $\leqslant_\alpha$ ordering, of the set $\{\beta \in K : A$ is true at $\beta\}$ $(= I(A)$ for the interpretation at issue). Now note that for each world there is another such ordering; because, intuitively, not all the worlds may lie on one straight line. In the models which I construct in this section, however, the following happens: if $\alpha \leqslant_\alpha \beta$ and $\beta \leqslant_\beta \gamma$, then $\alpha \leqslant_\alpha \gamma$. Indeed, think of $\leqslant_\alpha$ as determining a series α, β, δ, γ, ... and $\leqslant_\beta$ another series, say β, γ', δ', ... Then in *these* models, but not in general, the series β, γ', δ', ... is in fact the series β, γ, δ, ... So we can think of these models as having their worlds arranged on a set of parallel lines very far apart from each other, and on each line, the worlds are arranged like the numbers, 0, $\frac{1}{2}$, $\frac{3}{4}$, $\frac{7}{8}$, ... The reason I have explained this at length is to make clear that the tenability of the Thesis *for more complex sentences* has little to do with what happens in these special models.

Now let me explain how the construction works. We begin with a Stalnaker model $M = \langle K, \leqslant \rangle$, and a field F of propositions in M. These propositions are to be thought of as expressed by zero-degree sentences; sentences in which there are no arrows. Finally, we have a probability P which is defined at least for all of F. If A and B are in F, then $A \rightarrow B$ is a well-defined proposition in M, of course; but at this point we don't know what $P(A \rightarrow B)$ is, nor even whether it is defined.

We now make up the bigger model $M^* = \langle K^*, \leqslant^* \rangle$ where K^* is the set of all maps π of the natural numbers into K. The relation $\leqslant_\pi^*$, on K^*, where π is in K^*, is defined by:

$\rho \leqslant_\pi^* \sigma$ iff there is a number k such that $\rho(k+m) = \sigma(k+m)$ for all natural numbers m.

For obvious reasons, such a map ρ may be referred to as the sequence $\rho(1), \rho(2), \ldots$.

In this much bigger model M^*, the old model M reappears if we identify member α of M with the series $\alpha^* = (\alpha, \beta, \gamma, \ldots)$ which is the domain of $\leqslant_\alpha$ ordered by that relation.

Now we look at the field F^* of subsets of K formed this way: F^* is the family of all sets

$$A_1 \otimes \cdots \otimes A_n \otimes K^*: A_1, \ldots, A_n \text{ in } F$$

where $\otimes$ is the Cartesian product, and this notation is imprecise, but perspicuous for:

$$\{\pi: \pi(i) \in A_i \quad \text{for} \quad i = 1, \ldots, n\}.$$

This construction of F^* is such that if F is a (Borel) field, so is F^*.

What we have so far is what is called a *product* construction, and this has a familiar correlate extension for the probability; I shall use the same symbol 'P' for the new measure as well as the old:

$$P(A_1 \otimes \cdots \otimes A_n \otimes K^*) = P(A_1) \ldots P(A_n).$$

The sets $A \otimes K^*$ in F^* may be called *zero-degree propositions* in M^*. These are the reappearance of the old field F of course, and have the same probabilities as the correlate old propositions: $P(A \otimes K^*) = P(A)$. Henceforth it will be convenient to abbreviate '$A \otimes K^*$' to 'A' when this is not confusing.

In M^* we also have conditional propositions, of course, defined by

$$X \to Y = \{\pi \in K^*: \text{ for all } m, \text{ if } m \text{ is the first number such that } \pi_m \in X, \text{ then } \pi_m \in Y\}$$

where π_m is the series that results from π when you cut off the initial segment $\pi(1), \ldots, \pi(m-1)$.

The nice thing shown in the Appendix at this point is that F^* is closed under $\to$: if X and Y are in F^*, so is $X \to Y$. Hence in M^*, all propositions constructible from the zero-degree propositions (by $\to, \cap, \cup, -$), automatically receive a probability. Note also that if A and B are zero degree, then A, and also $A \to B$, is true at α in K iff it is true at α^*.

Now I shall show intuitively that if A and B are zero-degree, then $P(A \to B) = P(B/A)$. Choose at random a series π of worlds in K. This is itself a series formed by making a random selection of a world from K, again and again, ad infinitum. It is just like tossing a die forever. At some point, say at $\pi(m)$ we find a world in which A is true. Since previous selections do not influence this one in the least, it is itself nothing more nor less than a random selection from K. Hence the probability that this first A-world is a B-world, is just the probability that *any* world satisfies B given that it satisfies A. So that probability is $P(B/A)$. Hence the probability that $A \rightarrow B$ is true in the actual world (about which nothing is specified a priori *except* that it belongs to K) is just $P(B/A)$. So $P(A \rightarrow B) = P(B/A)$.

This proof is made precise in the Appendix, as well as similar, but longer, proofs that $P(A \rightarrow .B \rightarrow C)$ and $P(A \rightarrow B. \rightarrow C)$ conform to the Thesis as well. So at least for these simple conditionals, the Thesis is tenable in conjunction with Stalnaker's logic *and* Stalnaker's model theory.

University of Toronto

APPENDIX

For a summary of Stalnaker's semantics, see the end of Section 2 in the body of the text. In this Appendix, I shall use distinct symbols for set-theoretic operations and sentential connectives $(\cap, \cup, -; \&, \vee, \neg)$ except in the case of the arrow.

1. STALNAKER ALGEBRAS

David Lewis gave essentially the following simplified axiomatization[1] of Stalnaker's logic C2:

(A1) $\vdash A$ when A is a propositional tautology
(A2) $\vdash A \rightarrow A$
(A3) $\vdash (A \rightarrow B) \& (B \rightarrow A) \& (A \rightarrow C) . \supset B \rightarrow C$
(A4) $\vdash (A \vee B) \rightarrow A . \vee (A \vee B) \rightarrow B . \vee : (A \vee B) \rightarrow C \equiv A \rightarrow C . \& . B \rightarrow C$
(A5) $\vdash (A \rightarrow B) \supset (A \supset B)$
(A6) $\vdash A \rightarrow B . \vee . A \rightarrow \neg B$
(A7) $\vdash (A \& B) \supset (A \rightarrow B)$
(R1)   if $\vdash A$ and $\vdash A \supset B$ then $\vdash B$

(R2)   if $\vdash (A_1 \& \ldots \& A_n) \supset B$ then $\vdash (C \to A_1) \& \ldots \& (C \to A_n). \supset (C \to B)$

THEOREMS:

(1) $\vdash (A \to B) \& (A \to C). \equiv A \to (B \& C)$
(2) $\vdash (A \to B) \vee (A \to C). \equiv A \to (B \vee C).$

Almost all of this can be proved using (R2); there remains only the following part:

(2′) $\vdash A \to (B \vee C). \supset . (A \to B. \vee . A \to C)$

Suppose the consequent is false; then by (A6) and theorem (1), $A \to (\neg B \& \& \neg C)$ is true. So then if the antecedent is true, by (1) and (R2), $A \to B$ is true after all, contrary to supposition.

Now (A4) can be simplified: suppose that $(A \vee B) \to A$ and $(A \vee B) \to B$ are false; then by (A6) and (T1), $B \vee B \to (\overline{A \vee B})$ and so, via (A2) and (R2), $A \vee B \to C$ for all C. So $A \vee B$ is an impossibility and (A4) reduces to: if $A \vee B$ is an impossibility, so is A (and so is B mutatis mutandis). Hence (A4) can be replaced by:

(A4′) $\vdash (A \vee B) \to \overline{A \vee B}. \supset . A \to \bar{A}$

Also, (A5) and (A7) can together be replaced by

(A5′) $\vdash A \& (A \to B), \equiv (A \& B).$

Accordingly, a Stalnaker algebra of propositions is a field of propositions (sets) with unit K such that the following principles hold (I)–(IV) are same as in Section 6 in body of text):

(I) $(A \to B) \cap (A \to C) = (A \to . B \cap C)$
(II) $(A \to B) \cup (A \to C) = (A \to . B \cup C)$
(III) $A \cap (A \to B) = (A \cap B)$
(IV) $A \to A = K$
(V) $(A \to B) \cap (B \to A) \cap (A \to C) \subseteq (B \to C)$
(VI)   If $A \subseteq B$ then $B \to A \subseteq A \to A.$

(Here (VI) comes from (A4′).)

THEOREMS

(1)   If $A \subseteq B$ then $A \to B = K$.

For if $A \subseteq B$ then $B = A \cup B$; but $A \to A \cup B = A \to A \cup . A \to B = K$.

(2)   If $A \subseteq B$ then $C \to A \subseteq C \to B$.

For if $A \subseteq B$ then $B = A \cup B$ so $C \to B = C \to A \cup . C \to B$.

(3)   If $A_1 \cap ... \cap A_n \subseteq B$ then $C \to A_1 \cap ... \cap C \to A_n \subseteq C \to B$.

For $C \to A_1 \cap ... \cap C \to A_n = C \to . A_1 \cap ... \cap A_n$, and so by (2), $\subseteq C \to B$.
This recaptures algebraically the effect of (R2).

2. CHANCE SET-UPS AS STALNAKER MODELS

To produce a simple Stalnaker model, consider a probabilistic finite state machine with input. Let its present state be B_1, its possible states $B_1, ..., B_n$, and the possible inputs $I_1, ..., I_m$. There is a transition probability matrix $[p_{ij}]$ relevant given that the present state is B_1: the probability that the next state is B_j if one applies input I_i, equals p_{ij}: Since each input leads to some outcome, $\sum_j (p_{ij}) = 1$. In the usual picture, the rows sum to one:

	B_1	B_j	B_n
I_1	p_{11}	p_{1j}	p_{1n}
I_i		p_{ij}	
I_m			p_{mn}

We can also put this as follows:

$$P(B_j \text{ next} \mid B_1 \text{ now \& } I_i) = p_{ij}$$

Let the set of possible worlds (meaning, possible immediate futures) be the set of couples $\langle I_i, B_j \rangle$. So the actual world is the one in which the present state is B_1, the input I_i, and the next state B_j, and it is represented by the couple $\langle I_i, B_j \rangle$.

With respect to this set of worlds let the proposition I_k expressed by sentence I_k be $\{ \langle I_k, B_j \rangle : 1 \le j \le n \}$ and similarly let $B_k = \{ \langle I_i, B_k \rangle : 1 \le i \le m \}$. The set of worlds is linearly ordered in some way. The probabilities are as follows: for simplicity, let the input be chosen randomly, hence $P(I_i)$

should be $(1/m)$. So let $P(\{\langle I_i, B_j\rangle\})=(1/m)\,p_{ij}$.

$$P(I_k)=\sum_j (1/m)\,P_{kj}=(1/m)\sum_j p_{kj}=(1/m)$$
$$P(B_k)=\sum_i (1/m)\,p_{ik}=(1/m)\sum_i p_{ik}$$
$$P(I_k\cap B_j)=P(\{\langle I_k, B_j\rangle\})=(1/m)\,p_{kj}$$
$$P(B_j\mid I_k)=P(I_k\cap B_j)/P(I_k)=$$
$$=[(1/m)\,p_{kj}]/(1/m)=$$
$$=p_{kj}$$

all as it should be.

Now we consider the nearest ordering. It is linear, and since the set is finite, discrete. So there is for each world $\langle I_i, B_j\rangle$ a nearest I_k world, namely $\langle I_k, B_{kij}\rangle$ where kij is a number between 1 and n inclusive. The restriction on this is that if $i=k$ then $kij=j$. Apart from that, you would think, it could be anything.

$$(I_k\to B_1)=\{\langle I_i, B_j\rangle: kij=1\}$$
$$P(B_1\mid I_k)=P(I_k\to B_1)=\sum_{ij}\{P(\langle I_i, B_j\rangle): kij=1\}.$$

So
$$P_{k1}=\sum_{ij}(i/m)\,p_{ij}\delta(1, kij)$$

where $\delta(x, y)=1$ if $x=y$ and $=0$ otherwise.

For any two indices a and b, we have therefore the equation

(1) $p_{ab}=(1/m)\sum_{ij} p_{ij}\delta(b, aij)$

But that is very surprising!

There are now two questions: will there always be an ordering such that (1) is true? *and* when (1) is true, what is the ordering like? The answer to the first question is *NO*: in fact, it is difficult to cook up an example in which (1) *is* true. As one counterexample, let the matrix in question be

	B_1	B_2	B_3
I_1	$\frac{1}{5}$	$\frac{1}{10}$	$\frac{7}{10}$
I_2	$\frac{1}{3}$	$\frac{1}{9}$	$\frac{5}{9}$

and consider p_{11} in (1). The equation then becomes $\frac{3}{5} = \frac{1}{5} + ?$ and there just are no other entries summing to $\frac{2}{5}$. (But this problem can presumably be met by throwing in a large number of extra possible worlds, with fictional states, and a larger matrix; see the last section of this Appendix.)

3. THE MINIMAL LOGIC CE

I will define model structures with probability; in accordance with my own preferences, the algebra of propositions in the model structure is specified directly. The procedure of having a privileged family of propositions (as opposed to the family of *all* sets of possible worlds) is familiar from quantum logic, but has only recently become current in modal logic (for example, in S. K. Thomason).

3.1. *Frames*

Motivation: in a propositional logic of the usual stripe, the propositions (= sets of worlds) form a field but not a Borel field. Any field can be extended to a Borel field; and a function on the original field would not be a probability measure unless it could be extended to a probability measure on the generated Borel field. *Terminology*: a Borel field of subsets of V is a sigma-ring of such subsets, to which V itself belongs. A triple $\langle V, F, P \rangle$ I shall call a *probability space* exactly if P is a probability measure with Borel field F of subsets of V as its domain.

DEFINITION. A *frame* is a triple $\varphi = \langle V, F, P \rangle$ such that $V \neq A$ and $F \subseteq$ domain of P is a field and $\varphi' = \langle V, \text{domain of } P, P \rangle$ is a probability space.

Note that φ' is a frame also; I shall say that $\varphi \subseteq \varphi'$ just because their second members are so related, and the other members pairwise identical. A member X of F with $P(X) \neq 0$ will be called a *non-zero* set in φ (to be distinguished from *non-empty*).

DEFINITION. Frame $\varphi = \langle V, F, P \rangle$ is *full* exactly if every non-zero set X in φ is the union of a family X^* such that $P \mid X^*$ is *onto* $[0, P(X)]$.

The vertical bar denotes restriction of a function; a probability space is full exactly if every set of measure r has a subset of measure q, for each positive $q \leqslant r$.

THEOREM. Every frame can be homomorphically extended to a full frame (with measure preserved).

Proof. It is necessary only to consider frames which are probability spaces. If $\varphi = \langle V, F, P \rangle$ is thus, let $\varphi^* = \varphi \otimes [0, 1] = \langle V^*, F^*, P^* \rangle$ where $V^* = V \otimes [0, 1]$ the Cartesian product; the members of F^* are the sets $X \otimes E$ with X in F and E a Borel set on $[0, 1]$; $P^*(X \otimes E) = P(X) \cdot \mu(E)$ where μ is the usual 'length' measure on $[0, 1]$ with $\mu([a, b]) = |b - a| = \mu([a, b])$ and similarly for other kinds of intervals. This is a probability space – see any textbook under the heading 'Product measure'. The map $X \to X \otimes [0, 1]$ is a set homomorphism of F into F^* and preserves measure: $P(X) = P(X) \cdot 1 = P(X) \mu([0, 1]) = P^*(X \otimes [0, 1])$.

Finally, this is a full frame. For let r be any fraction between zero and one, and Borel set $E = \bigcup_{i=1}^{\infty} e_i$ where the e_i are disjoint intervals. If e_i has end points a and $a + m$ let re_i be the similar interval with end points a and $a + rm$. Then we argue

$$P^*\left(X \otimes \bigcup_{i=1}^{\infty} re_i\right) = \sum_{i=1}^{\infty} P^*(X \otimes re_i) =$$

$$= \sum_{i=1}^{\infty} P(X) \mu(re_i) =$$

$$= \sum_{i-1}^{\infty} P(X) r\mu(e_i) = rP^*(X \otimes E)$$

This suffices; any Borel set differs by measure zero from a countable union of half-open intervals.

Finally, I will call a frame $\langle V, F, P \rangle$ *finite* or *denumerable* if the field F is so.

3.2. *Proposition Algebras*

Let F be a field of subsets of V. I shall call $\langle V, F, \to \rangle$ a *proposition algebra* if $\to$ is a partial operation with domain $G \otimes F$, $G \subseteq F$ and range $\subseteq F$ such that (where defined)

(I) $(A \to B) \cap (A \to C) = (A \to .B \cap C)$

(II) $(A \to B) \cup (A \to C) = (A \to .B \cup C)$

(III) $A \cap (A \to B) = A \cap B$

(IV) $(A \to A) = V.$

Deducible properties of $\rightarrow$, where defined, are:

(tp0)   If $A \subseteq B$ then $(A \rightarrow B) = V$

(tp1) $A \rightarrow A = V$; from (p5)

(tp2) $(A \rightarrow B) = A \rightarrow (A \cap B)$; from (p1) and tp1)

(tp3) $(A \rightarrow B) = (A \rightarrow . B \cap C) \cup (A \rightarrow . B \cap \bar{C})$; from (p2)

(tp4) $(A \rightarrow B) \cup (A \rightarrow \bar{B}) = V$

If $\Rightarrow$ has properties (I)–(IV), and we define $A \rightarrow B$ to be $A \Rightarrow B$ when defined and to be $\bar{A} \cup B$ otherwise, for given sets A, B, then $\rightarrow$ also has properties (I)–(IV). This shows that the stipulation about the domain of $\rightarrow$ is rather innocuous.

The definition of proposition algebra is suggested by the properties of a model structure in which the sentences A receive propositions A and $(A \rightarrow B) = \{x : s(A, x) \in B\}$ where s is a (partial) operation subject to the restrictions

(s1) $s(A, x) \in A$

(s2)   if $x \in A$ then $s(A, x) = x$

but no others. I do not think all proposition algebras could be produced this way, since $\cap \{(A \rightarrow B_i) : i = 1, 2, \ldots\}$ might be empty while no member of that set is. But could any proposition algebra be homomorphically extended to one so produced? I don't know.

With reference to (III), define $[A, B] = (A \rightarrow B) - (A \cap B)$. Then

(q1) $[A, B \cap C] = [A, B] \cap [A, C]$

(q2) $[A, B \cup C] = [A, B] \cup [A, C]$

(q3) $[A, B] \subseteq \bar{A}$

(q4) $[A, B] \cup [A, \bar{B}] = \bar{A}$

(q5)   if $A \subseteq B$ then $[A, B] = \bar{A}$

(tq1) $[A, A] = \bar{A}$

(tq2) $[A, B] = [A, A \cap B]$

(tq3) $[A, B] = [A, B \cap C] \cup [A, B \cap \bar{C}]$

follow from (I)–(IV); and (q1), (q2), (q4), (q5) together imply (I)–(IV). (Note that (q4) implies (q3).)   ...

Hence I shall take (q1), (q2), (q4), (q5) as defining a proposition algebra when convenient.

3.3. *Models*

A model is a combination of frame and proposition algebra:
A *model* is a quadruple $M = \langle V, F, P, \rightarrow \rangle$ such that $\langle V, F, P \rangle$ is a frame and $\langle V, F, \rightarrow \rangle$ is a proposition algebra, and

(m1) $A \rightarrow B$ is defined at least if A and B are in F and A is non-zero

(m2) $P(A \rightarrow B) = P(A \cap B)/P(A)$ if $P(A) \neq 0$ for all A, B in F.

The proviso on (m2) is not operative if $\rightarrow$ is not defined except as required by (m1), but there is no need to be so stringent. I am willing to say even that $P(A \rightarrow B) = P(B/A)$ whenever $A \rightarrow B$ is defined, for I regard conditional probability as an undetermined concept when the antecedent is a zero set; $P(A/A)$ and $P(B/A)$, for example, may be any two numbers you like between 0 and 1 inclusive, or undefined if you like. The mathematical theory of probability certainly requires no choice among these options.

THEOREM. If φ is a denumerable full frame, then there is a model $M = \langle \varphi', \rightarrow \rangle$ such that $\varphi \leqslant \varphi'$.

The proof will be long and I shall begin with intuitive commentary and lemmas. Let $\varphi = \langle V, F, P \rangle$ be a denumerable full frame (recall that this means that F is denumerable). For A and B in F and A non-zero, I shall want to choose a set $[A, B] \subseteq \bar{A}$.

LEMMA 1. $\dfrac{P(A \cap B)}{P(A)} - P(A \cap B) \leqslant 1 - P(A)$ if $P(A) \neq 0$

For let $P(A \cap B) = y$ and $P(A) = x = y + m$ then it is necessary to show that $(y/x) - y \leqslant 1 - x$ hence that $f(y) = (x^2 - xy + y) \leqslant x$. Actually this function increases from x^2 when $y = 0$ to x when $y = x$:

$$x^2 - xy + y \leqslant x^2 - x(y + n) + (y + n)$$
$$\leqslant x^2 - xy - xn + y + n$$

because $0 \leqslant n - xn$ for $0 \leqslant x \leqslant 1$.

So it appears that for $[A, B]$ we can choose an appropriately large subset of $\bar{A}$.

Now let us go on to $[A, C]$, $[A, D]$, and so on for all the other sets in F. Well, in choosing $[A, B]$ we also chose $[A, \bar{B}]$, namely $\bar{A} - [A, B]$. To choose $[A, C]$ it will suffice to choose $[A, B \cap C]$ and $[A, \bar{B} \cap C]$; see (tq3). So we can choose $[A, B \cap C]$ and $[A, \bar{B} \cap C]$; then $[A, C]$ is determined and we choose $[A, B \cap C \cap D], (A, \bar{B} \cap C \cap D], [A, B \cap \bar{C} \cap D], [A, \bar{B} \cap \bar{C} \cap D]$ and that determines $[A, D]$, and so on. The question is only whether $[A, B \cap C]$, which is a subset of $[A, B]$ is not meant, ever, to have a probability greater than that of $[A, B]$. No:

LEMMA 2. $\dfrac{P(A \cap B \cap C)}{P(A)} - P(A \cap B \cap C) \leqslant \dfrac{P(A \cap B)}{P(A)} - P(A \cap B)$

$$\text{when } P(A) \neq 0$$

For let $P(A \cap B \cap C) = z$; $P(A \cap B) = y = z + m$; $P(A) = x = y + n$. Then

$$(z/x) - z \leqslant (y/x) - y$$
$$\leqslant (z + m/x) - (z + m)$$
$$\leqslant (z/x) + (m/x) - z - m$$

because $0 \leqslant (m/x) - m$ when $0 \leqslant x \leqslant 1$.

So it appears that $[A, B_1 \cap \ldots \cap B_{n+1}]$ can be chosen as an appropriately large subset of $[A, B_1 \cap \ldots \cap B_n]$.

After this long preamble – which, I hope, was *inhaltlich* rather than contentious – the proof should be transparent.

Proof of theorem. Let $\varphi = \langle V, F, P \rangle$ be a denumerable full frame and G the domain of P. Given that φ is full, G must be uncountable (having a cardinality at least as high as the range of P). Let F be enumerated as $A_1, \ldots, A_m, \ldots$.

The strategy of the proof is to define a series of fields $F_0 \subseteq F_1 \subseteq F_2 \subseteq \ldots$ and to define $\to$ first for F_0, then for F_1, and so on. Here F_0 must be finite and included in F, F_{m+1} is defined to be the least field containing F_m and every set $A \to B$ with A a non-zero set and A, B both in F_m, and A_{m+1}. The union $F^* = \bigcup_{i=1}^{\infty} F_i$ is then a field closed under $\to$.

The process of definition of $\to$ is exactly the same at each stage, since there are no principles of form 'if $R(A, B)$ then $R(A \to C, B \to C)$'; in this I fall short of Stalnaker's logic.

This process of definition consists simply in choosing subsets $[A, B]$ of $\bar{A}$ – this yields also $[A, \bar{B}]$, its relative complement; then $[A, B \cap C]$ of $[A, B]$ and $[A, \bar{B} \cap C]$ of $[A, B]$ – which yields also $[A, B \cap \bar{C}]$, $[A, \bar{B} \cap \bar{C}]$; and so on. In each case, one chooses a subset of appropriate measure, which is possible given the above lemmas and the reflection that the frame is full.

I will add only this *stipulation*: if choosing subset X of set Y in the above construction, then X must be Λ if the appropriate measure is zero, and X must be Y if that measure is $P(Y)$.

The quadruple $\langle V, F^*, P, \rightarrow \rangle$ is now clearly a model, provided only that $\langle V, F^*, \rightarrow \rangle$ is a proposition algebra. It suffices to check (q1), (q2), (q4), (q5). The operation $[A_i, A_j]$ for A_i non-zero is defined to be the union of the sets $[A_i, A_1^* \cap ... \cap A_{j-1}^* \cap A_j]$ where A_k^* is either A_k or $\bar{A}_k$. These are all disjoint sets, and subsets of $\bar{A}_i$. The definition guarantees (q1) and (q2) immediately. Also (q4) holds because the union of *all* the sets $[A_i, A_1^* \cap ... \cap A_{j-1}^* \cap A_j^*]$ covers $\bar{A}$.

Finally, if $A_i \subseteq A_j$ then

$$\frac{P(A_i \cap A_1^* \cap ... \cap A_{j-1}^* \cap \bar{A}_j)}{P(A_i)} = 0$$

so (because of my stipulation above) the sets of which $[A_i, A_j]$ is composed by union already cover $\bar{A}$. This establishes (q5).

3.4. *Logical System*

The logical system which corresponds to the proposition algebras is CE;

(A1)   Axiom schemata as for propositional calculus

(A2) $\vdash (A \rightarrow B) \equiv (A \rightarrow .A \mathbin{\&} B)$

(A3) $\vdash A \rightarrow (B \supset C) \supset .(A \rightarrow B) \supset (A \rightarrow C)$

(A4) $\vdash (A \rightarrow B) \supset (A \supset B)$

(A5) $\vdash (A \mathbin{\&} B) \supset (A \rightarrow B)$

(A6) $\vdash (A \rightarrow B) \vee (A \rightarrow \neg B)$

(R1) $A, A \supset B \vdash B$

(R2)   If $\vdash A \equiv B$ then $\vdash (C \rightarrow A) \equiv (C \rightarrow B)$

(R3)   If $\vdash A \equiv B$ then $\vdash (A \rightarrow C) \equiv (B \rightarrow C)$

(R4)   If $\vdash A \supset B$ then $\vdash A \rightarrow B$

Here (A2) and (R2) are redunbant. CE is sound for the described family of models.

Details. Let $M = \langle V, F, P, \rightarrow \rangle$ be a model and v a map of the sentences into F such that $v(A \,\&\, B) = v(A) \cap v(B)$, $v(\neg A) = v - v(A)$, and $v(A \rightarrow B) = {} = v(A) \rightarrow v(B)$ where defined, and $= v(A \supset B)$ otherwise. Then all the theorems of CE receive value V:

(ad A3) $v(A) \rightarrow v(B \supset C). \cap .v(A) \rightarrow v(B) = v(A) \rightarrow .v(B \supset C) \cap$
              $\cap v(B) \subseteq v(A) \rightarrow v(C)$ by (I) and (II); if $(A \rightarrow (-))$ is not defined, similarly for the horseshoe.

(ad A4) $v(A) \cap v(A \rightarrow B) \subseteq v(B)$ in either case; see (III)

(ad A5)   see (III)

(ad A6)   see (tp4)

(ad R2 and R3) if $v(A \equiv B) = V$ then $v(A) = v(B)$

(ad R4)   if $v(A \supset B) = V$ then $v(A) \subseteq v(B)$; see (tp0).

Completeness. This is proved by putting together the usual completeness proof for CE and Stalnaker's completeness proof for his probability semantics.[2] Let Σ be the set of all maximal consistent CE theories, $\Sigma(A) = \{\alpha \in \Sigma : A \in \alpha\}$ and define $\Sigma(A) \rightarrow \Sigma(B) = \Sigma(A \rightarrow B)$. Finally, for an arbitrary member α of Σ and any subset X of Σ, let $P(X) = 1$ if $\alpha \in X$ and $= 0$ otherwise. Then $\langle \Sigma, F = \{\Sigma(A) : A$ a sentencel$\}, P, \rightarrow \rangle$ is a model.

Details. That F is a field follows from $(A1)$. That $\langle \Sigma, F, \rightarrow \rangle$ is a proposition algebra:

(11) $\vdash A \rightarrow (B \,\&\, C). \equiv .(A \rightarrow B) \,\&\, (A \rightarrow C)$ via (A3), (R4)

(12) $\vdash A \rightarrow (B \vee C). \equiv .(A \rightarrow B) \vee (A \rightarrow C)$ via (A3) and (A6)

(13) $\vdash A \,\&\, (A \rightarrow B). \equiv .A \,\&\, B$ via (A4) and (A5)

(14) $\vdash (A \rightarrow B) \vee (A \rightarrow \neg B)$: see (A6)

(15)   if $\vdash A \supset B$ then $\vdash A \rightarrow B$; see (R4).

This yields conditions (I)–(IV). That P is a probability measure on the powerset of Σ is trivial. That F is closed under $\rightarrow$ is also trivial. So only condition (m2) remains. Suppose $P(\Sigma(A)) \neq 0$. Then it equals 1 and $A \in \alpha$. But then $A \rightarrow B \in \alpha$ iff B is in α too, via (13). above or via (A4) and (R5). So then $P(\Sigma(A) \rightarrow \Sigma(B)) = 1$ iff $P(\Sigma(A) \cap \Sigma(B)) = 1$; otherwise both are zero. So for $X, Y \in F$, we have $P(X \rightarrow Y) = P(X \cap Y)/P(X)$ when $P(X) \neq 0$.
This ends the proof.

4. STALNAKER BERNOULLI MODELS

In Section 1 above, Stalnaker algebras are equationally defined. They differ from proposition algebras in general by obeying two further principles; and also by having $\rightarrow$ everywhere defined (not a partial operation, as I have allowed in Section 3 above).

We begin with a probability space $\langle K, F, P \rangle$ as in the preceding section of this Appendix. From this we form the structure $M = \langle K^*, F^*, P \rangle$; the symbol '$P$' is used ambiguously between the original measure and the product measure. The construction of M is the usual product construction (for sequences of stochastically independent events, as in repeated tosses of a die). Hence: K^* is the set of denumerable sequences of members of K (here identified with maps π of the natural numbers into K, but also depicted as $\pi = \langle \pi(1), \pi(2), \pi(3), ... \rangle$). F^* is the family of sets *(generating sets)*

$$A_1 \otimes \cdots \otimes A_n \otimes K^* = \{\pi \in K^*: \pi(1) \in A, \& ... \& \pi(n) \in A_n\}$$

formed from sets $A_1, ..., A_n$ in F. Finally P is the product measure with domain F^* (which is a Borel field because F is) defined by

$$P(A_1 \otimes \cdots \otimes A_n \otimes K) = P(A_1) ... P(A_n).$$

If π is in K^*, let π_m be the sequence defined by:

$$\pi_m(k) = \pi(m+k-1)$$

so that $\pi_m = \langle \pi(m), \pi(m+1), ... \rangle$. Define the operation $\rightarrow$ on the powerset of K^* by

$$X \rightarrow Y = \{\pi \in K^*: \text{for all } m, \text{ if } m \text{ is the first natural number such that } \pi_m \text{ is in } X, \text{ then } \pi_m \text{ is in } Y\}$$

(For the case in which there is no X-world 'accessible to' π – i.e. no m such that π_m is in X – Stalnaker introduced an 'absurd world', but the same effect is gotten by the present definition; namely that π is in $X \rightarrow Y$ in that case for any Y.)

Now, F^* is closed under $\rightarrow$. For define $X(k) = K^{k-1} \otimes X$ with $X(1) = X$. Note that if X is a generating set, so is $X(k)$; moreover, $\bar{X}(k) = \overline{X(k)}$ and $\bigcup_{i=1}^{\infty} [X_i(k)] = [\bigcup_{i=1}^{\infty} X_i](k)$. So not only for the generating sets, but for all sets generated from them in the Borel way, we find that $X(k)$ is in F^* if

X is. As a special case, define $X(0) = K$. Then we have

$$X \Rightarrow Y = \bigcup_{k=1}^{\infty} \left[\bigcap_{i=1}^{k-1} \bar{X}(i) \cap (X \cap Y)(k) \right]$$

as the set of worlds π such that for some i, π is in $X(i)$ *and* π is in $(X \to Y)$. So $X \Rightarrow Y$ differs from $X \to Y$ only by leaving out the 'impossible antecedent' case. Therefore

$$X \to Y = (X \Rightarrow Y) \cup \left[\bigcap_{i=1}^{\infty} \bar{X}(i) \right]$$

All the operations thereby used to define $\to$ are such that a Borel field is closed under them; hence F^* is closed under $\to$.

I turn now to probabilities, and shall designate sets $A \otimes K^*$ with A in F as *zero-degree propositions*. Moreover, abbreviate '$A \otimes K$' to 'A' when convenient; the capital letters A, B, C, D are to stand only for zero-degree propositions and members of F (via this symbolic identification).

LEMMA 1. $P(A \to X) = P(A \Rightarrow X)$ when $P(A) \neq 0$. The reason is that $P(\bigcap_{i=1}^{\infty} \bar{A}(i)) = P(\bar{A})\dots P(\bar{A})\dots$ which equals zero unless $P(\bar{A}) = 1$.

LEMMA 2. (Fraction Lemma)[4] In any probability space $\langle W, G, P \rangle$, if $P(A) \neq 0$ then $\sum_{k=0}^{\infty} P(\bar{A})^k = 1/P(A)$.

For proof consider $s = P(A) \sum_{k=0}^{\infty} P(\bar{A})^k =$

$$= P(A) + P(\bar{A}) \, P(A) + P(\bar{A})^2 \, P(A) + \cdots =$$
$$= P(A \otimes K^*) + P(\bar{A} \otimes A \otimes K^*) + \cdots =$$
$$= P\left(A \otimes K^* \cup \left[\bigcup_{m=1}^{\infty} \bar{A}^m \otimes A \otimes K^* \right] \right)$$

because the sets in the third line are disjoint. We note that we can continue these equations as

$$= P(\{\pi \in K^* : \pi(m) \in A \text{ for some } A\})$$
$$= P(K^* - \{\pi \in K^* : \pi(m) \in \bar{A} \text{ for all } m\})$$
$$= 1 - P(\bar{A} \otimes \bar{A} \otimes \bar{A} \otimes \dots)$$
$$= 1$$

since we are given that $P(A) \neq 1$. But from the fact that $s = 1$, the lemma follows.

THEOREM. $P(A \rightarrow B) = P(B/A)$ if $P(A) \neq 0$, for all zero degree propositions A and B.

Intuitively: let m be the first number such that $\pi(m)$ is in A. The probability that $\pi(m)$ is in B is then $P(B/A)$, if B too is in F, the consequtive events $\pi(1), \pi(2), \ldots$ being independent. Now m could be 1, or 2, or 3,... the probabilities of these disjoint cases being $P(A)$, $P(\bar{A}) P(A)$, $P(\bar{A}) P(A) P(A)$, and so forth. Hence we calculate

$$P(A \Rightarrow B) = P(A) P(B/A) + P(\bar{A}) P(A) P(B/A) +$$
$$+ P(\bar{A})^2 P(A) P(B/A) + \cdots =$$
$$= P(A) \sum_{k=0}^{\infty} P(\bar{A})^k P(B/A) =$$
$$= P(B/A)$$

by the Fraction Lemma.

LEMMA 3. $P(A \rightarrow C. \rightarrow B) = P(A \rightarrow C. \Rightarrow B)$ if $P(A \rightarrow C) \neq 0$. This is a generalization of Lemma 1, the proof being similar and hinging on the point that $P(\bigcap_{m=1}^{\infty} \overline{(A \rightarrow C)} (m)) = P(Z) = 0$. Now the set Z introduced by definition contains the sequences π such that for each number m, there are one or more numbers $n > m$ such that $\pi(n) \in A$, while the first of these is not in C. So a fortiori, none of them is in C. So a somewhat bigger set than Z is

$$Z_1 = \{\pi: [\text{for all } m, \pi(m) \in \bar{A} \cup \bar{C}] \ \& \ [\text{for some } m, \pi(m) \ A]\}$$

and a still bigger set is

$$Z_2 = \{\pi: \text{for all } m, \pi(m) \in \bar{A} \cup \bar{C}\}.$$

Now $P(Z_2) = 0$ unless $P(\bar{A} \cup \bar{C}) = 1$. But if $P(\bar{A} \cup \bar{C}) = 1$ then $P(A \rightarrow \bar{C}) = 1$, so then $P(A \rightarrow C) = 0$ – which is ruled out – or else $P(A) = 0$. In both cases, however, we see that $P(Z_2) = 0$; hence also $P(Z) = 0$.

THEOREM. $P(A \rightarrow C. \rightarrow B) = P(B/A \rightarrow C)$ if $P(A \rightarrow C) \neq 0$ for zero-degree propositions A, B, C.

By the preceding lemma, we may concentrate on $P(A \rightarrow C. \Rightarrow B)$. Let the first number m such that π_m is in $A \rightarrow C$ be k; the probability that π_m is in B is entirely independent of the initial sequent $\pi(1), \ldots \pi(m-1)$; hence this is just $P(B/A \rightarrow C)$ – more precisely, in this case, it is $P(B \otimes K^*/(A \otimes K^*) \rightarrow (C \otimes K^*))$, the very probability that *any*, random, sequence in K^* is in B if it is in $A \rightarrow C$. The cases are $K = 1, 2, \ldots$; and these cases can be described for $k > 1$ as:

(a) $\pi(1), \ldots, \pi(k-2)$ are in $\bar{A} \cup \bar{C}$ [if any]

(b) $\pi(k-1)$ is in $A \cap \bar{C}$

(c) $\pi(k)$ is in $A \rightarrow C$

while for case $k = 1$, we leave out (a) and (b). So we have the calculation (omitting intersection signs where convenient, and so replacing also $\bar{A} \cup \bar{C}$ by $\overline{AC}$):

Let $r = P(B/A \rightarrow C)$

$$P(A \rightarrow C \Rightarrow B) = P(A \rightarrow C)\, r + P(\overline{AC})\, P(A \rightarrow C)\, r +$$
$$+ P(\overline{AC})\, P(A\bar{C})\, P(A \rightarrow C)\, r +$$
$$+ P(\overline{AC})^2\, P(A\bar{C})\, P(A \rightarrow C)\, r + \cdots =$$
$$= P(A \rightarrow C)\, r \left[1 + \left(\sum_{k=0}^{\infty} P(\overline{AC})^k \right) P(A\bar{C}) \right] =$$
$$= P(A \rightarrow C)\, r \left[1 + \frac{P(A\bar{C})}{P(AC)} \right]$$

by the Fraction Lemma for case $P(AC) \neq 0$; set this

$$= P(A \rightarrow C)\, rs$$

but then

$$s = \frac{P(AC) + P(A\bar{C})}{P(AC)} = \frac{P(A)}{P(AC)} = 1/P(C/A) =$$
$$= 1/P(A \rightarrow C)$$

by the preceding theorem. Hence

$$P(A \rightarrow C \Rightarrow B) = P(A \rightarrow C)\, rs = r = P(B/A \rightarrow C).$$

Now in the remaining case, $P(AC) = 0$. In that case $P(A \rightarrow C) = 0$ – which is ruled out – unless $P(A) = 0$. But if $P(A) = 0$ then $P(A \rightarrow C) = 1$ and

$P(B/A \to C) = P(B)$. However, in that case also $P(A \to C \Rightarrow B) = P(B)$, because, except for a set of measure zero, any sequence π we pick will be in $A \to C$. More precisely, if $P(X) = 1$, then

$$P(X \Rightarrow Y) = P\left(\bigcup_{k=1}^{\infty} \left[\bigcap_{i=0}^{k-1} \bar{X}(i) \cap (X \cap Y)(k) \right] \right)$$

which is the sum of the terms

$$P(X \cap Y(1)) + P(X(1) \cap (X \cap Y)(2)) + \cdots$$

of which all but the first has probability zero; and $P(X \cap Y(1)) = P(X \cap Y) = P(Y)$ because $P(X) = 1$.

So the theorem holds in all cases. Let me call $A \to C. \to B$ a *left* conditional and $C \to . A \to B$ a *right* conditional. To prove a corresponding result for right conditionals I need one more lemma.

LEMMA 4. (Independence Lemma) $P(\bar{A} \cap C \cap A \to B) = P(\bar{A}C) \, P(A \to B)$ for zero-degree propositions A, B, C.

Consider first $P(A) = 0$. Then the lemma holds for any C, because each side equals zero.

Consider next $P(A) \neq 0$. Abbreviate $\bar{A} \otimes \bar{A} \otimes \ldots$ as $\bar{A}^*$. Then $A \to B = \bar{A}^* \cup (A \Rightarrow B)$, so

$$P(\bar{A}C \cap (A \to B)) = P(\bar{A}C \cap \bar{A}^* \cup \bar{A}C \cap (A \Rightarrow B)) =$$
$$= P(\bar{A}C \cap (A \Rightarrow B))$$

because $P(\bar{A}^*)$ and hence $P(\bar{A}C \cap \bar{A}^*)$, equals zero. Now $\bar{A}C = \bar{A}C(1)$ while $(A \Rightarrow B) = AB(1) \cup \bar{A}(1) \, AB(2) \cup \bar{A}(1) \, \bar{A}(2) \, AB(3) \cup \ldots$. Hence $\bar{A}C \cap (A \Rightarrow B) = \Lambda \cup \bar{A}C(1) \, AB(2) \cup \bar{A}C(1) \, \bar{A}(2) \, AB(3) \cup \ldots$. Hence also

$$P(\bar{A}C \cap A \Rightarrow B) = P(\bar{A}C) \, P(AB) + P(\bar{A}C) \, P(\bar{A}) \, P(AB) +$$
$$+ P(\bar{A}C) \, P(\bar{A})^2 \, P(AB) + \cdots =$$
$$= P(\bar{A}C) \, P(AB) \sum_{k=0}^{\infty} P(\bar{A})^k =$$
$$= P(\bar{A}C) \, P(AB)/P(A) =$$
$$= P(\bar{A}C) \, P(B/A) =$$
$$= P(\bar{A}C) \, P(A \to B)$$

as required.

THEOREM. $P(C \to A \to B) = P(A \to B/C)$ if $P(C) \neq 0$, for zero-degree propositions A, B, C.

By Lemma 1, we can concentrate on $C \Rightarrow A \to B$. Consider first $P(A) = 0$. Then $P(A \to B) = 1$ as we have seen; so $P(A \to B/C) = 1$ also. But in addition, $P(C \Rightarrow A \to B) = 1$ because in general, if $P(Y) = 1$ then the probability that, for random π, the first m such that π_m is in C, is also in Y, equals 1. So consider henceforth that $P(A) \neq 0$. Now $C \Rightarrow A \to B$ is the union of the disjoint sets

$$X_k = \bar{C}(1) \dots \bar{C}(k-1) [C \cap (A \to B)] (k) =$$
$$= \bar{C}(1) \dots \bar{C}(k-1) \{[CAB(k) \cup C\bar{A}(k) AB(k+1) \cup$$
$$\cup C\bar{A}(k) \bar{A}(k+1) AB(k+2) \cup \dots] \cup C(k) \bar{A}^*(k)\}.$$

Because $P(A) \neq 0$, $\bar{A}^*(k)$ has probability zero, so that term can be ignored. Thus

$$P(X_k) = P(\bar{C})^{k-1} [P(CAB) + P(C\bar{A}) P(AB) +$$
$$+ P(C\bar{A}) P(\bar{A}) P(AB) + P(C\bar{A}) P(\bar{A})^2 P(AB) + \cdots] =$$
$$= P(\bar{C})^{k-1} [P(CAB) + P(C\bar{A}) P(AB) \sum_{m=0}^{\infty} P(\bar{A})^m] =$$
$$= P(\bar{C})^{k-1} [P(CAB) + P(C\bar{A}) P(AB)/P(A)] =$$
$$= P(\bar{C})^{k-1} [P(CAB) + P(C\bar{A}) P(A \to B)] =$$
$$= P(\bar{C})^{k-1} [P(CAB) + P(C\bar{A} \cap A \to B)]$$

(by the Independence Lemma),

$$= P(\bar{C})^{k-1} [P(CA \cap A \to B) + P(C\bar{A} \cap A \to B)]$$

(because $A \cap (A \to B) = AB$),

$$= P(\bar{C})^{k-1} [P(C \cap A \to B)].$$

Now we can finish the calculation as follows:

$$P(C \Rightarrow A \to B) = P\left(\bigcup_{k=1}^{\infty} X_k\right)$$
$$= \sum_{k=1}^{\infty} P(\bar{C})^{k-1} P(\cap A \to B) =$$
$$= P(C \cap A \to B)/P(C) =$$
$$= P(A \to B/C)$$

as required.

The conclusion is that any antecedently given probability assignment to the zero-degree propositions can be reflected in a Stalnaker model, in such a way that the first degree conditionals, and what I have called left and right (second degree) conditionals are in accord with the Stalnaker Thesis.

NOTES

[1] The research for this paper was supported by Canada Council grant S72-0810. I also wish to thank Yvon Gauthier, Ian Hacking, William Harper, David Lewis, Robert Meyer, and Richmond Thomason for helpful comments on earlier drafts of the Appendices. Further debts to Hacking and Lewis are detailed below; it will be clear that my reaction to Lewis' writings colours almost every section.

[2] I give the Stalnaker semantics essentially as simplified by Lewis in his [3].

[3] This refers to comments by Thomason at the end of [11], and my [12, 13]. See also the critical comments by Lewis on this in his [4, 5].

[4] See the Appendix to my [13].

[5] Lewis [6], p. 10.

[6] See also my discussion in [13].

[7] Lewis [4], p. 84

[8] I realize that on some views, the scientist does not make theoretical commitments, and does not believe his theories. So this criticism is based on my views on scientific theories; see [13].

[9] Stalnaker [7].

[10] Lewis [6], p. 5.

[11] Given in an unpublished paper circulated in April 1972.

[12] (*Added January 1974.*) I have just seen a proof by Stalnaker that no such result as Theorem 25 can hold for his C2.

Notes to Appendix

[1] In his 'Completeness and Decidability...'. I have modified the axioms and rules in some obvious ways to ensure continuity with the remainder of my discussion.

[2] In his 'Probability and Conditionals'; this proof is not affected by the apparent vulnerability of Stalnaker's paper to Lewis' triviality results.

[3] *Ibid.*

[4] This is the lemma for which I am indebted to Ian Hacking.

BIBLIOGRAPHY

[1] Chisholm, R., 'The Contrary-to-Fact Conditional', *Mind* 55 (1946), 289–302.

[2] Goodman, N., 'The Problem of Counterfactual Conditionals', *Journal of Philosophy* 44 (1947), 113–128.

[3] Lewis, D., 'Completeness and Decidability of Three Logics of Counterfactual Conditionals', *Theoria* 37 (1971), 74–85.

[4] Lewis, D., *Counterfactuals*, Blackwell, Oxford, 1973.

[5] Lewis, D., 'Counterfactuals and Comparative Possibility', *Journal of Philosophical Logic* 2 (1973), in No. 2–4.

[6] Lewis, D., 'Probabilities of Conditionals and Conditional Probabilities' (ditto'd), Princeton, March 1973.

[7] Stalnaker, R., 'Probability and Conditionality', *Philosophy of Science* **37** (1970), 64–80.

[8] Stalnaker, R., 'A Theory of Conditionals', in N. Rescher (ed.), *Studies in Logical Theory* (APQ Supplementary Volume), Blackwell, Oxford, 1968.

[9] Stalnaker, R. and Thomason, R. H., 'A Semantic Analysis of Conditional Logic', *Theoria* **36** (1970), 23–42.

[10] Teller, P., 'Conditionalization, Observation, and Change of Preference', this volume, p. 205.

[11] Thomason, R. H., 'A Fitch-Style Formulation of Conditional Logic', *Logique et Analyse* **13** (1970), 397–412.

[12] van Fraassen, B. C., 'Hidden Variables in Conditional Logic', *Theoria*, forthcoming.

[13] van Fraassen, B. C., 'Theories and Counterfactuals', forthcoming in a festschrift for W. Sellars (ed. by H-N. Castañeda).

DISCUSSION

Commentator: Giere: I should like you to fill out the motivations to the paper a little more. We have two paradigms for probability, the ticker tape paradigm and the sweepstake paradigm. In both cases it seems that we can say all that needs to be said using regular conditional probabilities. So why should we bother looking for a way of assigning probabilities to conditional statements?

Van Fraassen: Actually they arise in both. If you look upon conditionals the way Ramsey did, then iterated conditionals don't make sense, but there are many practical situations in which there are iterated conditionals which people do assert. The question arises whether we should assign probabilities to these iterated conditionals, and this is a question whether we should extend the probability calculus to these English sentences. In my own view we should extend probabilities to such sentences. For example, I may say "if this glass breaks when I drop it on the floor then these glasses will break if I throw them against that wall". Then we have a claim that is surely not a necessary truth and so one wants to know the probability of its being correct.

Giere: Still it is not clear why we cannot equally well regard the cases you mention as examples of complex systems for which we need never ask about the probability of an iterated conditional but rather simply about probabilities of various outcomes of trials on the entire complex system.

Van Fraassen: Yes there are some philosophers who claim that we get by in science with very little resources in this respect – if they are right, then the answer to the question doesn't carry a lot of practical weight. But the problem of explicating probability discourse in general remains, for philosophy of language and of logic, if not of science.

STALNAKER TO VAN FRAASSEN

Dear Bas:

When David Lewis brought in his artillary and fired his bombshell at Stalnaker's assumption, Stalnaker meekly capitulated. But Van Fraassen, chivalrously, has rushed to its defense, opening a second front by uncovering a different assumption which is involved, and diverting the fire to that one in the hope of saving the first. It is time for me to reenter the battle – this time on the other side.

I am not sure whether I am a "metaphysical realist" in the relevant sense, or whether the assumption made by Lewis (and by me) that you criticize is necessarily connected to such realism. But I am convinced by your argument that the assumption was essential to Lewis's argument, that it was essential to my argument using probability semantics to justify C2, and that the assumption is untenable. But I am not tempted to reembrace Stalnaker's assumption, since I can get the same trivialization result from it without "metaphysical realism," as long as I assume that the logic of conditionals is C2.

I list the following six theses for reference. *A, B* and *C* are any *propositions*. P is any *probability function*. In my original paper, I used probability functions which allowed non-trivial conditional probability values even when the absolute probability of the condition was zero. This complication is irrelevant to the current discussion, and so I will ignore it. It may be assumed that the conditional probability, $P(B/A)$, is defined only when $P(A) \neq 0$. A *subfunction*, P_A, is a function defined for any probability function P and proposition *A* such that $P(A) \neq 0$ as follows: $P_A(B) =_{df} P(B/A)$.

(1) If $P(A) \neq 0$, $P(A > B) = P(B/A)$.

(2) Any subfunction is a probability function.

(3) 'Metaphysical Realism': The proposition expressed by a conditional sentence is independent of the probability function defined on it. So, for example, the content of the sentence $B > C$ is the same in the context $P(B > C)$ as it is in the context $P_A(B > C)$.

(4) If $P(A \& C) \neq 0$, $P(A > B/C) = P(B/A \& C)$.

302

(5) The logic of the conditional corner is C2.

(6) For any probability function, there are at most two disjoint propositions which have non-zero probability.

The aim of my original paper was to derive (5) from (1), and so to provide an independent motivation for C2. My argument assumed (2) more or less explicitly, and (3) implicitly. (2) is not really an assumption, since it follows from elementary probability calculus, and depends on no special assumptions about conditional propositions. Lewis's argument derived (6) from (4). Since (4) follows from the premisses of my argument, this was sufficient to defeat my project. The argument generalizing (1) to (4) goes as follows:

1. $\quad P_C(A \ \& \ B) = P_C(A) \times P_C(B/A)$
2. $\quad\quad\quad\quad\quad = P_C(A) \times P_C(A > B)$
3. $\quad\quad\quad\quad\quad = P(A/C) \times P(A > B/C)$
4. $\quad P_C(A \ \& \ B) = P(A \ \& \ B/C)$
5. $\quad\quad\quad\quad\quad = P(A/C) \times P(B/A \ \& \ C)$
6. $\quad$ So, assuming $P(A/C) \neq 0$, $P(A > B/C) = P(B/A \ \& \ C)$.

How does thesis (3) enter into this argument? Without (3), the move from step 2 to step 3 will involve an equivocation, since $A > B$ need not express the same proposition in both places. So Lewis's argument, which makes no assumptions about the logic of conditionals, does depend on (3). But another argument gets exactly the same conclusion from (1) and (5), without (3), or (2).

Assumptions: (1), (5), and the denial of (6). The argument will show that a contradiction can be derived from these assumptions. By the denial of (6), there are at least three disjoint propositions that are assigned non-zero probability by some probability function. Call them $A \ \& \ B$, $A \ \& \ \bar{B}$, and $\bar{A}$.

I will make use of the following abbreviation:

$$C =_{\mathrm{df}} A \vee (\bar{A} \ \& \ (A > \bar{B})),$$

and the following lemmas which can be proved from the assumptions (1) and (5):

(7) If $P(\bar{X}) \neq 0$, then $P(X > Y/\bar{X}) = P(X > Y)$.

(8) $\bar{C}$ entails $C > \sim(A \ \& \ \bar{B})$.

Finally, since $\bar{A}$ entails $(\bar{A} \ \& \ (A > B)) \vee (\bar{A} \ \& \ (A > \bar{B}))$, and $P(\bar{A}) \neq 0$, it follows that either $P(\bar{A} \& (A > B)) \neq 0$ or else $P(\bar{A} \& (A > \bar{B})) \neq 0$. I will assume the former, but in case it is instead the latter, interchange B and $\bar{B}$ everywhere in the argument, including in the definition of C.

Now, from all these assumptions, it follows that

(9) The following propositions all have non-zero probability: $A \& B$, $A \& \bar{B}$, $\bar{A}$, C, $\bar{C}$.

Now for the inconsistency argument:

1.	$P(C > \sim(A \& \bar{B})/\bar{C}) = 1$	by (8)
2.	$P(\bar{C}) \neq 0$	by (9)
3.	$P(C > \sim(A \& \bar{B})) = 1$	from 1, 2, by (7)
4.	$P(C) \neq 0$	by (9)
5.	$P(\sim(A \& \bar{B})/C) = 1$	from 3, 4, by (1)
6.	$\dfrac{P(C \& \sim(A \& \bar{B}))}{P(C)} = 1$	from 5
7.	$\dfrac{P((A \& B) \vee (\bar{A} \& (A > \bar{B})))}{P((A \& B) \vee (A \& \bar{B}) \vee (\bar{A} \& (A > \bar{B})))} = 1$	from 6
8.	$P(A \& \bar{B}) = 0$	from 7

which contradicts (9).

Note that if we had substituted B for $\bar{B}$ everywhere, the conclusion would have been $P(A \& B) = 0$, which also contradicts (9).

As you say, Stalnaker's logic is one thing, his models are another. The above argument makes no assumptions about models, and so applies to the logic, however interpreted. But if one considers things semantically for a minute, one can see a connection between the above argument and your shielding effect (Notes, III). (My argument was worked out before I went through your Notes III, so the connection was unexpected.)

You define the shielding effect in terms of normal models, but as you point out, the notion is more general. Say that a proposition X *encloses* a proposition Y iff for every possible world $i \in \bar{X}$, $s(X, i) \in \bar{Y}$. Then the following is the generalization of the result you state on page 4 of the Notes: If thesis (1) holds, then there is no pair of propositions X and Y such that $P(X) < 1$, $P(Y) > 0$, and X encloses Y. But, for any probability function defined on any model which assigns non-zero probability to at least three disjoint propositions, there will be a pair of propositions meeting those conditions. The above argument can be taken to show this, since $P(C) < 1$, $P(A \& \bar{B}) > 0$, and C encloses $(A \& \bar{B})$.

So where does this result leave me? Since I want to keep (5), I can avoid disaster only by rejecting (1). But it is not only to avoid disaster that I take this option since I think there are good intuitive arguments against (1). In fact, I am as taken with the distinction between the probability of the conditional and the conditional probability as I once was with their supposed identity.

Where does the result leave you? It does answer, I think, the question you ask on page 297 of your paper. It is not tenable (although it is consistent) to accept the thesis for the C2 conditional. But you don't like C2 that much anyway, so you can just keep (1) and reject (5). But I think this would be a mistake.

Suppose there is a general theoretical proposition T which entails a conditional $A > B$, or perhaps it entails the theoretical claim that $P(A > B) = r$. Since T might be a theory with diverse consequences, the evidence for or against T – and so the evidence which bears on $A > B$, or $P(A > B) = r$ – could be anything. I see no reason to rule out a priori the possibility that A itself be part of the evidence for or against T. If it were, then the conditional might not be stochastically independent of its antecedent (which as you point out the thesis requires). Consider a concrete example: H_1 and H_2 are propositions describing the outcomes of two flips of a coin. Say H_1 says that flip 1 comes up heads, and H_2 says that flip 2 comes up heads. Suppose (as is very plausible) that it is known on the basis of general theoretical considerations that the flips are causally independent of each other, but it is unknown what the bias of the coin is. Suppose it was drawn at random from a bag of coins half of which are biased for heads, half biased to the same degree for tails. On the basis of the causal independence, we know that the truth value of H_2 would have been the same as it in fact was whatever the truthvalue of H_1, and this holds for all possible worlds compatible with the general theoretical assumption of causal independence. Hence, although we may not know the value of $P(H_2)$, we know that $P(H_1 > H_2) = P(H_2)$. But H_1 still is *evidence* for H_2, since it is evidence for the proposition that the chosen coin is biased for heads. Thus, in this example, $P(H_2/H_1) > P(H_2)$, and so $P(H_2/H_1) > P(H_1 > H_2)$. This result seems intuitively plausible to me, and it does not depend on any assumptions about the logic of conditionals.

I see that (1) is formally compatible with your logic of conditionals,

CE, and with any probability distribution defined on all the zero degree sentences. But I am not convinced that any kind of intuitively satisfactory semantics can be given for CE which explains why the thesis should be true. Whatever the truth conditions for conditional propositions (assuming they have *truth* conditions), the effect of the thesis is to impose the following requirement: for any A with probability between 0 and 1 exclusive, and for any B, the ratio of the measure of worlds in which $A > B$ is true to the measure of worlds in which it is false ll be the same in the A worlds as it is in the $\bar{A}$ worlds. But why shou.d this be true? One does need some intuitive explanation.

  Best.

Yours,

BOB STALNAKER

VAN FRAASSEN TO STALNAKER

February, 1974

Dear Bob:

That was a very exciting letter! I was very pleased to see your exposition, in precise form, of exactly where the 'metaphysical realism' premiss enters David's reductio. And your subsequent argument doing without that premiss fills *me* with admiration – though not pleasurably so. Also, you are right: the crucial question is whether conditionals should be stochastically independent of their antecedents.

But, I really doubt that you could convince me of the contrary by *first degree* examples. In your last page you apparently define

H_2 is causally independent of H_1 iff $P(H_1 > H_2) = P(H_2)$
H_2 is stochastically independent of H_1 iff $p(H_2/H_1) = P(H_2)$

and say that causal independence does not imply stochastic independence. Of course, my doubt is this: have you got your fingers on a correct symbolization of 'causal independence'?

Suppose that S is the thesis that the coin is fair, and T the thesis that the coin is biased 3 to 1, either in favor of heads or in favor of tails, each with probability $\frac{1}{2}$. Then – your symbolism – $P(H_1) = P(H_2) = \frac{1}{2}$ on either hypothesis. Hence to bring out the difference, to represent adequately such information as T conveys, we cannot use the sample space of single tosses. Let K be the set of all countable sequences of tosses, and let

P_1 be the product measure based on $P_1(H) = P_1(\sim H) = \frac{1}{2}$
P_2 that based on $P_2(H) = \frac{3}{4}$
P_3 that based on $P_3(H) = \frac{1}{4}$
$P_4 = \frac{1}{2}P_2 + \frac{1}{2}P_3$
H_i is true in d in K iff d(i) is H.

So $P_1(H_1 \& H_2) = \frac{1}{4}$ but $P_4(H_1 \& H_2) = \frac{10}{32}$. Now let α be a random member of K: what is the probability that the nearest H_1-world to α is an H_2-world? Since α is *random*, and we know nothing about what is near what, is the answer $P(H_2/H_1)$? And isn't the question the same as: what is the probability that $H_1 > H_2$ is true in α?

(I know, this assumes that we have *no* prior knowledge of what 'nearest' amounts to. But the problem gives no information about the nearest ordering, does it? And for the first-degree conditional, it can make no difference that nearest-orderings have certain abstract properties, which keep the thesis from being true in general in Stalnaker models.)

I respectfully submit that you took $P(H_1 > H_2)$ to be a measure of causal rather than stochastic dependence because you were thinking of $H_1 > H_2$ as meaning something like: if we brought it about that H_1 we would *thereby* bring it about that H_2 – rather than, if H_1 were the case then H_2 would be the case.

Well, I hope to resist the temptation to start working on this again. I had one more idea, a kind of generalization of Lewis' counterfactual probability. Remember how in my 'hidden variables' paper, each Lewis model is a family of Stalnaker models with the same worlds? Suppose we have a measure P on this family F and define

$$P_\alpha(A) = P(\{w \in F: \text{in } w, A \text{ is true in } \alpha\})$$
$$P_\alpha(A/B) = P(\{w \in F, A \text{ is true in } \alpha \text{ in } w: B \text{ is true in } \alpha \text{ in } w\}).$$

Then I figure that if you average over α, you should have a probability assignment. But this too is just a hunch. As I say, I want to resist the temptation this time around.

Best regards,

BAS

THE UNIVERSITY OF WESTERN ONTARIO
SERIES IN PHILOSOPHY OF SCIENCE

A Series of Books on Philosophy of Science, Methodology, and Epistemology
published in connection with
the University of Western Ontario Philosophy of Science Programme

Managing Editor:

J. J. LEACH

Editorial Board:

J. BUB, R. E. BUTTS, W. HARPER, J. HINTIKKA, D. J. HOCKNEY,

C. A. HOOKER, J. NICHOLAS, G. PEARCE

1. J. LEACH, R. BUTTS, and G. PEARCE (eds.), *Science, Decision and Value.* Proceedings of the Fifth University of Western Ontario Philosophy Colloquium, 1969. 1973, vii + 213 pp.

2. C. A. HOOKER (ed.), *Contemporary Research in the Foundations and Philosophy of Quantum Theory.* Proceedings of a Conference held at the University of Western Ontario, London, Canada. 1973, xx + 385 pp.

3. J. BUB, *The Interpretation of Quantum Mechanics.* 1974, ix + 155 pp.

4. D. HOCKNEY, W. HARPER, and B. FREED (eds.), *Contemporary Research in Philosophical Logic and Linguistic Semantics.* Proceedings of a Conference held at the University of Western Ontario, London, Canada. 1975, vii + 332 pp.

5. C. A. HOOKER (ed.), *The Logic-Algebraic Approach to Quantum Mechanics.* 1975, xv + 607 pp.

309